Positron Emission Tomography

Birendra Kishore Das
Editor

Positron Emission Tomography

A Guide for Clinicians

Springer

Editor
Birendra Kishore Das, MD, ANM, FAR, FAMS
Director
Utkal Institute of Medical Sciences
Bhubaneswar
India

ISBN 978-81-322-2998-8 ISBN 978-81-322-2098-5 (eBook)
DOI 10.1007/978-81-322-2098-5
Springer New Delhi Heidelberg New York Dordrecht London

Printed on acid-free paper

Springer is part of Springer Science+Business Media (www.springer.com)

Foreword

While it is true that certain types of cancers such as those of the pancreas and brain are still conditions with a grave prognosis, there are many other common cancers for which appropriate treatment can be curative for a substantial number of patients. This has been possible with the development of new technology in diagnosis, chemotherapeutic drugs, and delivery of precise radiation therapy. This hopeful aspect of oncology is often insufficiently appreciated even by the medical profession.

There can be no doubt that knowledge of these new developments and their role in the management of cancer is becoming of increasing importance to doctors throughout the world. One such new development is the role of PET and PET-CT in the overall management of cancer. Many potentially curable cancer patients are deprived of this opportunity due to lack of knowledge about this new technology and its versatile applications.

Professor B. K. Das, who is a specialist of international repute, has now put together with the help of top nuclear physicians of India a unique book that organizes this large body of information about PET and its applications by distilling the key facts, which the treating physicians can effectively use. This book has culminated from the author's vast experience including his training in radiology, radiotherapy, and nuclear medicine in the 1970s in Germany and subsequent practice of nuclear medicine both in Germany and India. In his interaction with thousands of patients, Dr Das has realized that many treating colleagues are not acquainted with the vast potential of PET technology in the routine management of cancer patients.

This book can serve as a valuable resource for general practitioners, cancer specialists, and health care givers to have access in a very comprehensive manner to knowledge pertaining to this evolving technology and its practical application in routine management of especially cancer patients.

OH, USA — Joseph C. Mantil, MD, PhD

Preface

Positron Emission Tomography (PET) has become a buzzword in the oncology community. In the history of medical sciences, no high tech–based imaging modality has ever become so popular and widespread in such a short time as the PET technology. In spite of very high cost (the costliest imaging equipment ever!), PET centers are coming up rapidly in many cities of India.

PET imaging is mostly used in Oncology (85 %), Neurology (10 %), and Cardiology (5 %) as also in some other situations like infection imaging (cause of fever in cases of "Fever of Unknown Origin," etc.). The main focus, of course, is in the management of cancer patients. PET (PET-CT) is not only very sensitive as it can detect changes in abnormal biochemical processes at cellular levels (as it happens during the development of cancer), but in one go all such areas can be detected in a whole-body scan. It can show response to therapy, eradication of the disease, or recurrence of the cancer during the follow-up period. It can also contribute significantly in identifying exact areas where radiotherapy is to be targeted, avoiding unnecessary radiation exposure to surrounding tissue.

The purpose of this book is to provide basic information about this technology, its clinical application, and practical guidance for referring physicians. The chapters have been contributed by experts from top institutions of the country who have tried to provide necessary information in a most comprehensive manner.

Because most of the new PET equipment comes as hybrid machines with CT or MRI, two small chapters have been included at the end of the book to provide basic information about these two technologies.

I hope that this book will be helpful for those who wish to use this new technology for the rational management of cancer patients and in other appropriate areas.

I wish to express my gratitude to all who have contributed various chapters and encouraged to bring out this first edition.

I thank the staff of the department who have helped in many ways in compiling the chapters.

My special thanks go to my wife Dr. Gouri Das, who has contributed significantly going through all the chapters and suggesting changes necessary to fulfill the objectives of the book.

Bhubaneswar, India — Birendra Kishore Das, MD, ANM, FAR, FAMS

Contents

Contributors

C.S. Bal, MD, DSc Prof. & Head, Department of Nuclear Medicine and PET, All India Institute of Medical Sciences, New Delhi, India

Sandip Basu, MD Professor, Radiation Medicine Centre, Bhabha Atomic Research Centre, Tata Memorial Hospital Annexe, Mumbai, India

Birendra Kishore Das, MD, ANM, FAR, FAMS Director and Chief of Nuclear Medicine, Utkal Institute of Medical Sciences, Bhubaneswar, India

Vandana Dhingra, MD Consultant, Radiation Medicine Centre, Bhabha Atomic Research Centre, Tata Memorial Hospital Annexe, Mumbai, India

Prathamesh Joshi, DRM, DNB Consultant, Radiation Medicine Centre, Bhabha Atomic Research Centre, Tata Memorial Hospital Annexe, Mumbai, India

B.R. Mittal, MD, FICNM, FAMS Prof. & Head, Department of Nuclear Medicine and PET, Postgraduate Institute of Medical Sciences and Research, Chandigarh, India

Raveena Mubalsha Senior Resident, Department of Nuclear Medicine and PET, Sanjay Gandhi Postgraduate Institute of Medical Sciences, Lucknow, India

Rahul Parghane, MD Senior Resident, Department of Nuclear Medicine and PET, Postgraduate Institute of Medical Sciences and Research, Chandigarh, India

Alok Pawaskar, DRM, DNB Consultant, Department of Nuclear Medicine, Radiation Medicine Centre, Tata Memorial Hospital Annexe, Mumbai, India

P.K. Pradhan, MD, DNB Professor, Department of Nuclear Medicine and PET, Sanjay Gandhi Postgraduate Institute of Medical Sciences, Lucknow, India

J. Mohan Roop, MD Junior Resident, Department of Nuclear Medicine and PET, Postgraduate Institute of Medical Sciences and Research, Chandigarh, India

Gowri Sankar, MD Asst.Prof., Department of Nuclear Medicine, Sanjay Gandhi Postgraduate Institute of Medical Sciences, Lucknow, India

Sunita Sonavane, DRM, DNB Radiation Medicine Centre, Bhabha Atomic Research Centre, Tata Memorial Hospital Annexe, Mumbai, India

Madhvi Tripathy, MD, DNB Asst. Prof., Department of Nuclear Medicine and PET, All India Institute of Medical Sciences, New Delhi, India

1 Positron Emission Tomography: An Overview

Birendra Kishore Das

In this era of evidence-based medicine, both anatomical as well as functional imaging is necessary. Computed Tomography (CT) and MRI provide excellent structural details of diseased organs, and conventional nuclear medicine procedures complement the anatomical and structural details by providing mostly quantitative functional information. Positron Emission Tomography (PET) has added a new dimension to the existing modalities.

PET deals with biochemistry and metabolic changes that occur at molecular level. Hence, PET differs fundamentally from other imaging modalities. CT imaging is based on tissue density, whereas MRI conveys anatomic information based on proton density and proton relaxation dynamics. CT and MRI are useful in clinical diagnosis only when the disease process has caused significant anatomic alterations. However, in most diseased conditions, chemical changes precede anatomic changes that can be detected by PET technology. Thus, PET can provide earliest and unique information about ongoing disease process long before anatomic or structural changes take place. There is no other modality available at present that can replace PET technology. Although PET produces cross-sectional images like that obtained in MRI or CT, they represent circulation, function and metabolism, and not anatomical structure. PET is extremely sensitive, measuring quantitatively the concentration of tracers in nano- to picomolar range. Thus, PET enables merger of biochemistry and biology in medicine, giving birth to molecular medicine that focuses on identifying the molecular errors of disease leading to developing molecular corrections, including gene therapy. Molecular imaging with PET has been playing a role in examining the biological nature of a diseased condition and its characterization to guide selection and evaluation of treatment.

This is the reason why PET facilities (Fig. 1.1) have been growing exponentially in the United States and elsewhere in the world, including even in developing countries like India where the majority of the patients are extremely poor. PET centers are coming up rapidly, although mostly in metropolitan cities. In coming years, it is expected that the price structure may become favorable so that such facilities are made available in many smaller cities throughout the country.

How the Scan Is Done?

To conduct the scan, a short-lived radioactive tracer is injected into the body of the patient (usually into blood circulation). The tracer is chemically incorporated into a biologically active molecule. There is a waiting period while the active molecule becomes concentrated in tissues

B.K. Das, MD, ANM, FAR, FAMS
Director and Chief of Nuclear Medicine,
Utkal Institute of Medical Sciences,
Jaydev Vihar, Bhubaneswar, India
e-mail: bkishoredas@yahoo.com

B.K. Das (ed.), *Positron Emission Tomography: A Guide for Clinicians*,
DOI 10.1007/978-81-322-2098-5_1, © Springer India 2015

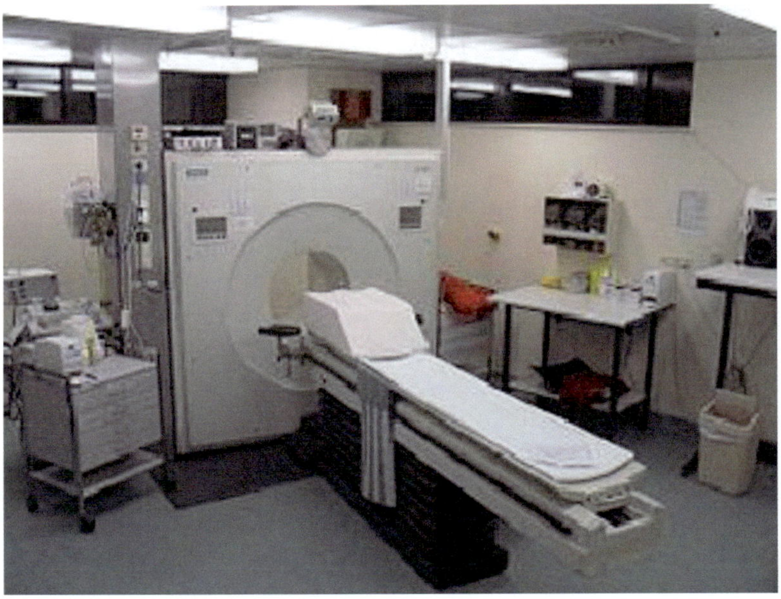

Fig. 1.1 View of a typical positron emission tomography (PET) machine

of interest; then the patient is placed in the imaging scanner. The molecule that is most commonly used for this purpose is fluorodeoxyglucose (FDG), a sugar, for which the waiting period is typically an hour. During the scan, a record of tissue concentration is made as the tracer decays.

As the radioisotope undergoes positron emission decay (also known as positive beta decay), it emits a positron, a particle with the opposite charge of an electron. After traveling up to a few millimeters, the positron encounters and annihilates with an electron, producing a pair of annihilation (gamma) photons moving in opposite directions. These are detected when they reach a scintillator in the scanning device, creating a burst of light which is detected by photomultiplier tubes or silicon avalanche photodiodes (Si APD). The technique depends on simultaneous or coincident detection of the pair of photons; photons which do not arrive in pairs (i.e., within a timing window of few nanoseconds) are ignored.

The most significant fraction of electron–positron decays result in two 511 keV gamma photons being emitted at almost 180° to each other; hence, it is possible to localize their source along a straight line of coincidence (also called the line of response or LOR). In practice, the LOR has a finite width as the emitted photons are not exactly 180° apart. If the recovery time of detectors is about 1 ps rather than about 10 ns, it is possible to localize the event to a segment of a cord, whose length is determined by the detector timing resolution. As the timing resolution improves, the signal-to-noise ratio (SNR) of the image will improve, requiring fewer events to achieve the same image quality. This technology is available on most new systems.

Image Reconstruction Using Coincidence Statistics

Using statistics collected from tens-of-thousands of coincidence events, a set of simultaneous equations for the total activity of each parcel of tissue along many LORs can be solved by a number of techniques, and thus a map of radio activities as a function of location for parcels or bits of tissue (also called voxels) may be constructed and plotted. The resulting map shows the tissues in which the molecular probe has become concentrated, and can be interpreted by a nuclear medicine physician in the context of the patient's diagnosis and treatment plan (Figs. 1.2 and 1.3).

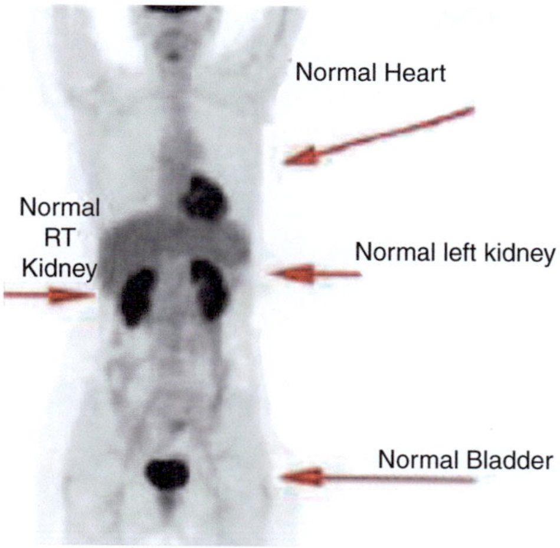

Fig. 1.2 Normal PET Scan

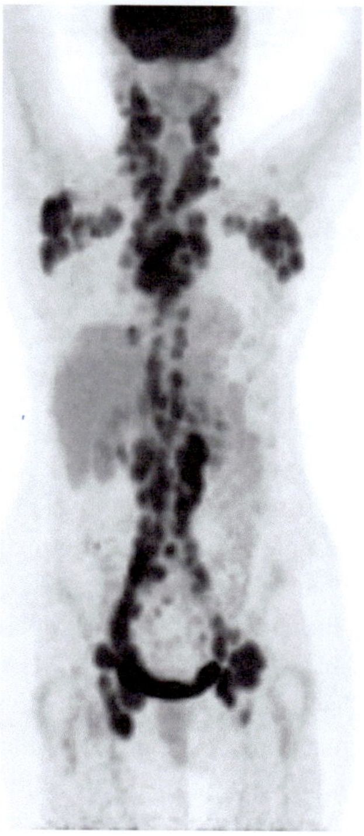

Fig. 1.3 PET scan showing malignant lymph nodes

Combination of PET with CT and MRI

PET scans are increasingly read alongside CT or MRI scans, the combination ("co-registration") giving both anatomic and metabolic information (i.e., what the structure is, and what it is doing biochemically). Because PET imaging is most useful in combination with anatomical imaging, such as CT, modern PET scanners are now available with integrated high-end multi-detector-row CT scanners. Because the two scans can be performed in immediate sequence during the same session, with the patient not changing position between the two types of scans, the two sets of images are more-precisely registered, so that areas of abnormality on the PET imaging can be more perfectly correlated with anatomy on the CT images. This is very useful in showing detailed views of moving organs or structures with higher anatomical variation, which is more common outside the brain. The same procedure is also done with MRI in place of CT (Fig. 1.4).

Radioisotopes

Radio nuclides used in PET scanning are typically isotopes with short half lives such as carbon-11 (~20 min), nitrogen-13 (~10 min), oxygen-15 (~2 min), and fluorine-18 (~110 min). These radio nuclides are incorporated either into compounds normally used by the body such as glucose (or glucose analogues), water or ammonia, or into molecules that bind to receptors or other sites of drug action. Such labeled compounds are known as radiotracers. It is important to note that PET technology can be used to trace the biologic pathway of any compound in living humans (and many other species as well), provided it can be radio labeled with a PET isotope. Thus, the specific processes that can be probed with PET are virtually limitless, and radiotracers for new target molecules and processes are being synthesized all the time. There are already dozens in clinical use and hundreds applied in research. Due to the short half lives of most radioisotopes, the radiotracers must be produced using a cyclotron and radiochemistry laboratory that is in close proximity to the PET imaging facility. This adds actually to the most part of the high initial and maintenance cost of the facility.

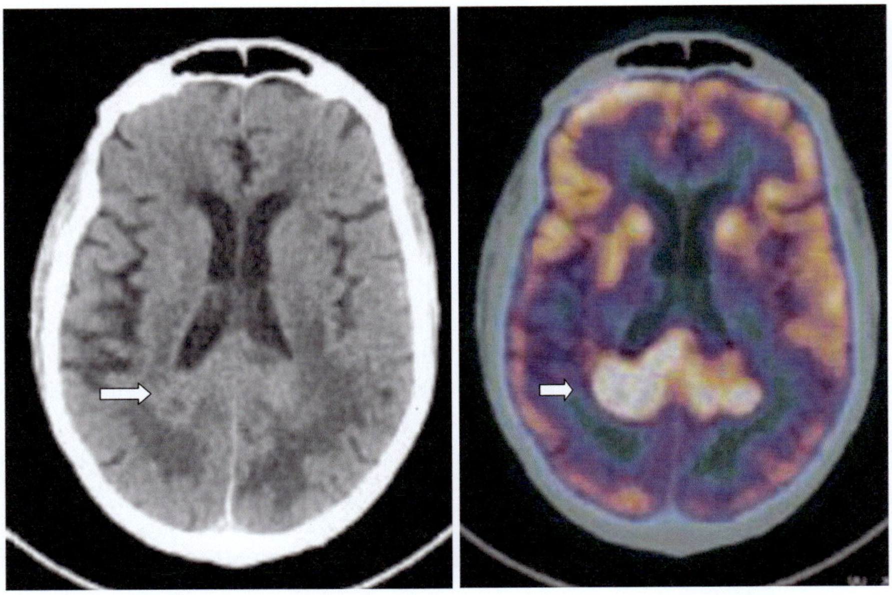

Fig. 1.4 Transaxial plain CT (*left image*) and fused F-18 FDG PET-CT (*right image*) of a 51-year-male with a mass involving the corpus callosum posteriorly on either side of midline (see *arrow*)

Limitations

As mentioned above, limitations to the widespread use of PET arise from the high cost of the cyclotron needed to produce the short-lived radio nuclides for PET scanning and the need for specially adapted on-site chemical synthesis apparatus to produce the radiopharmaceuticals.

However, the minimization of radiation dose to the subject is an attractive feature of the use of short-lived radio nuclides. Besides its established role as a diagnostic technique, PET has an expanding role as a method to assess the response to therapy, in particular, cancer therapy where the risk to the patient from lack of knowledge about disease progress is much greater than the risk from the test radiation.

Applications of PET

PET is both a medical and research tool. It is used heavily in clinical oncology (medical imaging of tumors and the search for metastases), and for clinical diagnosis of certain diffuse brain diseases such as those causing various types of dementias. PET is also an important research tool to map normal human brain and heart function. Some of the main uses are as follows:

In Oncology

PET scanning with the tracer fluorine-18 (F-18) fluorodeoxyglucose (FDG), called FDG-PET, is widely used in clinical oncology. This tracer is a glucose analog that is taken up by glucose-using cells and phosphorylated by hexokinase (whose mitochondrial form is greatly elevated in rapidly-growing malignant tumors). A typical dose of FDG used in an oncological scan is 200–400 MBq for an adult human. Because the oxygen atom that is replaced by F-18 to generate FDG is required for the next step in glucose metabolism in all cells, no further reactions occur in FDG. Furthermore, most tissues (with the notable exception of liver and kidneys) cannot remove the phosphate added by hexokinase. This means that FDG is trapped in any cell which takes it up, until it decays, since phosphorylated sugars, due to their ionic charge, cannot exit from the cell (Fig. 1.5).

This results in intense radiolabeling of tissues with high glucose uptake, such as the brain, the liver, and most cancers. As a result, FDG-PET can be used for diagnosis, staging, and monitoring treatment of cancers, particularly in Hodgkin's disease, non-Hodgkin's lymphoma, and lung cancer. Many other types of solid tumors will be found to be very highly labeled on a case-by-case basis—a fact which becomes especially useful in searching for tumor metastasis, or for recurrence after a known highly-active primary tumor has been treated.

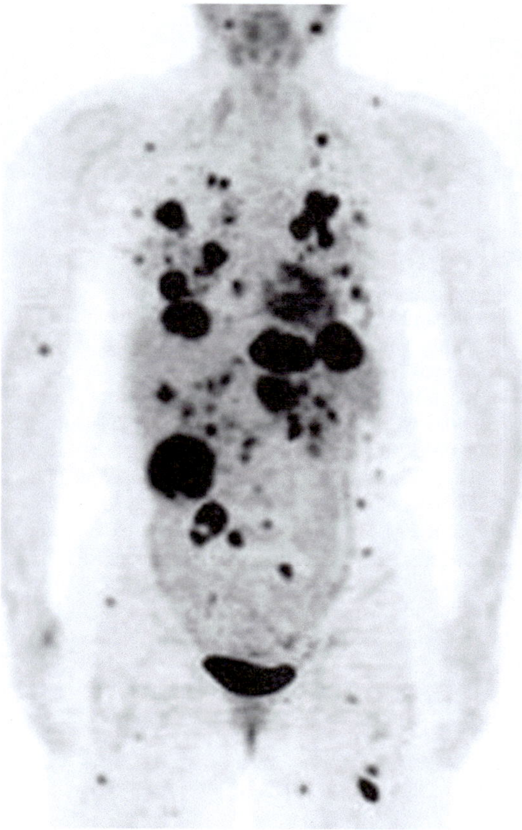

Fig. 1.5 Image of a whole body FDG-PET study showing extensive metastases throughout the body

In Neurology

PET neuroimaging is based on an assumption that areas of high radioactivity are associated with brain activity. What is actually measured indirectly is the flow of blood to different parts of the brain, which is generally believed to be correlated, and has been measured using the tracer oxygen-15. However, because of its 2-min half-life, O-15 must be piped directly from a medical cyclotron for such uses, and this is difficult. In practice, since the brain is normally a rapid user of glucose, and since brain pathologies such as Alzheimer's disease greatly decreases brain metabolism of both glucose and oxygen in tandem, standard FDG-PET of the brain, which measures regional glucose use, is successfully used to differentiate Alzheimer's disease from other dementing processes, and also to make early diagnosis of Alzheimer's disease. PET imaging with FDG can also be used for localization of seizure focus: A seizure focus will appear as hypometabolic during an interictal scan and as hypermetabolic focus during an attack. Several radiotracers (i.e., radioligands) have been developed for PET that are ligands for specific neuroreceptor subtypes such as ^{11}C raclopride and ^{18}F-fallypride for dopamine D2/D3 receptors, ^{11}CMcN 5652 and ^{11}CDASB for serotonin transporters, or enzyme substrates (e.g., 6-FDOPA for the AADC enzyme). These agents permit the visualization of neuroreceptor pools in the context of several neuropsychiatric and neurologic illnesses (Fig. 1.6).

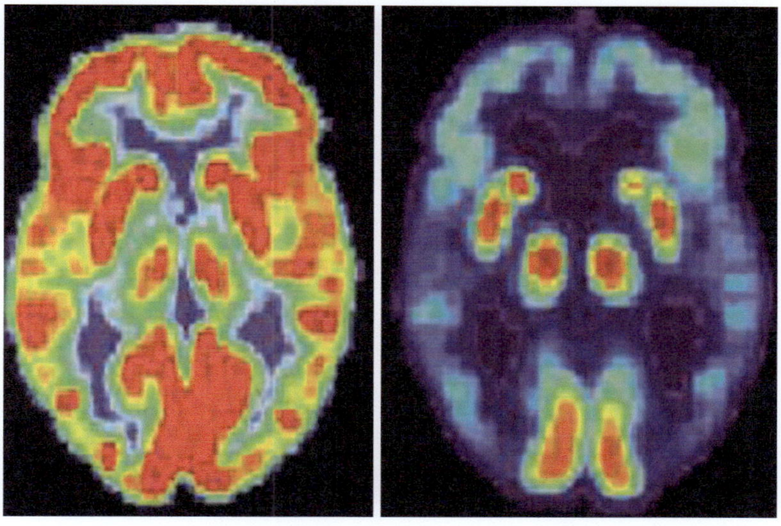

Fig. 1.6 PET image of normal brain (*left image*) and brain of an Alzheimer's patient (*right image*)

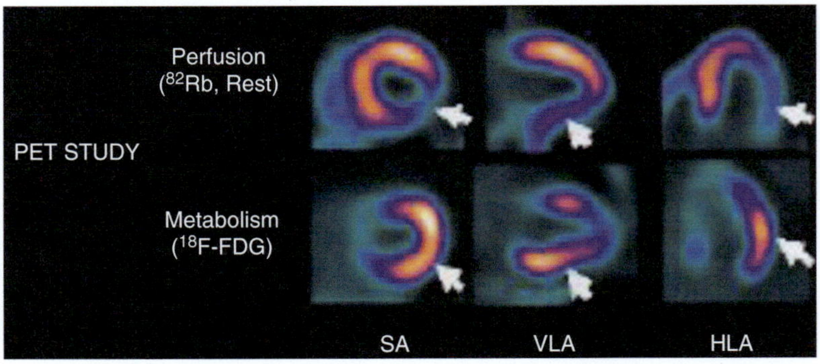

Fig. 1.7 PET study showing 'Hibernating Myocardium' (*arrow*) which can be salvaged by revascularization. *SA* short axis, *VLA* vertical long axis, *HLA* horizontal long axis

In Cardiology

In clinical cardiology, FDG-PET can identify so-called "hibernating myocardium," which can be activated by revascularization. It is also increasingly used for myocardial viability studies. Recently, a role has been suggested for FDG-PET imaging of atherosclerosis to detect patients at risk of stroke (Fig. 1.7).

PET is an evolving technology. Many more uses are expected to develop in the future, particularly in drug research and molecular medicine.

For Further Reading

1. Coleman RE, Tesar RD. A perspective on clinical PET imaging. Clin Positron Imaging. 2000;2(6):297–9.
2. DeGrado TR, Turkington TG, Williams JJ, Stearns CW, Hoffman JM, Coleman RE. Performance characteristics of a whole-body PET scanner. J Nucl Med. 1994;35:1398–406.
3. Ell PJ. The contribution of PET-CT to improved patient management. Br J Radiol. 2006;79:32–6. Q.
4. Rohren EM, Turkington TG, Coleman RE. Clinical applications of PET in oncology. Radiology. 2004;231:305–32.
5. Townsend DW, Carney JP, Yap JT, Hall NC. PET-CT today and tomorrow. J Nucl Med. 2004;45:4S–14.

2 Development of Positron Emission Tomography (PET): A Historical Perspective

Birendra Kishore Das

PET imaging was developed more than 30 years ago. It was too expensive and difficult a modality and was used only for research. During the late 1980s and the 1990s, the clinical applications of PET became apparent. PET imaging is now a routine clinical procedure. The number of patients having PET scans is increasing rapidly, and the projections are that the number of patients who will have PET imaging will continue to increase in future (Fig. 2.1).

Many eminent individuals who have made major contributions to the field of nuclear medicine have been involved in the development of PET. Some of these names include Brownell, Anger, Phelps, Muehllehner, Ter-Pogossian, Hoffman, Mullani, and others. Although many individuals attempted to do PET imaging in the 1950s and 1960s, Phelps et al., working at the Mallinckrodt Institute of Radiology in the early 1970s, developed the initial PET scanner that led to PET imaging as we know it today. They demonstrated the quantitative nature of PET imaging and the ability to study multiple physiologic parameters using PET imaging. The original scanner was a single slice device with a resolution of 17 mm. The initial detectors were made of sodium iodide, and the detectors were in a hexagonal array around the patient. The banks of detectors scanned across the patient, the gantry rotated, and the banks of detectors scanned back across the patient. Although this system was crude by the standards of today, it did demonstrate the power of PET. A lot of improvements in scanner technology have occurred since that time, and improvements are continuing to be made. The commercial scanners today have a 15–30 cm axial field of view and a 4–5 mm intrinsic resolution.

Muehllehner et al. attempted to perform coincidence imaging using opposed gamma cameras in the mid 1970s. The group was working at Searle Radiographics. Anger had proposed using two gamma cameras for coincidence imaging as early as 1959. The electronics of the gamma cameras were too slow and the available computers were too limited for the technology to be effective. The electronics of gamma cameras have changed dramatically since the mid 1970s and more sophisticated computers have become available. The thickness of the crystals in gamma cameras was also increased so that more counts could be obtained from radiation emanating from the patient.

In between, marked improvements in camera-based PET have occurred since the introduction of camera-based PET. The counts in an image are 5–10 times greater today than they were with the systems introduced earlier.

While PET instrumentation has continued to improve since its development in the early 1970s, the other major development that facilitated clinical PET was the introduction of fluorine-18 2-fluoro-2-deoxyglucose (FDG) by Al Wolf and

B.K. Das, MD, ANM, FAR, FAMS
Director and Chief of Nuclear Medicine,
Utkal Institute of Medical Sciences,
Jaydev Vihar, Bhubaneswar, India
e-mail: bkishoredas@yahoo.com

B.K. Das (ed.), *Positron Emission Tomography: A Guide for Clinicians*,
DOI 10.1007/978-81-322-2098-5_2, © Springer India 2015

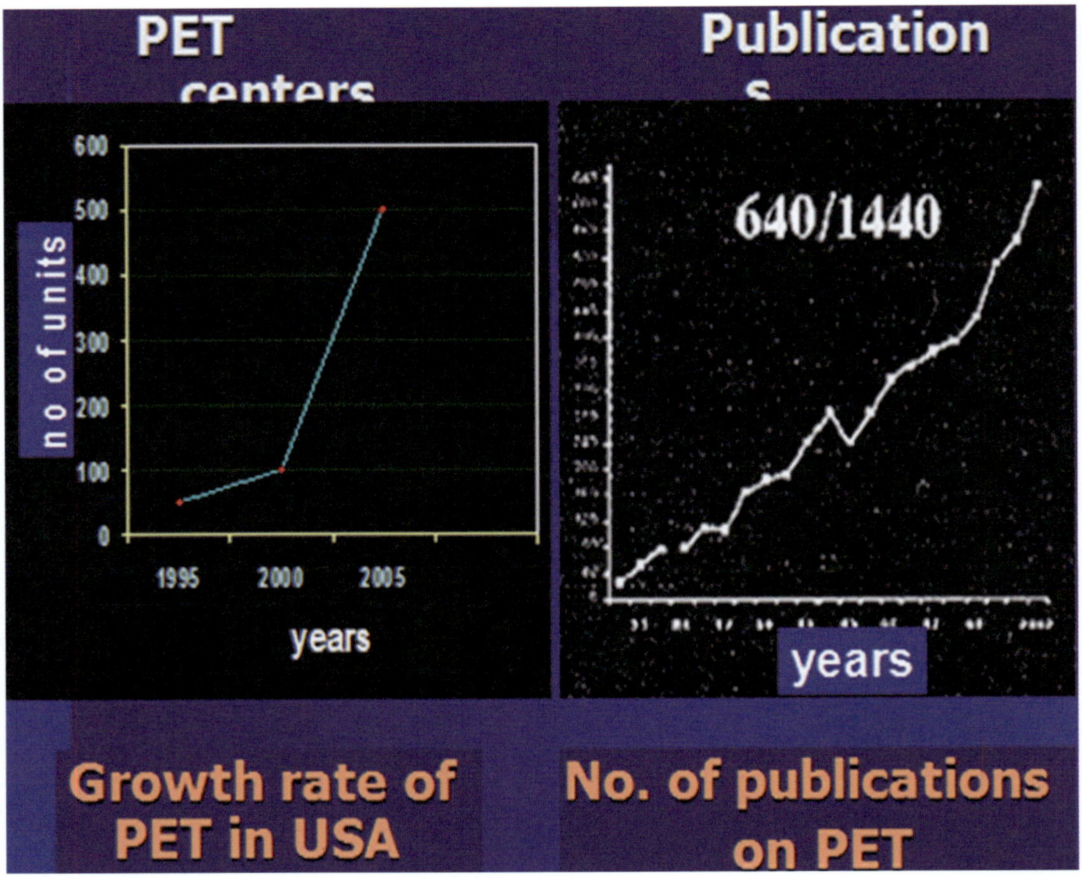

Fig. 2.1 Number of PET centers in the United States showing exponential growth, which is similar in many countries, including India in recent times (*left*). Growth in number of scientific publications related to PET (*right*)

colleagues at Brookhaven National Laboratory. Carbon-14 2-deoxyglucose was used for many years to quantify glucose metabolic rate using autoradiography. Louis Sokoloff and colleagues at the NIH demonstrated the information available from using this radiopharmaceutical in characterizing lesions in experimental models in the brains of animals. With the introduction of FDG, the data from the experimental models using carbon-14 2-deoxyglucose could now be extrapolated to patient studies. The combination of quantitative PET imaging and FDG started the road to the clinical applications of PET. FDG is handled by the body like glucose. It crosses cell membranes through the glucose transporters and is phosphorylated to FDG-6 phosphate by hexokinase. Once phosphorylated, it is no longer metabolized by most tissues. The major exception occurs in the liver. The high concentration of phosphatase enzymes in the liver results in dephosphorylation and clearance of the FDG from the liver. More than 70 years ago, biochemists demonstrated that tumors use more glucose than normal tissue. The glucose transporters and hexokinase enzymes are increased in tumors compared to normal tissue, and thus there is greater accumulation of FDG in tumors than normal tissue.

Issues Related to Regulatory Agencies

Wider clinical applications of PET were apparent in the early 1990s. These applications included determination of myocardial viability, evaluation of refractory seizure disorder, evaluation of dementia, and evaluation of brain tumors. Attempts to obtain reimbursement for these studies were met with resistance from third-party payers and the regulatory agencies. Third-party payers would ask if the radiopharmaceutical FDG was FDA-approved, and FDA approval had not been obtained. Asking the FDA to develop a New Drug Application (NDA), the process was extremely difficult. Because the drug was not proprietary and because there was no pharmaceutical firm backing its development, no industry was available to support obtaining an NDA. The FDA had guidelines for developing drugs, but they had no flexibility in approving drugs that would be made and used locally. At that time, the individual PET centers were making the fluorine-18 in their cyclotrons and synthesizing the radiopharmaceutical on site. The FDA procedures were designed for pharmaceutical manufacturing and required current good manufacturing practice (cGMP) standards. Very few, if any, PET centers in the United States could meet the standards required by the FDA. However, through the efforts of the Institute for Clinical PET and a group of dedicated individuals, Methodist Medical Center in Peoria, Illinois, did receive an NDA for FDG. The NDA was for the limited application of the evaluation of refractory seizure disorder.

Because of the uniqueness of PET radiotracers and because of the inability of the FDA to have flexibility in regulating PET radiopharmaceuticals, the US Congress made changes in the Food and Drug Administration through the Food and Drug Administration Modernization and Accountability Act (FDAMA) that was passed in November 1997. In this Act, it was stated that PET radiopharmaceuticals listed in the US Pharmacopoeia would have the equivalence of FDA approval until a new regulatory mechanism was developed for PET radiopharmaceuticals. The Act stated that the PET community and the FDA would have 2 years to develop a regulatory mechanism for PET radiopharmaceuticals. The PET community would then have 2 years to comply with these new regulatory guidelines. The PET community worked since then diligently with the FDA in developing appropriate regulatory guidelines. Finally, FDA approved FDG as a general oncologic imaging agent, as also an agent that can be used in identifying myocardial viability, and an agent that can be used for identifying sites of abnormalities in the brains of patients who have refractory seizure disorder and are being considered for surgery. Nitrogen-13 ammonia has been approved as a myocardial perfusion agent, and F-18 fluoride as a bone-imaging agent. Generator-produced rubidium-82 has been approved as a myocardial perfusion imaging agent. The FDA has been in the process of evaluating O-15 water as a perfusion-imaging agent, F-18 fluorodopa as an agent for evaluating movement disorders, and FDG as an agent for evaluating dementia.

After FDAMA was passed by Congress in 1997, the Health Care Financing Administration (HCFA) approved FDG PET imaging for two indications: the evaluation of indeterminate solitary pulmonary nodules and the initial staging of non-small cell lung cancer. These indications were approved for all devices approved by the FDA for marketing as PET scanners. Thus, dedicated and camera-based PET scanners were covered. In January 1999, a Town Hall Meeting was held at the HCFA to discuss further indications for PET imaging and added following indications for PET imaging: recurrent colorectal cancer with rising CEA, recurrent malignant melanoma, and staging and restaging of Hodgkin and non-Hodgkin lymphoma.

In July 2000, a document was submitted to the HCFA as an application for broad coverage of PET imaging. The document included more than 600 articles on approximately 25,000 patients who had FDG PET scans. This document demonstrated that PET had a sensitivity of 84 % and specificity of 88 % in all malignancies. In those studies that evaluated change in management, 32 % of the patients had their management changed based on the result of the PET scan. Among the 8,004 patients who had both PET and CT scans, PET was 19 % more sensitive and 13 % more specific than CT scanning. The PET community met with the HCFA in November 2000 to discuss the application for broad coverage. The HCFA published a Decision Memorandum on December 15, 2000. In this document, the

HCFA noted they would be covering the following indications: diagnosis, staging, and restaging of six malignancies: non-small cell lung cancer, esophageal cancer, colorectal cancer, lymphoma, melanoma, and head and neck cancer as also presurgical evaluation of refractory seizure disorder and myocardial viability after an inconclusive SPECT. The document also noted that the indication for breast cancer and dementia were being referred to the Medicare Coverage Advisory Committee.

Representatives of the PET Community met with the Imaging Subcommittee of the Medicare Coverage Advisory Committee in June 2001. At that meeting, the Imaging Subcommittee recommended coverage of recurrent metastatic breast cancer.

With the coverage of PET scans by the HCFA and other third-party payers, the use of PET has increased tremendously.

Optimal Conditions for PET Imaging

For the FDG PET scan to be most effective, the patient has to be well hydrated with low blood sugar and low insulin levels. For the nondiabetic patient, this can be achieved with drinking water but having no caloric intake for 4 h prior to the administration of the FDG. If patients are not diabetic and have eaten more recently than 4 h prior to FDG administration, the study can be delayed for at least 3 h after caloric intake.

For diabetic patients, the blood sugar level should be as low as reasonably achievable. PET scan is performed if the blood sugar is less than 200. The patient should not have any subcutaneous regular insulin for 4 h prior to the FDG administration. If insulin levels are high, either from subcutaneous administration or endogenous release after a glucose load, the administered FDG will go into muscle and will not be available to tumor for accumulation. Thus, it is important that the patient is fasting and has not had any subcutaneous regular insulin within 4 h of the FDG administration.

Usually, patients have the FDG administered at dose of 140 μCi/kg with a minimum adult dose of 10 mCi and maximum adult dose of 20 mCi. The pediatric dose is 140 μCi/kg. The delay time from injection to imaging is a minimum of 30 min for brain imaging and 45–60 min for whole body imaging. A further delay in imaging for whole body studies may improve the target-to-background ratio and improve lesion detection. A possible limitation of delayed imaging is that standardized uptake values have generally been obtained with a delay of 45 min to 1 h. A further delay in imaging will increase the SUV value, and the values differentiating malignant lesions from benign lesions have not been determined for the further delay.

For brain imaging, a 3D acquisition mode is used. A 6–8-min acquisition is obtained, which results in excellent image quality. For whole body imaging, a 2D-acquisition mode is typically used. However, with better scatter correction algorithms, 3D acquisitions are also routinely used. For the 15-cm axial field-of-view scanner, 5–6 bed positions are needed to acquire images from the base of the skull through the midthigh. Images of the lower extremity are only obtained if the patient has known or suspected disease in the lower extremity. These images are obtained in a 2D mode and emission scans are obtained. Attenuation correction is not needed for the extremities.

PET as an Evolving Technology

Continuous changes in instrumentation and radio pharmaceuticals are occurring. New detector materials are being used that provide higher sensitivity for the annihilation radiation and provide better photon energy determinations. These characteristics result in shorter imaging times and the ability to use 3D acquisition because of the separation of the non-scatter from non-scattered events. Thus, 3D imaging has been done more reliably with these new detectors that include LSO and GSO.

Another major advance in the instrumentation of PET includes the combination of PET and CT and MRI. The addition of CT/MRI to PET imaging has several advantages. The attenuation scan using CT is basically noise free and can be done faster than using the germanium-68 rods that have typically been used for transmission imaging. The CT scan provides excellent anatomic information, which is easily registered with the PET information. As reconstruction algorithms improved, the anatomic information from the CT scan could be incorporated into the reconstruction of the PET data. The CT scan also provides an excellent way to incorporate the PET information into the radiation therapy planning. Intensity modulated radiation therapy is being increasingly used in the treatment of cancer, and the PET information provides more information than the CT scan that is currently used in radiation therapy planning. The role of PET and PET CT in radiation therapy planning has been evaluated at many medical centers and well documented.

Several new radio pharmaceuticals are on the horizon. For the radio pharmaceuticals to be clinically applicable, they need to be labeled with fluorine-18. The 110-min half-life of F-18 permits widespread distribution of these radio pharmaceuticals. The potential radio pharmaceuticals of interest include F-18, fluoro-L-thymidine. This radio pharmaceutical measures cell replications, and it is now being used widely. F-18 fluorodopa is being used for the evaluation of movement disorders such as Parkinson's Disease. F-18 fluorocholine is incorporated into cell membranes, and this radio pharmaceutical shows promise in detecting prostate cancer. FDG is not accumulated in most primary prostate cancers or metastatic lesions, and fluorocholine has demonstrated superiority to FDG in detecting prostate cancer and metastases.

Concluding Remarks

PET imaging is an elegant technology in nuclear medicine and is becoming an important modality in the practice of nuclear medicine. The use of PET will continue to increase in the next several years, and will become the major study performed in a nuclear medicine service. The oncologic applications of PET will continue to predominate the type of studies performed, but other applications such as evaluation of dementia and infection are likely to become prevalent. PET imaging will have an important role in the future of nuclear medicine.

For Further Reading

1. Ter-Pogossian MM. The origins of positron emission tomography. Semin Nucl Med. 1992;22(3):140–9. Review.
2. Wagner Jr HN. A brief history of positron emission tomography (PET). Semin Nucl Med. 1992;22(3): 140–9. Review.

3 Planning of A PET Facility in India: Regulatory Aspects

Birendra Kishore Das

The ideal situation for establishing a PET facility is to have both the scanner as well as the cyclotron required for onsite production of positron emitters. However, this situation may prevail mainly in large medical institutions with sufficient funding/investment. Most of the centers may not be able to afford an onsite cyclotron, which requires huge investment and running expenses. In this chapter, we will discuss mainly a facility as the representative model for FDG whole body scan with the FDG purchased from an off-site vendor. Depending on the scanner, 5–20 mCi will be needed per patient.

With a single state-of-the-art scanner, approximately 15–20 patients could be scanned per 8-h day. If additional capacity is needed, all of the numbers can be multiplied by the number of scanners. Depending on the scanner used, between 20 and 60 min should be allocated for a scan. A typical scan will be approximately 35 min. Since the uptake period is 45–60 min and some additional time is needed with the patient prior to injection, it may be advisable to have more patient prep rooms in order to keep the scanner busy. In an 8-h day, approximately two scans per hour or sixteen scans can be acquired. Space for the prototypical PET suite then will consist of more than two prep rooms, one scanner room, one scan control room, and a reading room. Each of these spaces will now be considered separately.

The scan room, of course, requires sufficient space for the scanner. Around the scanner, there must be sufficient space to bring a patient in comfortably so that they can be transferred easily onto the scanner bed. There may also be occasions when anesthesiology is required; so sufficient space for anesthesia equipment should also be available. In the scanner room, a good amount of storage will be required for such things as sheets, needles, restraint bands, etc. A sink and small hot cell with a dose calibrator are also required.

Each prep room requires enough space for the patient to comfortably recline in a chair and the technologist and perhaps a second person to sit in the room. Injections will take place in this room, as well as histories or any other interaction with the physician prior to the scan. A small amount of storage is needed.

The scan control room is where the scanner will be operated and where images will be processed and evaluated for quality control. For this, two computer systems are required—one to primarily operate the scanner and the other to perform all other processing duties. The control room also functions as the central space for the PET center. This is where the technologists will spend their time when not specifically with a patient and where physicians and others interested in the study will meet. Because of this, the control room

B.K. Das, MD, ANM, FAR, FAMS
Director and Chief of Nuclear Medicine,
Utkal Institute of Medical Sciences,
Jaydev Vihar, Bhubaneswar, India
e-mail: bkishoredas@yahoo.com

B.K. Das (ed.), *Positron Emission Tomography: A Guide for Clinicians*,
DOI 10.1007/978-81-322-2098-5_3,

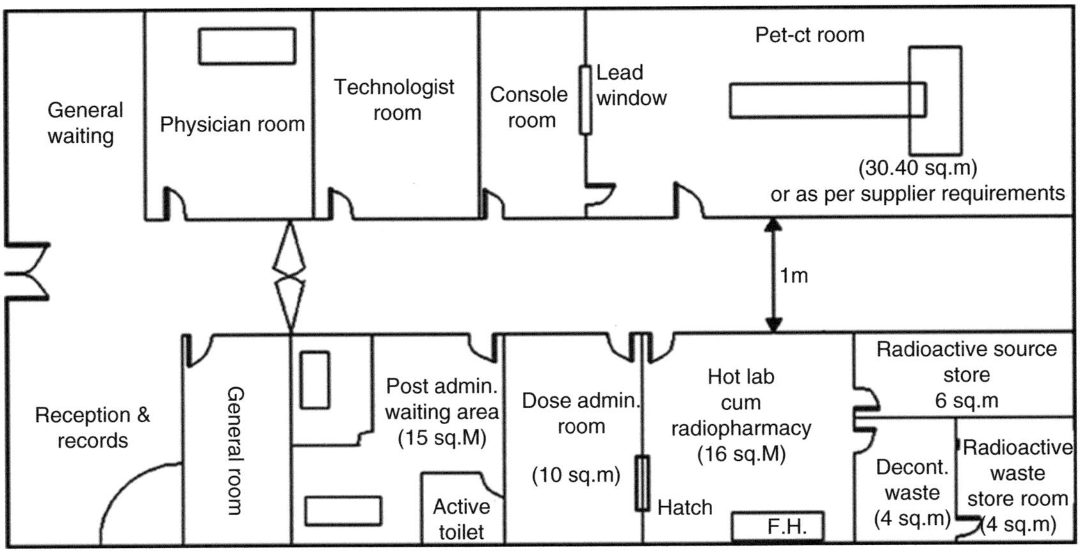

Fig. 3.1 A typical layout of a positron emission tomography (PET) facility

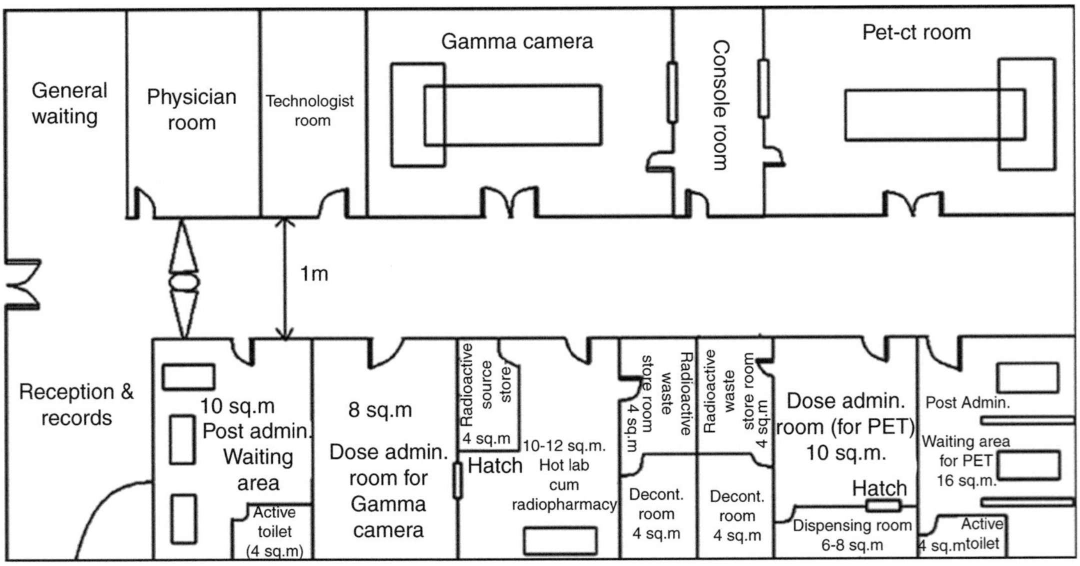

Fig. 3.2 Layout plan of a nuclear medicine facility having both gamma camera and positron emission tomography (PET) facility

needs much more space than the minimum required to house two computer systems.

The scan interpretation room—reading room—requires enough space for a computer system and two or three seats with easy viewing access to the computer monitor. Enough space may be kept in view of the fact that results may have to be explained in this room to referring physicians. In addition to the computer system, medical records pertinent to the PET scanning will be stored in this room.

Typical layout plans as suggested by AERB are given below (Figs. 3.1 and 3.2)

Several different models of PET scanners can be considered for the PET center. However, for all practical purposes, any new center will opt for a PET-CT machine. PET-CT scanners have the advantage of the very fast CT acquisition, and hence little time is required for transmission imaging. Transmission scanning with ^{68}Ge rod sources requires approximately 15 min for the whole body. This is compared with the CT scan that requires only 30 s. This timesaving translates into approximately one more patient able to be scanned per hour.

Traditionally, the PET scanner has been made with either NaI (now obsolete) or BGO detector material. BGO has been the most common material because it has greater stopping power than sodium iodide and it can count faster. Newer scanners are beginning to use new detector materials, including LSO, GSO, and YSO. These new detector materials promise better energy resolution and better count rate performance. The energy resolution improvement is used to reduce scatter; however, the practical improvement is not very significant. The count rate performance, on the other hand, is likely to impact imaging. With the newer detector materials, the timing of an event is more precisely determined. This allows the coincidence timing window to be shortened and thereby fewer random events are recorded. This would allow a scanner to perform higher count rate studies in 3D mode. So, if the scanner is planned for uses other than FDG whole body imaging, such as cardiac imaging, scanners with the new detector materials should be preferred.

The personnel needed to operate a PET center include technologists, a scheduling person, and the physician reading the images. Depending upon the workload, at least four technologists may be required to operate the PET-CT facility. One technologist will be dedicated to operating the scanner. His duties will be to set up the scanner, make sure the acquisition is proceeding correctly, process the data afterward, and make sure all studies are archived. Two technologists are required to handle patients in the prep rooms. The fourth technologist will act as the Supervisor helping out as needed and performing all of the scheduling and administrative tasks. The physician who would read PET scans may need a part-time radiologist to read the CT part of the PET-CT scans.

Because cyclotron-produced radionuclides have higher energy than the other routinely used radio nuclides for diagnosis, it is necessary to know about the regulatory requirement and radiation safety precautions that have to be taken for the installation of this modality. The various stages of approval of PET-CT facility by the Atomic Energy Regulatory Board (AERB) have to be followed for planning the facility.

Plan Approval by AERB

Two copies each of the proposed layout plan, site plan, and elevation drawing of the facility indicating the floor, nature of occupancy around, above and below, has to be submitted in 'B3' size paper (353 × 500 mm^2) along with the application form AERB/RSD/NMF/SLA (downloadable from www.aerb.gov.in) has to be submitted to the Head, Radiological Safety Division (RSD), AERB. The plan has to clearly indicate the dimension of each of the rooms associated with the facility of the NM department. When the user has to plan the laboratory, it is required that the arrangement of the various rooms associated with the facility has to follow the principle of low active area to high active area, that is, entrance of the facility should have reception/general waiting area, and at the end, hot laboratory cum radiopharmacy/radioactive waste storage area is to be planned. The typical layout plans for the facility can be downloaded from the website and may be referred to design the PET-CT facility alone or PET-CT facility along with the gamma camera facility with respect to the arrangement/allocation of rooms and area requirement.

Submission of Regulatory Consent Form

After approval of the layout plan, the user has to submit the details of the completion of the construction work as per the approved plan, installation of equipments, procurement of radiologic protection accessories, enrollment of radiation workers in Personal Monitoring Services, and availability of qualified staff as per AERB Safety Code shall be intimated to Head, RSD, AERB, by submitting the Regulatory Consent form no. AERB/444-NM/RC-FORM (can be downloaded from www.aerb.gov.in). The Radiological Safety Officer (RSO) as per the qualification mentioned in the AERB Safety Code AERB/SC/MED-IV (Rev-1 2001) has to be nominated by the employer for the NM facility by submitting the application form No. AERB/441/RSOM-II/III (downloadable from www.aerb.gov.in).

Radiation Protection Aspects of PET-CT Facilities

PET-CT uses relatively large activities of high-energy photon-emitting radioisotopes. This, coupled with the current dose limits for members of the public, can result in a shielding requirement. Even modest reductions in the radiation levels at 511 keV require significant amounts of shielding. A number of accessory equipments are required to ensure appropriate level of shielding. Some of such equipments are listed below:

- Automated dose dispenser to reduce the dose to the technologist while dispensing the unit dose for patient administration.
- Dose calibrator with thick lead shielding to reduce technologist exposure during the dose assay.
- Well counters with external shields to reduce background from the stored doses.
- Sources in the scanners and calibration sources.
- Tungsten syringe shields to reduce finger dose during injection.
- Remotely actuated syringes that keep the syringe totally enclosed in a shield while the operator delivers the dose by pushing on an extension rod.
- Extra thick L-block table-top shields (5 cm of lead compared with 1.2 cm of lead in standard NM applications)
- Syringe carriers for transporting the dose from one room to another.
- Apart from the above, shielding accessories following radiation monitoring devices are also mandatory:
- GM-based survey meter/ionization chamber-based survey meter
- Contamination monitor
- Pocket dosimeters
- Isotope dose calibrator.

As in NM facility, radioisotopes should be stored, used, and transported safely and securely all the time. The radiopharmaceuticals should be prepared, handled, administered to the patients, and disposed of in a safe manner, taking into account adequate radiation protection measures. Any unusual event that has resulted or has the potential to result in overexposure to the workers or public should be reported to the AERB. Annual Safety Status report of the facility has to be submitted to AERB in the prescribed format at the end of each calendar year.

To meet the growing requirement of PET studies in India, both the provider of the facility and the involved authorities of the government have to cooperate and help each other for the benefit of the patients.

For Further Reading

1. Tandon P. Regulatory requirements for designing PET-CT facility in India. Indian J Nucl Med 2010; 25:39–43
2. Brunetti J, Caggiano A, Rosenbluth B, Vialotti C. Technical aspects of positron emission tomography/computed tomography fusion planning. Semin Nucl Med 2008;38:129–36
3. Guidelines to Set up a Medical Cyclotron Facility. Available at: www.aerb.gov.in

4 PET in Comparison with Other Modalities

Prathamesh Joshi and Sandip Basu

Introduction

Cancer is one of the leading causes of morbidity and mortality worldwide. Various imaging modalities like computed tomography (CT), ultrasonography (USG), magnetic resonance imaging (MRI) based on anatomic information play a vital role for the management of cancer patients; however, they have a notable limitation of inability to identify morphologically normal but functionally abnormal tissues. This limitation is overcome by functional imaging modalities such as Positron Emission Tomography (PET) and Single Photon Emission Computed Tomography (SPECT). Fusion imaging in form of PET-CT has integrated the functional and morphological information in one investigation and has emerged as one of the most useful modality in practice of oncology. It has changed dramatically the management of numerous cancers and its impact has been felt worldwide.

In this chapter, we present the basics of PET-CT imaging along with a brief review of other functional imaging modalities. F-18 fluoro-2-deoxy-D-glucose (FDG) is the most frequently used PET radiopharmaceutical today and is the workhorse of PET-CT imaging. Hence the chapter primarily focuses on FDG with mention of other clinically relevant tracers as required.

Apart from oncology, FDG PET-CT has an established role in neurological diseases, myocardial viability assessment and is evolving as a promising modality for infection and inflammation imaging. However, description of these applications is beyond the scope of this chapter.

Definitions of Equipments

(a) *PET-CT scanner:* An integrated device containing both a CT scanner and a PET scanner with a single patient table and therefore capable of obtaining a CT scan, a PET scan, or both. Structure of PET-CT scanner is shown in Fig. 4.1. An example of current generation of PET-CT scanner is shown in Fig. 4.2.

(b) *PET-CT fusion*: is the combined display of registered PET and CT image sets. Superimposed data typically are displayed with the PET data color coded to the CT data in gray scale. An example is shown in Fig. 4.3.

P. Joshi, DRM, DNB
Consultant, Radiation Medicine Centre,
Bhabha Atomic Research Centre, Tata Memorial Hospital Annexe, Mumbai, India

S. Basu, MD (✉)
Professor, Radiation Medicine Centre,
Bhabha Atomic Research Centre,
Tata Memorial Hospital Annexe, Parel,
Mumbai, India
e-mail: drsanb@yahoo.com

B.K. Das (ed.), *Positron Emission Tomography: A Guide for Clinicians*,
DOI 10.1007/978-81-322-2098-5_4, © Springer India 2015

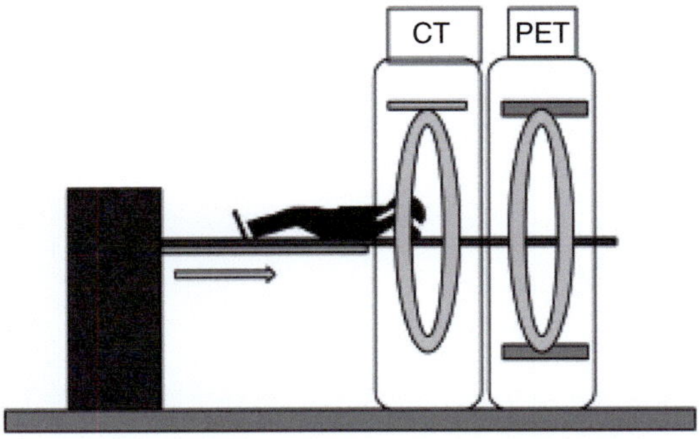

Fig. 4.1 Schematic diagram of PET-CT scanner. The gantries of CT and PET are sequentially arranged with a common patient table

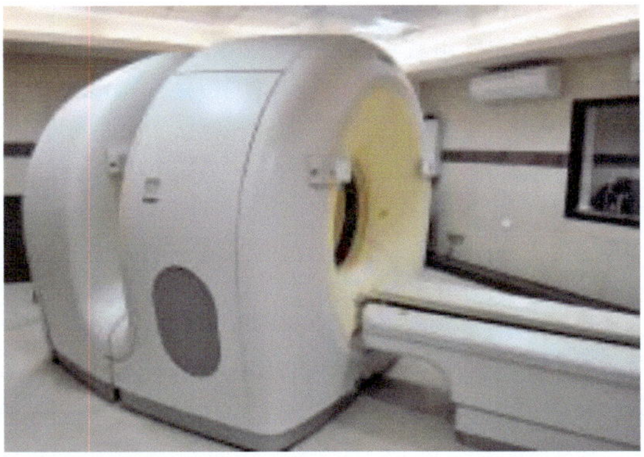

Fig. 4.2 Prototype of current generation of Time of Flight PET-CT (GEC)

Principles of FDG PET-CT Imaging

In the 1930s, Otto Warburg (1883–1970) reported that anaerobic metabolism of glucose is a fundamental property of all tumors. The phenomena known as the "Warburg effect" was described by Warburg during his lifetime of work into cellular metabolism and respiration, for which he was awarded the Nobel Prize in 1931. He also demonstrated a relationship between the degree of anaerobic metabolism and tumor growth rate. Today, the "Warburg effect" forms the basis of tumor imaging with FDG PET-CT.

FDG is a radio pharmaceutical analog of glucose that is taken up by metabolically active tumor cells facilitated by (glucose transporters GLUT), which is similar to that used by glucose (Fig. 4.4). The rate of uptake of FDG by the tumor cells is proportional to their metabolic activity. Like glucose, it undergoes phosphorylation to form FDG-6-phosphate; however, unlike glucose, it does not undergo further metabolism, thereby becoming trapped in metabolically active cells. This trapped radiotracer FDG in the tumor sites is detected by the PET detectors and is represented in the PET images as "hot spots".

Scanning Technique of FDG PET-CT

Patient Preparation

Patients are required to fast for approximately 4–6 h prior to PET-CT. They can have plain water during this period. Before intravenous injection of FDG, the blood glucose level is measured; a level of less than 150 mg/dL is desirable. Good control of blood glucose is essential because the uptake of FDG into cells is competitively inhibited by glucose, as they use a common transport mechanism (glucose transporters GLUT) facilitating transport into both normal and tumor cells. Patients are also instructed to avoid any kind of strenuous activity prior to the examination and following injection of the radioisotope to avoid physiologic muscle uptake of FDG. FDG can be administered orally if intravenous access is not available. The same delay time before imaging used with intravenous injection can be employed with oral administration.

Intravenous fluids containing dextrose or parenteral feedings should be withheld for 4–6 h. When intravenous contrast material is to be used as part of CT protocol, patients should be screened for a history of iodinated contrast material allergy and renal disease. Intravenous contrast should not be administered when the serum creatinine level is above 2.0 mg/dL. For CT scan of the abdomen or pelvis, an intraluminal gastrointestinal contrast agent may be administered to provide adequate visualization of the gastrointestinal tract unless it is medically

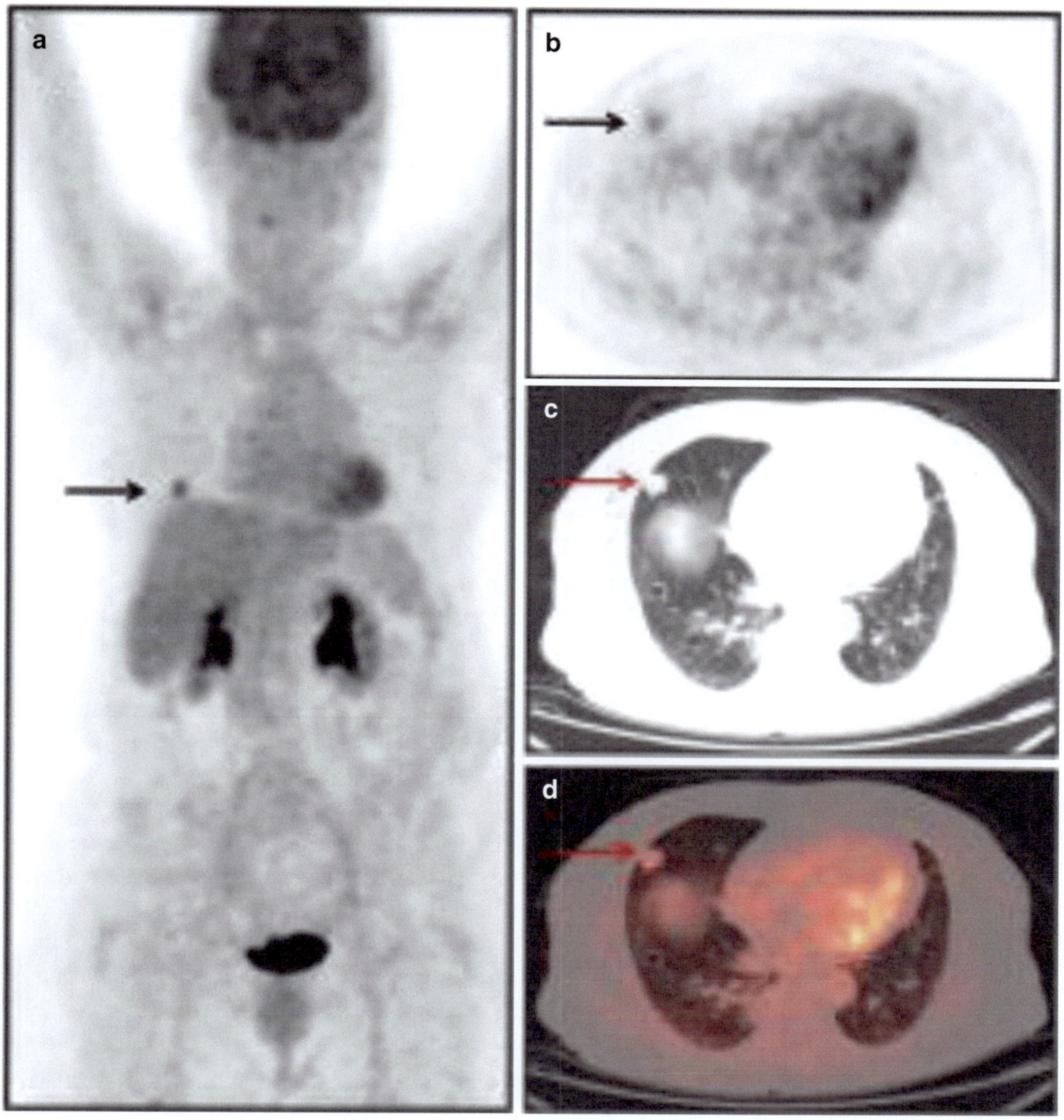

Fig. 4.3 Maximum intensity projection (MIP) image of PET-CT. This image is 3D reconstruction of whole body distribution of FDG (**a**). The PET image showing the lesion (*arrow*) (**b**) shows focal uptake in right side of chest (**c**), CT image reveals 1.5 × 1.1 cm nodule in right lung. Fused PET-CT image localizes the uptake to right lung nodule, suggesting neoplastic pathology (**d**). Biopsy confirmed it to be non-small-cell lung carcinoma (*arrow*)

contraindicated or unnecessary for the clinical indication.

As both PET and CT procedures involve use of radiation, pregnancy should be ruled out before injection of FDG in female patients. Nursing mothers are advised that breastfeeding should be stopped for 6–12 h.

Scan Acquisition

The PET-CT scan is typically acquired 45–60 min after FDG injection. The acquisition consists of PET emission scan and CT transmission scan. For most oncology indications, skull base-to-mid thigh tumor imaging is sufficient. Extended

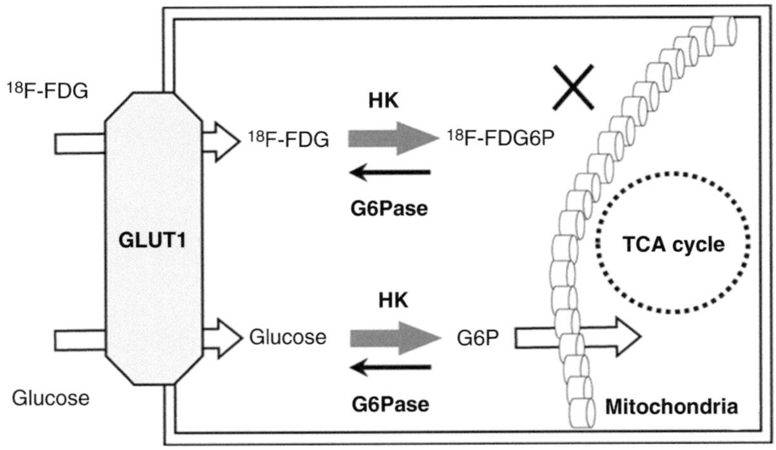

Fig. 4.4 Mechanism of metabolic trapping of FDG

whole-body examinations are performed in tumors that show a high probability of metastases in the head, skull, brain, and in the lower extremity.

Radiation Exposure

The radiation dose of FDG is approximately 6–8 mSv for an administered activity of 10 mCi. The effective CT dose could range from 1 to 20 mSv depending upon the CT protocol used.

Quantitative Analysis of Metabolism

Semiquantitative estimation of tumor glucose metabolism by use of the standardized uptake value (SUV) is based on relative lesion radioactivity measured on images corrected for attenuation and normalized for the injected dose and body weight, lean body mass, or body surface area.

SUV Is Defined by the Following Equation

SUV = Mean ROI activity (mCi/mL) × body weight (grams)/Activity administered (mCi), where ROI is a user-specified region of interest.

Other semiquantitative parameters include ratio of FDG uptake in a lesion to FDG uptake in internal reference regions, such as the blood pool, mediastinum, liver, and cerebellum.

Physiological Uptake Pattern of FDG and Variations

The distribution of FDG represents glucose metabolism in various organs. A physiological and variable FDG accumulation can be observed to a certain degree in most viable tissue: brain, myocardium (in which the FDG accumulation can be high in the fasting state), breast, liver, spleen, stomach, intestine, kidneys, urine, skeletal muscle, lymphatic tissue, bone marrow, salivary glands, thymus, uterus, ovaries, and testicles. The accumulation of FDG in urinary bladder is due to its excretion in urine. Figure 4.5a is example of normal, physiological FDG uptake in various organs.

The commonly seen variants of physiological uptake are brown adipose tissue uptake (Fig. 4.5b, arrows), increased skeletal muscle uptake due to muscular activity or food intake before the scan or in FDG uptake phase. Vocal cord uptake is seen if patient is talking (or a pediatric patient crying) in the post injection uptake phase. Myocardial FDG uptake is variable and can be very intense sometimes in the fasting state (Fig. 4.5c, arrow).

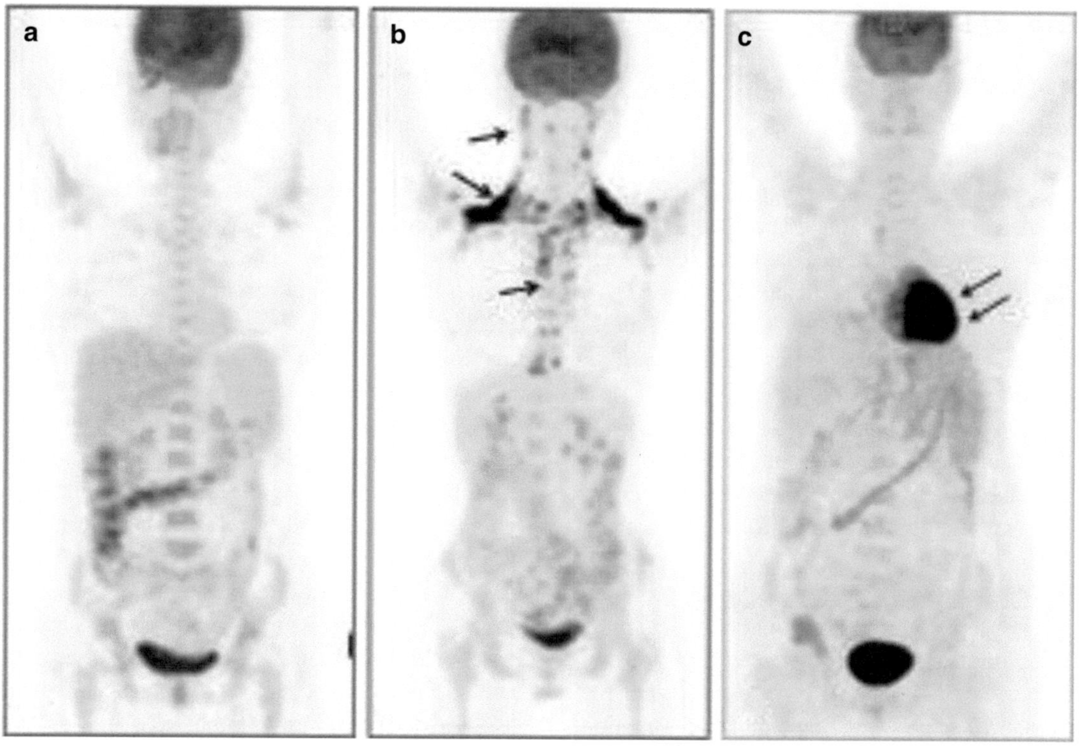

Fig. 4.5 (**a**) Physiological distribution of FDG in brain, bowel, kidneys, liver, urinary bladder, and bone marrow. (**b**) *Arrows* represent physiological uptake in brown adipose tissue (BAT) in cervical, supraclavicular, and paraspinal region. Without CT correlation such uptake can be misinterpreted as lymph nodal uptake. (**c**) *Arrow* shows intense FDG uptake in myocardium. Myocardial FDG uptake is variable and can be very intense sometimes in the fasting state

Table 4.1 Incidental FDG uptake in some tissues and possible interpretation

Incidental uptake	Suggested investigations	Comment
Focal uptake in thyroid	USG, TC Scan, FNA	Upto 1/3rd correlation (27–50 % in different studies) of focal thyroid uptakes are due to underlying neoplastic disease
Diffuse uptake in thyroid	Thyroid function tests, TC Scan, Anti-thyroid antibodies	Thyroiditis
Focal uptake in pituitary region	MRI and pituitary hormone evaluation	Functioning and non-functioning, micro as well as macro adenomas of pituitary may show focal tracer uptake Other causes are granulomatous diseases and very rarely metastases
Ovary in post-menopausal age	USG with tumor	Could be physiological if marker correlation seen in pre-menopausal age However in post-menopausal age malignancy should be suspected

Owing to its whole body imaging ability, occasionally incidental FDG uptake is seen in various tissues, which can be unrelated to the pathology under evaluation and such uptake may eventually diagnose second primary neoplasm or a benign pathology that may warrant therapeutic intervention. The list of such conditions is ever-growing and few of such incidentally seen uptakes that need further investigations are listed in Table 4.1.

Uses of FDG PET-CT

Because of the complementary information of metabolic activity and anatomical details supplied by PET and CT, respectively, together they possess unique ability of diagnosis, staging, and monitoring of malignancy. This ability of PET-CT to guide management decisions has lead to a new era in "personalized medicine" (Table 4.2).

The uses of FDG PET-CT can be broadly divided in two categories—*established and investigational.*

FDG PET-CT has established its utility in

1. Evaluating the extent of disease in known malignancies (staging/restaging).
2. Evaluating disease response to chemotherapy or radiotherapy.
3. Detecting tumor recurrence in the presence of elevated tumor markers but no clinical or morphological evidence of disease.
4. Searching for an unknown primary when metastatic disease is the first clinical presentation.

At present, FDG PET-CT is an investigational modality for

1. Selecting tumor region for biopsy guidance.
2. Radiotherapy planning.

Staging Malignancies

Accuracy in staging permits selection of the most appropriate treatment. Identification of metastases that are not seen with conventional imaging (upstaging) avoids the high morbidity and cost of treatments that cannot benefit the patient and permits more effective choices. Likewise, metabolic reclassification of malignant lesions to benign (downstaging) allows patients with false-positive anatomic findings but limited disease to receive potentially curative treatment.

PET-CT has been shown to alter stage of the malignancy in up to 39 % of patients [4]. Upstaging (25 %) was observed more frequently than downstaging (14 %) in this particular series. This has direct implications on type of therapy (surgical vs. medical), intent of therapy (curative vs. palliative), and economic burden of treatment.

FDG PET-CT is indicated in staging of non-small-cell lung carcinoma (NSCLC), esophageal carcinoma, Hodgkin's disease, non-Hodgkin's lymphoma, locally advanced breast cancer, and head, neck cancers. In colorectal carcinoma, FDG PET-CT is indicated for exclusion of disease elsewhere, in patients who are undergoing surgical resection of hepatic or pulmonary metastases with curative intent and characterization of equivocal lesions detected by conventional imaging.

Table 4.3 represents superior performance of PET-CT in staging as compared to conventional imaging (especially CT) in head and neck, lung, esophageal carcinomas and lymphomas. Examples of PET-CT for staging purpose are shown in Fig. 4.6.

Monitoring Response to Therapy

Once the diagnosis and stage of malignancy is established, the next challenge for clinicians is monitoring response to the therapy. This is extremely important considering the economic

Table 4.2 Illustrates characteristics of FDG PET-CT that are exploited for its use in specific clinical indications

No.	Characteristics	Used clinically in
1.	Exquisite sensitivity for GLUT expressing tumor cells	1. Identification of tumor in morphologically normal tissue 2. Differentiation of post therapy residual active disease from post therapy fibrosis
2.	Ability to image whole body status	Detection of unsuspected distant metastasis
3.	Quantification of metabolic activity in pathology	1. Response evaluation to therapy 2. Differentiation of benign Vs malignant lesions
4.	Correlation of metabolic activity with tumor biology and growth rate	1. Prognostication 2. Guiding tissue sampling from 'active' areas of tumor and increase diagnostic yield of biopsy 3. Radiotherapy planning

Table 4.3 Some examples of superior performance of PET-CT in staging as compared to conventional imaging (especially CT) in head and neck, lung, esophageal carcinomas and lymphoma

Sr No	Carcinoma	Sensitivity of PET/CT (%)	Specificity of PET/CT (%)	Sensitivity, of CT (%)	Specificity of CT (%)	Comment
1	Head & neck carcinoma – lymph nodal staging	94	84	78	84	PET/CT alters the initial clinical staging and TNM category o in 14–57 % of the patients when compared with CT-based evaluation alone,
2	Non small cell lung carcinoma					1. PET/CT is a more accurate method for staging of NSCLC; it changes the course of management in up to 52 % of cases and has a major role in reducing the number of futile thoracotomies
	(a) Operability	79	–	60	–	2. The high NPV of ^{18}F-FDG PET/CT (up to 97 %) for mediastinal disease has led to the recommendation to omit mediastinoscopy in patients with negative mediastinal ^{18}F-FDG PET/CT. However, special attention should be paid to central tumors, which have a high incidence of occult N2 disease. If PET/CT is positive, then mediastinoscopy is necessary to exclude a false-positive result
	(b) Mediastinal nodal metastasis	85	84	70	69	
3	Lymphoma staging (Hodgkin's and high grade NHL)	86	96	81	41	In patients with HL or high-grade NHL, the sensitivity and specificity of ^{18}F-FDG PET/CT for lymph node involvement are 94 and 100 %, respectively, while for organ involvement they are 88 and 100 %
4	Esophageal carcinoma-					1. PET/CT is superior imaging modality as compared to CT & endoscopic ultrasound for detection of distant metastases and it changes staging and management in around 1/3 rd of cases. Recent review article on this subject concluded that FDG PET/CT is the 'Gold standard' for evaluation of distant metastases in esophageal carcinoma.
	Loco-regional and distant nodal metastasis	43	98	46	90	2. In M-staging, the addition of PET/CT results in up-staging of 15–20 % and down-staging of 5–7 % of the patients. 3. In addition, synchronous primary tumors are identified in 5.5 % of patients, of which 75 % are not identified by conventional imaging

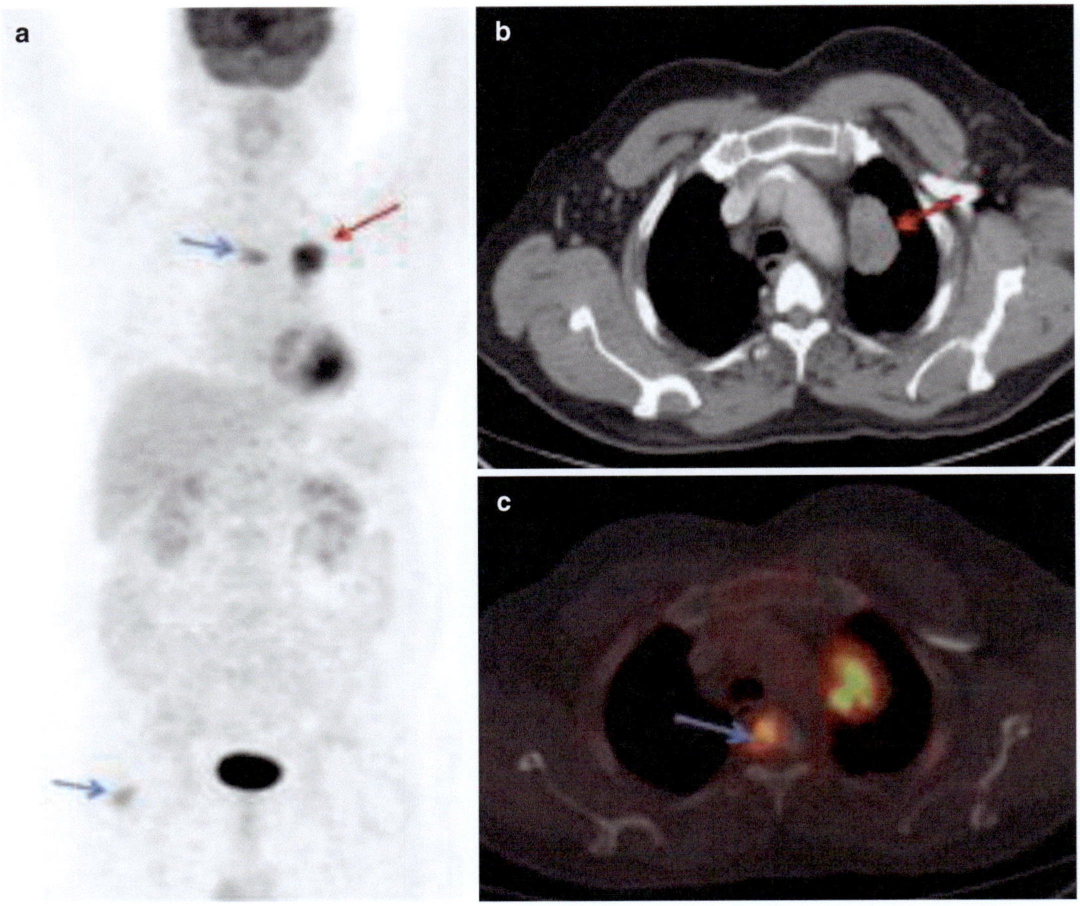

Fig. 4.6 Staging PET-CT of 41-year-old male patient, recently diagnosed case of left lung NSCLC (**a**). Conventional imaging reported the carcinoma as stage II neoplasm (**b**). In addition to hypermetabolic left lung upper lobe primary (*red arrow*), PET-CT showed metastatic FDG avid lesions in mid-dorsal vertebra and neck of right femur (*blue arrows*), upstaging it to stage IV disease (**c**)

burden of treatment and morbidity associated with cancer treatments. Conventional anatomical imaging with CT and MRI are limited in assessing response to therapy because

1. Inability to distinguish between residual fibrosis and residual active tumor
2. Anatomical regression usually lags behind metabolic cell death.

One of the strengths of FDG PET-CT is its ability to characterize metabolic activity of a lesion. This property is being used to detect early metabolic changes in patients who have received systemic or local anticancer treatment and categorize them as responders versus non-responders.

Therapy response PET-CT imaging can be divided into two categories: early and late prediction.

Early Prediction

The goal of imaging is to predict response early during therapy. In this case, the cancer should be one in which viable alternative therapies are available if the first-line therapy is ineffective. If there is a lack of response, the initial therapy can be changed early during the course of therapy. PET is particularly valuable as an early response to therapy often does not result in change on conventional imaging modalities.

Table 4.4 General guidelines for post-therapy scans

Intervention	Time interval after which which PET-CT should be done
Post-biopsy	1 week
Post-surgery	6 weeks
Post-chemotherapy	4–6 weeks or schedule scan just before next cycle
Post-radiotherapy	2–6 months
Colony stimulating factor (G-CSF) (see Fig. 4.7)	Short acting agent 1 week. Long acting-several weeks

Late Prediction

The goal of imaging is to evaluate response after the completion of therapy and predict future outcome. PET is particularly valuable as conventional imaging techniques are often not able to differentiate tumor from scar tissue.

It is useful to know the tumor SUV before the initiation of therapy to assess tumor grade and evaluate treatment response following radiation therapy or chemotherapy. It is important to standardize the time interval between injection of the radiotracer and the PET study because SUV variability with time has been well documented. Another important consideration during evaluation of response is timing of PET-CT scan. Treatment-related inflammatory response can alter the FDG PET-CT findings and lead to misinterpretation. Table 4.4 shows general guidelines for post-therapy scans.

The two neoplasms where FDG PET-CT has played a crucial role in response monitoring and has evolved as the cornerstone of management are gastrointestinal stromal tumors (GISTs) and lymphomas.

In lymphoma (Hodgkin's (HL) and high-grade NHLs), interim PET-CT is shown to be an important predictor of prognosis. Early interim ^{18}F-FDG PET-CT results (after two to four cycles) correlate well with event-free survival in HL and high-grade NHL. In high-grade NHL, the event-free survival at 2 years and 5 years has been reported to be 82 and 88.8 %, respectively, for negative interim PET patients in compassion to 43 and 16.2 %, respectively, for positive interim PET patients (Fig. 4.8).

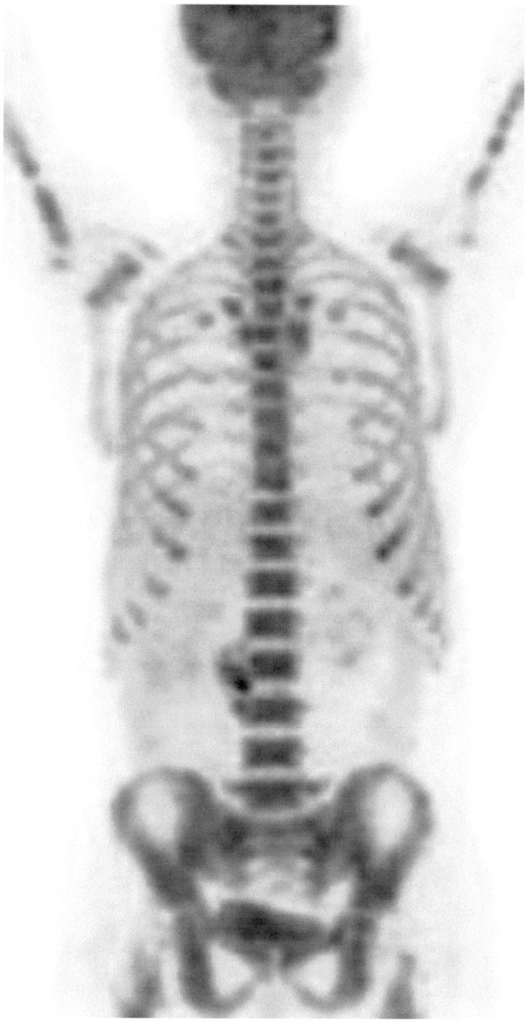

Fig. 4.7 Generalized increase in bone marrow FDG uptake due to G-CSF stimulation of marrow

Chemotherapy can be curative in a majority of patients of Hodgkin's and high-grade NHLs. A major problem in these patients is that they are often left with residual masses at the end of chemotherapy, and this may represent persistent disease or fibrosis. Residual post-therapy masses are seen in up to 85 % of the cases of HL and up to 40 % of the cases of NHL in conventional imaging on CT. Knowledge of persistent disease is critical since alternate treatments are available, including radiation, other chemotherapies, and even high-dose therapy with stem cell rescue. FDG PET has been shown to be more accurate at

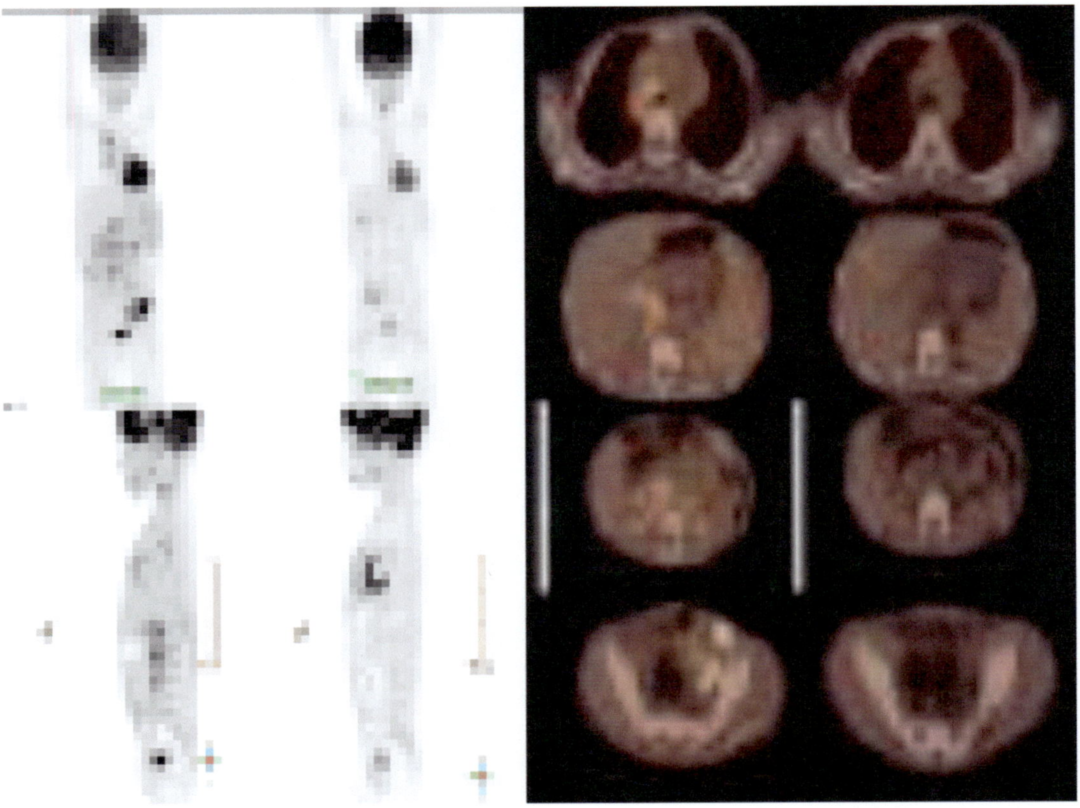

Fig. 4.8 Demonstrates interim PET-CT evaluation in a case of NHL (after 2 # of chemotherapy), which has produced major impact in the clinical management of lymphoma

end of chemotherapy response assessment as compared to CT. Figure 4.9 shows example of complete metabolic response to chemotherapy in a case of NHL.

The two advances that have revolutionized the management of GISTs are imatinib mesylate for targeted treatment and FDG PET-CT for accurate response monitoring. There is now convincing evidence that serial PET study is more sensitive and reliable for determining treatment response to imatinib mesylate in patients of GIST, when compared with only conventional CT monitoring. The findings of detection of disease recurrence on discontinuing imatinib and acquired resistance to imatinib provide insight into the issue of therapeutic endpoint definition.

In head-to-head comparison, PET is shown to be superior to CT in response assessment to imatinib in GIST patients. PET accurately diagnosed tumor response in 85 % of patients at 1 month and 100 % at 3 and 6 months. CT was found to be accurate in 44 % of patients at 1 month, 60 % at 3 months, and 57 % at 6 months.

Other malignancies where PET-CT has shown promise for this application are esophageal, colorectal, lung, and breast carcinoma.

Detecting Tumor Recurrence

PET-CT plays vital role in establishing and localizing disease sites as a cause for elevated serum markers in colorectal (Fig. 4.10), ovarian, and germ–cell tumors, as well as for restaging of proven recurrences in breast, cervix carcinomas, and melanomas. In thyroid carcinoma, FDG PET-CT is indicated in Thyroglobulin (Tg) Elevation but Negative Iodine Scintigraphy (TENIS Syndrome) to identify source of Tg. To highlight the importance of PET and PET-CT in

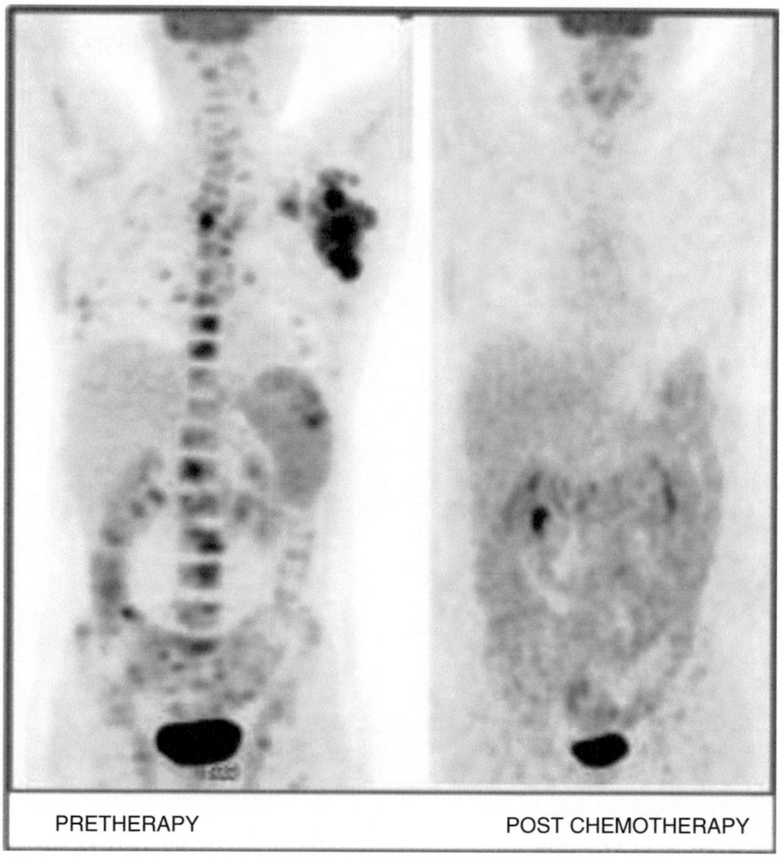

Fig. 4.9 A 30-year-old lady revealed a stage IV NHL on pretherapy MIP image of PET-CT with involvement of supradiaphragmatic lymph nodes, spleen, and bone marrow. Post chemotherapy with 6 cycles of R-CHOP regimen, PET-CT shows complete resolution of metabolically active disease

this scenario, example of colorectal carcinoma is described briefly in current chapter.

In colorectal carcinoma, serial measurements of carcinoembryonic antigen (CEA) are used to monitor the patients for recurrence. Imaging is warranted to detect and localize recurrence in patients with risen/rising CEA. Barium studies have been reported to detect local recurrence with accuracy in the range of 80 %, but are only 49 % sensitive for overall recurrence; hence, these are used relatively infrequently. CT has been the conventional imaging modality used to localize recurrence with an accuracy of 25–73 %, but it fails to demonstrate hepatic metastases in up to 7 % of patients and underestimates the number of lobes involved in up to 33 % of patients. In addition, metastases to the peritoneum, mesentery, and lymph nodes are commonly missed on CT, and the differentiation of postsurgical changes from local tumor recurrence is often equivocal.

FDG PET and PET-CT is known to be imaging modality of choice in this setting, especially when conventional imaging is negative or equivocal. A meta-analysis of 11 clinical reports and 577 patients determined that the sensitivity and specificity of FDG PET for detecting recurrent colorectal cancer were 97 and 76 %, respectively. A comprehensive review of the PET literature (2,244 patients studied) has reported a weighted average for FDG PET sensitivity and specificity of 94 and 87 %, respectively, compared to 79 and 73 % for CT (Fig. 4.11).

Carcinoma of Unknown Primary (CUP)

CUP is defined as histologically confirmed metastatic cancer with no identifiable primary site after a complete history and physical examination

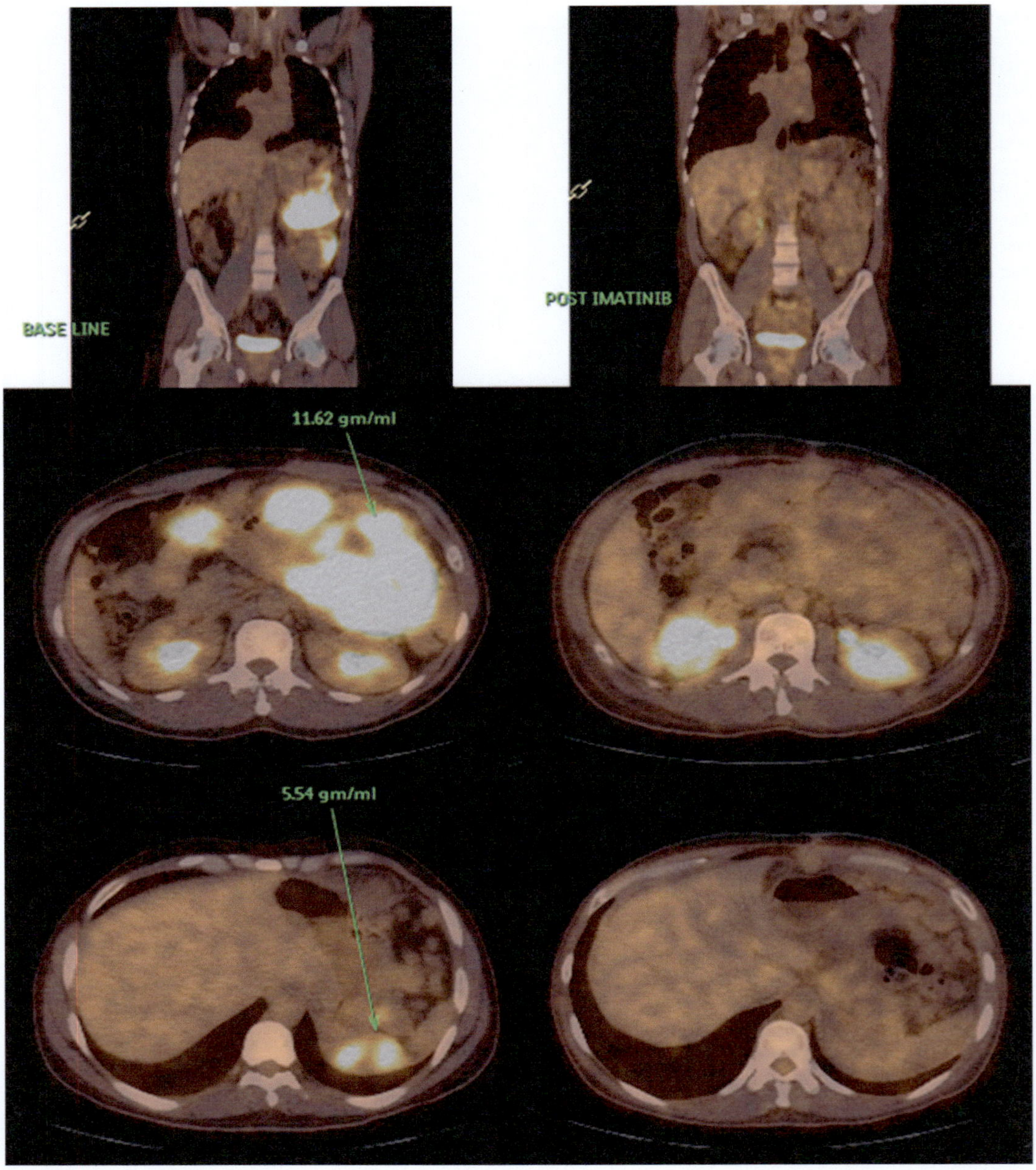

Fig. 4.10 Shows the utility of FDG PET-CT to assess metabolic response to imatinib mesylate in patients of GIST early in the course of therapy

(including a pelvic and rectal examination), complete blood count, biochemistry, urinalysis, occult fecal blood test, serum tumor markers, histopathology review of biopsy material with the use of immunohistochemistry, mammography in females, computed tomography (CT) of the chest, abdomen and pelvis, or any other relevant test. Because of its ability to image whole body and excellent sensitivity for majority of tumors, FDG PET-CT is a valuable tool in localizing a primary tumor in patients with CUP. FDG PET-CT can also identify additional sites of metastases that can alter the patient's management, which probably improves survival time and at the same time can serve as a guide for biopsies. FDG PET-CT is also useful for treatment monitoring following a

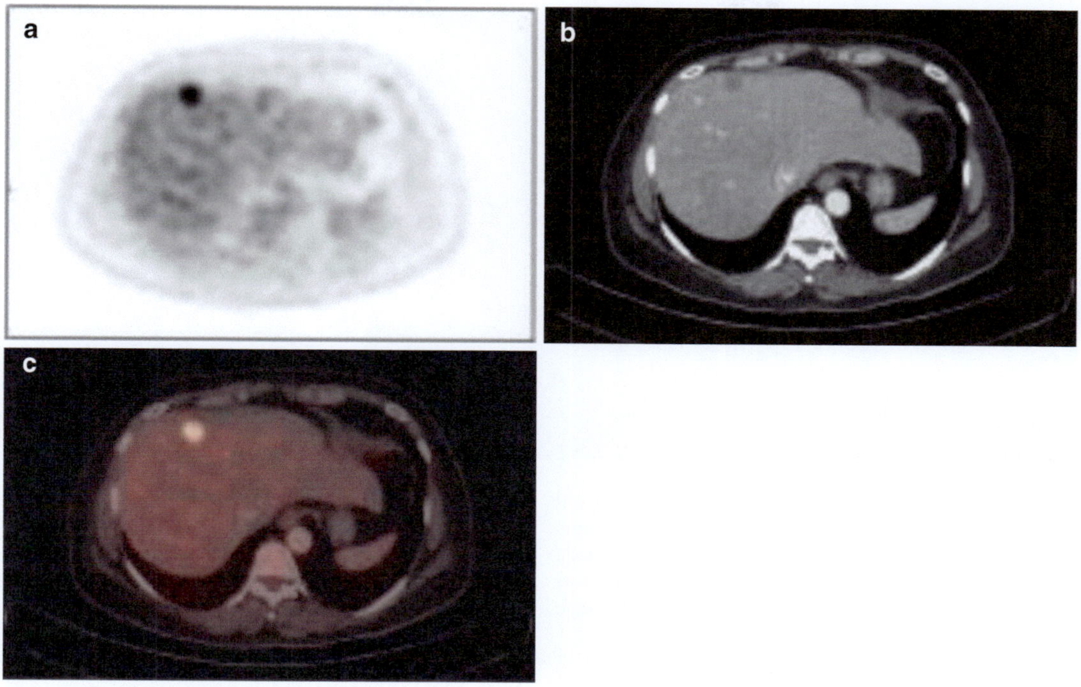

Fig. 4.11 The status of 40-year-old female post colorectal tumor resection, with rising CEA, Focal FDG uptake in liver on PET (**a**). Subcentimeter hypoenhancing lesion in right lobe of liver on CT (**b**). Fused FDG PET-CT image demonstrating focal FDG uptake in the CT demonstrable hypoenhancing area, suggesting hepatic recurrence as the cause of rising CEA (**c**)

therapeutic intervention. Various reviews have proposed that FDG PET should be the first imaging procedure performed in the case of cervical metastases since it is unlikely that other imaging modalities will identify a primary tumor that cannot be detected by FDG PET. If the FDG PET is positive, biopsies can be obtained from the suspected area. If the FDG PET is negative, panendoscopy with biopsy sampling can be performed.

As per published literature, performance of FDG PET-CT in CUP shows that

1. Conventional radiological imaging identifies about 20–27 % of the primary tumors after intensive investigation. FDG-PET and PET-CT detects the primary tumor in 30–60 % of patients with negative results in the conventional diagnostic procedures.
2. If the primary is not located by FDG-PET, it is usually not detected in further follow-up because of the high sensitivity and specificity of FDG-PET in general.
3. PET-CT PET findings-based investigations to diagnose primary tumor are shown to be more cost effective as compared to conventional approach, with approximately 50 % cost reduction.

Hence, there is convincing basis for the use of PET-CT at earlier point in patient investigation, rather than after exhausting other modalities. Figure 4.12 highlights the application of FDG PET-CT in CUP.

Miscellaneous

Apart from the indications described above, a few others deserve to be mentioned

Characterization of Solitary Pulmonary Nodule (SPN)

The prevalence of non-calcified pulmonary nodules in smokers or ex-smokers is 23–69 %, but only 1.4–2.7 % have malignant nodules. The Fleischner Society has published imaging guidelines for patients with pulmonary nodules that are smaller than 8 mm.

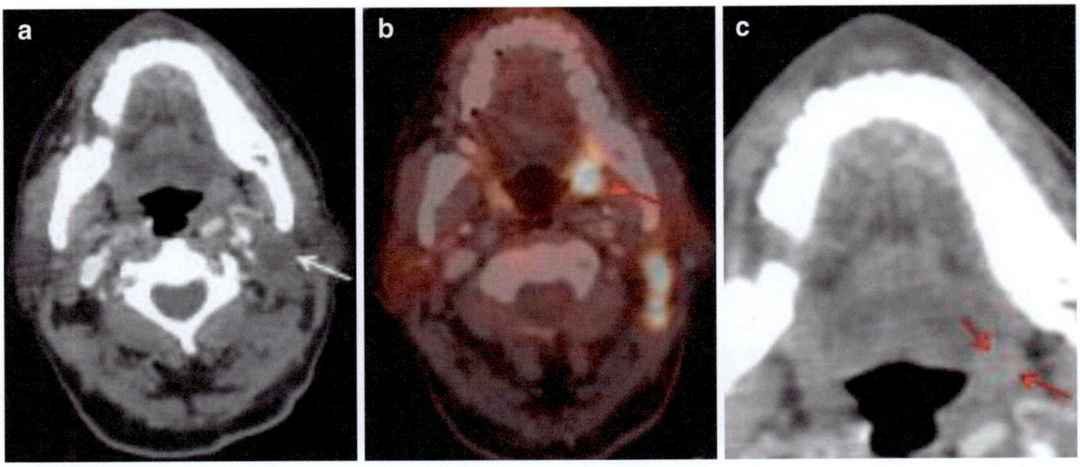

Fig. 4.12 Contrast enhanced CT (CECT) of 39-year old male diagnosed with metastatic squamous cell carcinoma of unknown primary origin. *Arrow* represents metastatic left level II cervical node (**a**). Conventional imaging was unable to locate primary neoplasm, Fused PET-CT image revealed an intense uptake in left gingivobuccal sulcus (GBS) suspicious of primary (**b**), Guided by PET findings, careful review of CECT reveals subtle enhancing lesion in left GBS, which later on confirmed as the primary site on histopathology (**c**)

PET-CT is more accurate than CT alone for characterizing pulmonary nodules, resulting in fewer equivocal findings, and higher specificity. Low-to-intermediate risk nodules of 8 mm should be evaluated by PET-CT, whereas high-risk nodules should be biopsied or excised.

The sensitivity of PET-CT for lung cancer is >90 %, so low-to-intermediate risk nodules that are metabolically inactive can be followed radiographically to ensure stability or resolution. The specificity of PET-CT is approximately 80 % because inflammatory nodules can be metabolically active. PET positive nodules require further investigation.

PET-CT with FDG has been shown to be cost-effective for characterizing lung nodules of 8 mm when risk estimates based on clinical versus morphological parameters are conflicting, or when indeterminate nodules are found in high-risk individuals.

Differentiating CNS Lymphoma from Toxoplasmosis in Immunocompromized Patients (Especially in HIV Patients)

Both toxoplasmosis and lymphoma are frequent CNS complications of AIDS, and it is not always possible to distinguish between the two at CT and MRI. FDG PET is considered very useful in this setting. CNS lymphoma is highly metabolically active, whereas toxoplasmosis is not. Quantitative assessment has shown that the standardized uptake values (SUV) of toxoplasmosis are significantly lower than those of lymphoma, with virtually no overlap between the uptake values of the two conditions.

Investigational Roles of PET-CT in Oncology

Because FDG uptake acts as the surrogate marker of tumor metabolism, visual analysis and/or quantitative analysis of FDG uptake (using SUV) can determine differential tumor cell activity in a lesion. This property is used in RT planning for "dose painting" of tumor and give a "boost" to those areas that show higher activity. Same principle is used in PET-CT guided biopsy to obtain the tissue from most active area of a neoplastic lesion and to improve the biopsy yield.

Limitations of FDG PET-CT

False-Positive Results

FDG, though extremely sensitive for detecting tumor, is not specific for neoplastic process. False-positive findings are most commonly associated with the uptake of FDG in infectious or inflammatory tissue.

Careful history, clinical examination and correlation with recent investigations along with knowledge of benign processes that concentrate FDG is essential to reduce false-positive results. Granulomatous diseases (especially in developing countries), post radiation inflammation and post chemotherapy "flare" are few of the common causes of misinterpretation.

False-Negative Results

The inherent spatial resolution limit of PET-CT scanner and tumors with low avidity for FDG are the two main reasons of false-negative results.

The limited reconstructed spatial resolution of 4–10 mm in available commercial systems means negative scan findings cannot exclude the presence of a smaller tumor or microscopic neoplastic disease. Hence, a negative result when clinical/biochemical suspicion of active disease is high, should be dealt with caution.

Tumors with a low metabolic rate (e.g., bronchoalveolar carcinoma and mucinous adenocarcinoma) may show insignificant uptake of FDG, and certain tumors are known to have poor avidity for FDG (well-differentiated thyroid cancer, prostate carcinoma, and hepatocellular cancer). FDG PET is not useful in the evaluation of possible cerebral metastases. High levels of FDG are normally present in the cerebral cortex and substantially limit the utility of FDG PET in this regard. Hence negative PET-CT for brain involvement should not be used to rule out brain metastases in certain tumors that have propensity of secondaries in brain, and relevant investigations like MRI should be performed in these scenarios.

Non-FDG Tracers in PET-CT

Although 18 F-FDG-PET-CT imaging provides high sensitivity and acceptable specificity in oncological imaging, it is recognized that FDG is not a "specific" radiotracer for imaging malignant disease. Highly "tumor-specific" and "tumor cell signal-specific" PET radiopharmaceuticals are being developed to meet the growing demand of radioisotope-based molecular imaging technology. In the last 15 years, many alternative PET tracers have been proposed and evaluated in preclinical and clinical studies to characterize the tumor biology more appropriately.

Table 4.5 provides an overview of various new PET agents that are finding applications in oncological practice. Out of these agents, 18 F—Na Fluoride (for bone imaging) and 11C—Choline (for prostate carcinoma imaging) have earned the FDA approval for routine use.

Future Directions

Apart from established functional imaging modalities like PET-CT and SPECT/CT, other functional imaging modalities that may play a role in oncology are PET/MRI, functional MRI, and contrast-enhanced ultrasound.

PET/MRI

PET/MRI is a new hybrid modality that is increasingly being used in clinical settings, although both clinical uses and technical optimization are still under evaluation. Like PET-CT, FDG is the radiopharmaceutical employed in PET/MRI modality. PET/MRI appeared in the

Table 4.5 Overview of various new PET agents that are finding applications in oncological practice

Biochemical process	PET agent	Uses
Bone metabolism	18 F-Na fluoride	Diagnosis of bone metastases
Cell membrane	11C-Choline, 18 F-Choline	Prostate carcinoma proliferation
Somatostatin	68 Ga- DOTATOC receptor imaging	Neuroendocrine tumor – staging, therapy response monitoring and planning Peptide receptor radionuclide therapy
Hormone receptors on breast	18 F-Estradiol	In vivo imaging of estrogen receptors

clinical setting in 2007. Current evidence shows overall good correlation between PET-CT and PET/MRI in lesion detection across most publications and oncologic diseases. In some tumors, PET/MRI may provide comprehensive TNM staging plus potentially improved N staging in the abdomen (i.e., pancreatic, colorectal, and gynecologic cancers). However, there is lack of clearly defined prospective studies with a sufficient number of patients to validate these findings.

Functional MRI

Current clinical techniques are based on microcirculation imaging using extracellular low-molecular-weight contrast agents such as gadopentetate dimeglumine and analogues. The temporal evolution of the enhancement visualizes the angiogenic properties of lesions with regard to vascular density and permeability, heterogeneity and changes during therapy. Stationary organ systems can be imaged more readily such as the central nervous system, bone marrow, the musculoskeletal system, the breast, and pelvic regions. Imaging of moving organs such as the lung, liver, and kidney can be burdened by motion-induced artifacts.

Contrast-Enhanced Ultrasound

Over the past decade, molecularly-targeted contrast-enhanced ultrasound has attracted significant attention in preclinical research of cancer diagnostic and therapy. Potential applications for ultrasound molecular imaging run the gamut from early detection and characterization of malignancies to monitoring treatment responses and guiding therapies. To date, the majority of preclinical studies have employed the use of contrast microbubbles, which are gas–liquid emulsions of several micrometers in size that are confined to the intravascular space. Targeted microbubbles have binding ligands on their surfaces that come in contact with specific molecular targets on vascular endothelial cells and accumulate at tissue sites that overexpress those molecular targets. At present, it is shown to be effective in characterization of focal liver lesions, including screening surveillance for HCC in high-risk patients.

Concluding Remarks

PET-CT represents fusion of functional information of PET with anatomical information of CT. FDG PET-CT has become integral part of oncology practice with applications in staging, therapy response monitoring, recurrence detection, and therapy planning. The field of PET-CT is expanding with advent of newer radiopharmaceuticals, and in future these novel agents are likely to play an important role in cancer patient management.

For Further Reading

1. Basu S, Alavi A. Unparalleled contribution of 18 F-FDG PET to medicine over 3 decades. J Nucl Med. 2008;49:17N–21N. 37N.
2. Basu S, Mohandas KM, Peshwe H, Asopa R, Vyawahare M. FDG-PET and PET-CT in the clinical management of gastrointestinal stromal tumor. Nucl Med Commun. 2008;29:1026–39.
3. Boellaard R, O'Doherty MJ, Weber WA, Mottaghy FM, Lonsdale MN, Stroobants SG, et al. FDG PET and PET-CT: EANM procedure guidelines for tumour PET imaging: version 1.0. Eur J Nucl Med Mol Imaging. 2010;37:181–200.
4. Delbeke D, Coleman RE, Guiberteau MJ, et al. Procedure guidelines for tumor imaging with FDG-PET-CT 1.0. J Nucl Med. 2006;47:885–94.
5. Valk PE, Bailey DL, Townsend DW, Maisey MN. Positron emission tomography: basic science and clinical practice. London: Springer London Ltd; 2006. XIV, p. 475.

5 Technical Considerations During PET Imaging

B.R. Mittal and J. Mohan Roop

Positron emission tomography (PET) imaging has emerged as the most important component of medical imaging in the management of oncology patients. It is a functional imaging technique in which radioactive tracer distribution in tissues is projected as a computer-generated image. The image is produced through the detection of annihilation photons emitted by radionuclides, which decay by emission of positrons. PET imaging with F-18 FDG depends on alteration in glucose metabolism of tissues for detection of disease. It differs from conventional imaging techniques that rely on morphologic changes for disease detection. As glucose metabolism is increased in many malignancies, F-18 FDG PET is a sensitive method for detecting, staging, and monitoring the effects of therapy in various malignancies.

Dual-modality PET-CT imaging has been widely used and received medical attention since its introduction in 1998. In this, patients can be examined with both CT and PET in a one-stop shop method, which is the major advantage of this technique. Complementary anatomic and functional imaging, such as that performed with CT and PET, allows for improved diagnosis and thus better patient care in clinical oncology. A whole-body survey is the standard mode of acquisition. CT images are used for attenuation correction of the PET data and as an anatomic reference of the tracer uptake patterns imaged in PET. It is very important to be aware of the technical aspects and the interpretation of the reconstructed PET-CT images for maximum benefit to the patient.

B.R. Mittal, MD, FICNM, FAMS (✉)
Prof. & Head, Department of Nuclear Medicine and PET, Postgraduate Institute of Medical Sciences and Research, Chandigarh, India
e-mail: brmittal@yahoo.com

J.M. Roop, MD
Junior Resident, Department of Nuclear Medicine and PET, Postgraduate Institute of Medical Sciences and Research, Chandigarh, India
e-mail: j.mohanroop@gmail.com

Prerequisites and Technique

Patient Preparation

Proper patient preparation is necessary to minimize tracer uptake in normal tissues, while maintaining uptake in target tissues (disease sites).

(i) Fasting for 4–6 h is necessary before the injection of F-18 FDG.
(ii) Blood glucose level should be measured to ensure that glucose level is <200 mg/dL
(iii) Strenuous exercise should be avoided for at least 24 h before PET scan.
(iv) Patient should be kept warm and relaxed to reduce the brown fat and skeletal muscle uptake during the 40–60 min (uptake phase) waiting period, post injection of radiotracer.
(v) Intravenous line for F-18 FDG administration should be secured.
(vi) Remove any metallic objects.

B.K. Das (ed.), *Positron Emission Tomography: A Guide for Clinicians*,
DOI 10.1007/978-81-322-2098-5_5, © Springer India 2015

(vii) Patient should void before start of the scan.
(viii) If oral CT contrast agents are to be given, patients are asked to drink up to 1,500 mL of oral contrast solution during the F-18 FDG uptake phase.
(ix) Depending on scanner characteristics, such as PET detector material and acquisition mode, the dose of F-18 FDG injected may vary between 300 and 700 MBq.
(x) For imaging of the kidneys or pelvic region, Foley catheter in the bladder to reduce urine activity may be considered. Delayed imaging with diuretic administration with good hydration and frequent voiding is also helpful

Relevant Patient History

Relevant patient history and data should be collected before performing F-18 FDG PET scan, which includes

(i) Clinical history: complaints, type, site, and date of diagnosis of tumor
(ii) Surgical history: type and date of previous surgery or biopsy
(iii) Current therapy: chemo/radiotherapy, bone marrow–stimulating factors like granulocyte colony-stimulating factor (G-CSF)
(iv) Pregnancy and breastfeeding
(v) History of diabetes and recent infection
(vi) History of previous trauma, accidents, or falls
(vii) Reports of radiologic investigations if done and review of the patient's film (CT, MRI)
(viii) History of claustrophobia and patient's ability to lie still for the duration of the acquisition

Patient Positioning

Patients must be positioned comfortably on the examination pallet with their arms raised above their heads, which is standard practice in CT. For optimal imaging of the body, this is the ideal position, if that position can be tolerated by the patient. Arms along the side may lead to the production of beam-hardening artifacts over the torso. However, the arms should be positioned along the side for optimal imaging of the head and neck. Patients should be supported with adequate positioning aids like knee rest, head and neck support to limit involuntary motion, which may cause misalignment during the combined examination.

Scout Scan

A topogram or scout scan is an x-ray image overview scan of the patient acquired at the start of PET-CT examination. It is acquired during continuous table motion, with the x-ray tube/detector assembly typically locked in either frontal or lateral position, producing an anatomic overview image. This topogram is used to define the axial examination range of the PET-CT study. Truncation of the anatomy is predicted by observing the visual markers for the measured transverse field of view of the CT and the PET in the topogram image. Patients should be repositioned before the CT scan to avoid truncation artifact.

CT Acquisition

The CT can be performed either for attenuation correction/anatomical localization or as an optimized diagnostic CT scan. If a CT is done solely for attenuation and scatter correction and co-localization, the acquisition parameters (tube current, voltage, slice thickness) should be adjusted in order to minimize the radiation exposure for the patient. Most PET-CT users acquire a single continuous spiral CT scan.

The CT scan acquisition is delayed by 20–50s if intravenous contrast agents are administered. A breath-hold command is usually given to match the anatomy more closely to the physiology of the patient, which is acquired during the free-breathing emission scan.

PET Acquisition

Immediately after completion of the CT scan, the table is advanced to the field of view of the PET, where emission scanning is acquired in the caudocranial direction, starting at the thighs. It is acquired in this fashion to limit artifacts from the FDG metabolite excretion into the urinary system. Many facilities acquire PET images at 60 or 90 min after F-18 FDG administration. The emission image acquisition time is based on the administered activity, patient body weight, and sensitivity of the PET scanner and can vary from 2 to 5 min or longer per bed position for body imaging. A second set of images to assess the change in uptake over time are also acquired in some facilities. When two studies are being compared by use of semi-quantitative parameters, especially the SUV, the F-18 FDG uptake time should be kept constant.

Data Processing and Image Reconstruction

CT image reconstruction occurs in parallel with emission acquisition. The CT data acquired is usually reconstructed by use of filtered back projection or a similar algorithm. Separate CT reconstructions for the PET attenuation correction and for the diagnostic CT are performed, depending on the CT protocol and the diagnostic question. These reconstructions differ in their slice thickness, slice overlap, filter, etc. An algorithm is used for the calculation of the attenuation correction factors (ACF) from the CT transmission data. Emission images are reconstructed consecutively with the completion of the emission acquisition and using the available ACFs. The emission data should be corrected for detector efficiency (normalization), system dead time, random coincidences, scatter, attenuation, and sampling non-uniformity. Scanners without septa acquire data in the 3-dimensional (3-D) mode only, whereas scanners with retractable septa can acquire data in both 2-dimensional (2-D) and 3-D modes. Datasets acquired in the 3-D mode can either be rebinned into 2-D data and reconstructed with a 2-D algorithm or can be reconstructed with a fully 3-D algorithm. Iterative reconstruction approaches are now widely available, largely replacing the direct, filtered back-projection methods, which were used previously. Reconstructed 3-D volume data set is available in transaxial, coronal and sagittal slices, but the maximum intensity projections should also be available.

Image Interpretation and Pitfalls

During interpretation of PET images, the following points should be taken into consideration:

(i) Correlation with clinical history of the patient
(ii) Clinical correlation with any other data from previous biochemical and morphological examinations
(iii) Physiological distribution of F-18 FDG should be known. Physiologic uptake of F-18 FDG can be seen to some extent in every viable tissue, including the brain, myocardium, breast, liver, spleen, stomach, intestines, kidneys and urine, muscle, lymphoid tissue (e.g., tonsils), bone marrow, salivary glands, thymus, uterus, ovaries, testes, and brown adipose tissue
(iv) Anatomical localization of the abnormal F-18 FDG uptake and its intensity
(v) Semi-quantitative (quantitative) value (if available)
(vi) Brain parenchyma shows a high F-18 FDG uptake. FDG PET is therefore only of limited value for the detection of brain metastases. As a result, F-18 FDG PET is routinely not indicated in detection or exclusion of brain metastases.
(vii) An increased FDG uptake is not only observed in neoplastic lesions but also in granulation tissue, infections, and other inflammatory processes.
(viii) Tumors showing low to absent F-18 FDG uptake (genitourinary cancers, bronchoalveolar cell lung cancer, mucinous adenocarcinomas, well-differentiated neuroendocrine tumors, low-grade sarcomas, low-grade lymphoma, differentiated thyroid cancer, hepatocellular carcinoma)
(ix) Various factors causing alteration of F-18 FDG distribution, those increasing the F-18 FDG uptake (higher tumor grade, inflammation, tumor hypoxia, acute radiation, acute chemotherapy, recent surgery) and those leading to reduction in uptake (necrosis, low-grade or low-cellularity tumor, mucinous tumors, bronchoalveolar carcinomas, hyperglycemia, presence of scar tissue, chronic radiation).

(x) Causes of false-negative results (small size of the lesion, low metabolic rate, concomitant drug use interfering with the uptake, physiological uptake masking cancer lesions, osteoblastic skeletal metastases)

(xi) Causes of false-positive results (artifacts, sites of physiological uptake, muscular activity, post-therapy uptakes: bone marrow and spleen (after G-CSF), thymus in young patients, benign neoplasms, inflammatory process).

(xii) Patient motion during acquisition, if any, can produce artifacts in the reconstructed images and thus should be considered while interpreting the data. Non-attenuated corrected data should be viewed to rule out artifacts due to metallic objects.

Additional Points

An interval of at least 10 days (ideally several weeks) is generally considered between the last chemotherapy treatment and PET. The effects of growth factors (G-CSF) on F-18 FDG biodistribution usually lasts less than 2 weeks after the final administration. It is presumed that the effects of radiotherapy are somewhat longer lasting. So, it is best to wait for about 3 months after the end of treatment before conducting F-18 FDG PET scan.

Reporting

The final report should include following essential elements:

(i) Clinical history–includes brief clinical and treatment history and their respective dates, indication for PET study.

(ii) Technique/Procedure–it should include radiopharmaceutical name, dose/activity, route of administration, uptake time (i.e., from radiopharmaceutical injection to imaging), precise body region scanned, and the CT technique (including whether oral or intravenous contrast was used) followed.

(iii) Comparison studies–Whether comparison was made with previous PET or PET-CT studies along with its date.

(iv) Findings–should include the location, extent and intensity of abnormal F-18 FDG uptake and the relevant morphologic findings related to PET abnormalities on the CT images. Incidental PET findings and incidental CT findings should also be included.

(v) Impression–Whenever possible, a precise diagnosis should be given, otherwise differential diagnosis should be provided. If necessary, follow-up and additional diagnostic studies required to clarify or confirm the impression should be recommended.

For Further Reading

1. Kluetz PG, Meltzer CC, Villemagne VL, Kinahan PE, Chander S, Martinelli MA, et al. Combined PET-CT imaging in oncology. Impact on patient management. Clin Positron Imaging. 2000;3:223–30.
2. Townsend DW, Carney JP, Yap JT, Hall NC. PET-CT today and tomorrow. J Nucl Med. 2004;45 Suppl 1:4S–14.
3. Wahl RL, Quint LE, Cieslak RD, Aisen AM, Koeppe RA, Meyer CR. "Anatometabolic" tumor imaging: fusion of FDG PET with CT or MRI to localize foci of increased activity. J Nucl Med. 1993;34:1190–7.

Applications in Oncology: An Overview

6

Birendra Kishore Das

Introduction

Treatment of patients with cancer requires appropriate histological diagnosis and, more importantly, staging of the disease using the least invasive and most sensitive and specific procedures. Over the last decades, several studies have demonstrated immense contribution of nuclear medicine to the management of cancer. Positron emission tomography (PET) scanning has added significant clinical value in evaluating cancer patients. PET is an imaging modality that is used to provide a three-dimensional image of functional changes in the body. PET is commonly used to detect and stage different types of cancer. Accurate information about diagnosis and staging of disease is critical for planning the most appropriate treatment strategy (Figs. 6.1 and 6.2). PET has also been used to monitor therapy. The rationale for this is that the early detection of disease that is not responding to treatment could allow for a change to a more effective treatment strategy. PET scanning has, therefore, been used as a tool in diagnosis, staging, restaging, and possibly follow-up of cancer patients. Another potential contribution of PET scan is monitoring delivery of target cancer therapy.

B.K. Das, MD, ANM, FAR, FAMS
Director and Chief of Nuclear Medicine,
Utkal Institute of Medical Sciences, Jaydev Vihar,
Bhubaneswar, India
e-mail: bkishoredas@yahoo.com

PET imaging changes the intended patient management strategy in about 40 %. In some studies, it is even more than 50 %. Results were consistent across all cancer types. In as many as 90 % of cases, referring physicians indicated that the scan results allowed them to avoid additional imaging tests or procedures, indicating that PET can significantly reduce the number of testing procedures and result in substantial healthcare savings. PET imaging allowed physicians to avoid costly biopsy surgeries in as many as 70 % of cases.

PET has been found to be a cost-effective modality specially in following situations:

- The differential diagnosis of solitary pulmonary nodules
- The staging of non-small cell lung cancer
- The restaging of colorectal carcinoma after recurrence
- The restaging of Hodgkin's and non-Hodgkin's lymphoma

Cost savings in lung and colorectal cancer primarily result from avoiding costly surgical procedures in cases where no reasonable chance of cure exists.

Role of PET in Lung Cancer

Squamous cell and small cell carcinomas (SCLC) occur predominately in smokers and arise from the proximal bronchial tree. Adenocarcinomas are more often peripherally located in the lung.

B.K. Das (ed.), *Positron Emission Tomography: A Guide for Clinicians*,
DOI 10.1007/978-81-322-2098-5_6, © Springer India 2015

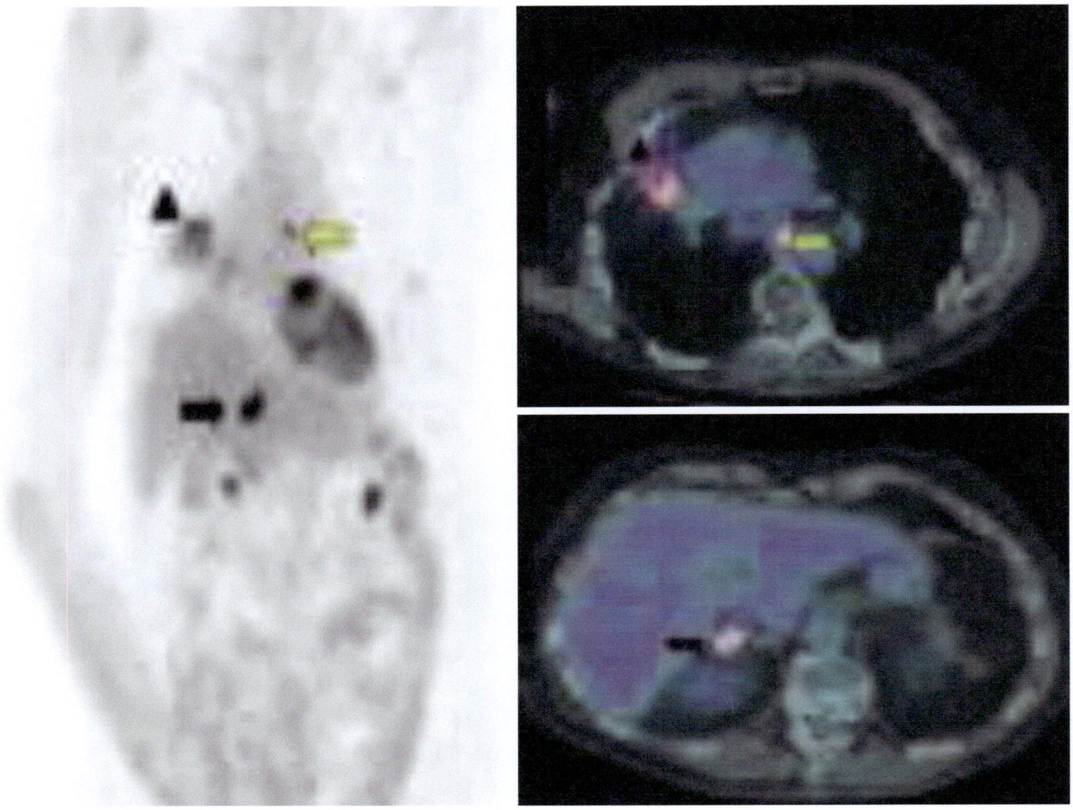

Fig. 6.1 Initial staging of non-small cell lung cancer. MIP image (*left panel*) and fused images (*right panel*) of 18 F-FDG PET-CT showed the primary tumor (*arrowhead*) with mediastinal nodal involvement (*yellow arrow*) and extra-thoracic right adrenal metastasis (*black arrow*)

SCLC is of neuroectodermal origin and has the poorest prognosis, usually having spread systemically at the time of initial diagnosis.

FDG-PET imaging is recommended for the evaluation of indeterminate pulmonary nodules greater in size than the resolution of most imaging systems (5–15 mm). Because of the high negative predictive value of FDG-PET, FDG-negative lesions can be followed at 6-monthly intervals radiographically. FDG-positive lesions need to be biopsied because of the high rate of false-positive active granulomatous processes. FDG-PET should also be performed for staging non-small cell lung cancer (NSCLC), because it can detect metastatic lymph nodes that do not meet CT size criteria for malignancy. Using whole-body imaging, PET can reliably detect unsuspected distant metastases. Unlike CT, FDG-PET imaging can differentiate scar tissue from recurrent tumor and therefore has applications for monitoring therapy and in the evaluation of recurrent disease.

FDG-PET scan complements conventional radiological studies in the evaluation of focal pulmonary nodules, lung cancer staging and tumor recurrence, treatment response, and prognosis. Increased glucose metabolism by malignant cells allows physiologic differentiation of benign and malignant lung lesions. In practice, nodules with low FDG uptake can be followed radiographically, and nodules with increased uptake should be evaluated with biopsy or resection. FDG-PET scanning has also shown tremendous promise in the evaluation of mediastinal node involvement and previously unidentified distant metastasis. However, when pathologic confirmation of lung cancer is required, minimally invasive techniques, such as bronchoscopy, thoracoscopy, and

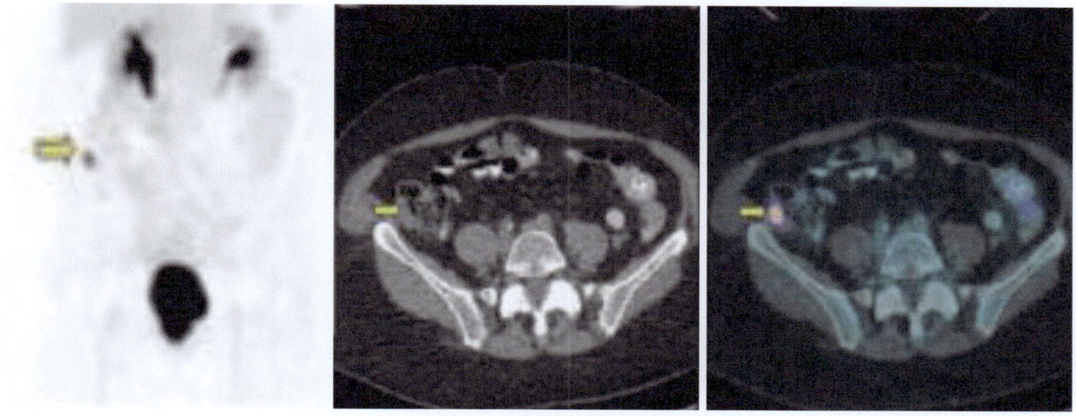

Fig. 6.2 Staging of non-small cell lung cancer. FDG image (*left side*), CT images (*middle*), and fused images (*right image*) in PET-CT show apart from the primary tumor (not shown here) incidental right iliac fossa a small focal uptake (*arrow*) is noted, which cross-correlated to a small soft tissue lesion in the cecum and turned out to be a synchronous primary adenocarcinoma

anterior mediastinoscopy among other methods are valuable and simple ways to obtain tissue.

Anatomic (CT, MRI) and physiologic imaging (PET) techniques should be considered complementary rather than competitive imaging methods. This problem does not exist anymore with the routine use of hybrid machines like PET-CT.

PET is also better than CT for the differentiation of recurrent tumor versus scar in patients who have already been treated. Several studies have shown that FDG-PET can be useful in predicting and assessing the response to radiation therapy as well as chemotherapy.

Role of PET in Colorectal Cancer

The main indications for FDG-PET whole-body imaging in patients with suspected, recurrent, or metastatic colorectal carcinoma are as follows:

(a) When there is a rising serum CEA level in the absence of a known source
(b) To increase the specificity of structural imaging when an equivocal lesion is detected
(c) As a screening method for the entire body in the preoperative staging before curative resection of recurrent disease
(d) To differentiate post-therapy changes versus persistent/recurrent viable tumor
(e) To monitor response to therapy

It may be interesting to know that the US Health Care Financing Administration has approved reimbursement of FDG-PET imaging by Medicare for diagnosing, staging, and restaging colorectal carcinoma since 2001.

Detection and Staging of Recurrent Colorectal Carcinoma

Serial measurements of CEA and CT have been conventionally used in the follow-up and detection of metastases. A large number of patients present with isolated liver metastases as their first recurrence, and 20 % of these die with metastases exclusive to the liver. Hepatic resection may result in a cure in up to 25 % of these patients, but the size and number of hepatic metastases and the presence of extrahepatic metastases all adversely affect prognosis. The presence of extrahepatic metastases is thought to represent a contraindication to hepatic resection. Serial CEA determinations are used to monitor patients for recurrences with a sensitivity of 59 % and specificity of 84 %, and CT has been the conventional imaging modality used to localize recurrence. However, CT fails to demonstrate hepatic metastases in many cases and underestimates the number of lobes involved. Extrahepatic abdominal metastases are commonly missed on CT, and

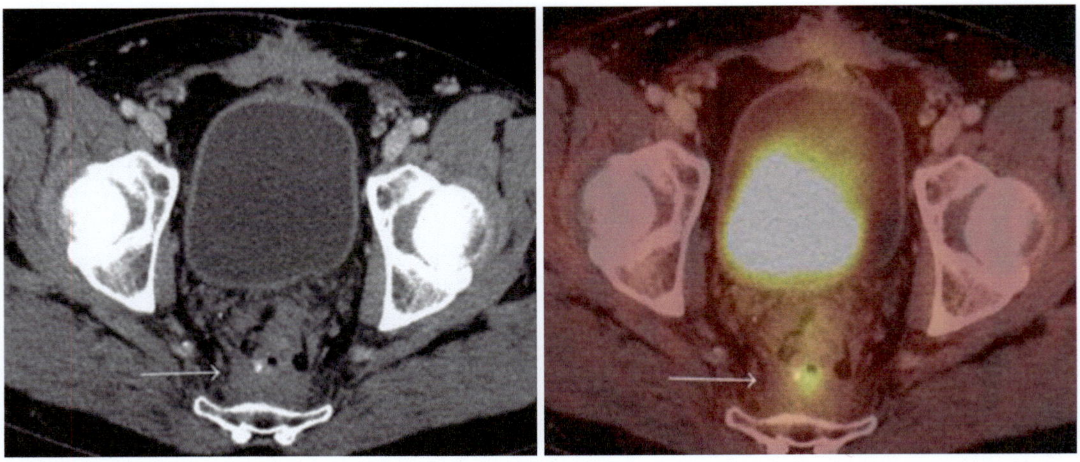

Fig. 6.3 Doubtful recurrence of rectal cancer seen in CT image (*left*) (*arrow*). Confirmation of malignancy in PET-CT image (*right*) (*arrow*)

the differentiation of postsurgical changes from tumor recurrence is problematic. Superior mesenteric arterial CT portography is more sensitive than CT for detection of hepatic metastases but has a high rate of false-positive findings, lowering the positive predictive value.

Functional imaging with SPECT used in the detection and staging of malignancies has well-known limitations. Radioimmunoscintigraphy is limited by difficulties with antigen modulation and variable depiction of tumor and nontumor cells, as well as by physiologic hepatic and bowel activity. Moreover, due to slow blood pool clearance, images are not acquired for several days after injection, a major disadvantage.

Numerous studies have demonstrated a strong role for FDG-PET in identifying recurrences of colorectal carcinoma (Fig. 6.3). For detection of recurrent colon carcinoma, FDG-PET-CT has been found to be more sensitive than CT at all anatomic sites, including the lung. One-third of PET-positive metastases in the extrahepatic abdomen and pelvis are CT negative. Whole-body PET is especially useful for detecting distant metastatic disease, including abdominal nodal disease, pulmonary metastases (indeterminate lung nodules), and differentiating postsurgical scarring from recurrent disease, all of which are problematic for CT. For differentiation of post-therapy scar from local recurrence, PET is clearly more accurate (90–100 %) than CT (48–65 %), and CT is often equivocal. For hepatic metastases, FDG-PET has a higher accuracy (92 %) than CT (78 %) and CT portography (80 %) despite the slightly higher sensitivity of CT portograph. In patients with unexplained CEA elevation and no abnormal findings on conventional evaluation, including CT, the sensitivity of PET for the detection of recurrent disease has been found to be 93–100 %. In multiple studies, including detecting recurrence, FDG-PET imaging led to a change in management in an estimated 32 % of the patients.

Monitoring Therapy of Colorectal Carcinoma

Increased FDG uptake immediately following radiation may be due to inflammatory changes, and is not always associated with residual tumor. Although the time course of postirradiation FDG activity has not been studied systematically, it is generally accepted that FDG activity present 6 months after completion of radiation therapy most likely represents tumor recurrence. There are many reports suggesting that the response to chemotherapy in patients with hepatic metastases can be predicted with PET. Responders may be discriminated from nonresponders after 4–5 weeks of chemotherapy with fluorouracil by measuring FDG uptake before and during therapy.

Regional therapy to the liver by chemoembolization can also be monitored with FDG-PET imaging. FDG uptake decreases in responding lesions. The presence of residual uptake in some lesions can help in guiding further regional therapy.

Role of PET in Lymphoma

Hodgkin's disease (HD) and non-Hodgkin's lymphoma (NHL) belong to those kind of malignancies that are potentially curable. The extent of the disease is the most important factor influencing relapse-free and total survival of patients. Conventional methods for staging like CT and ^{67}Gallium imaging have limitations. Although ^{67}Gallium plays a role in the evaluation of the presence of viable tumor in residual post-therapy masses, it is not superior to CT in initial staging of untreated lymphoma. FDG-PET imaging is recommended for staging lymphoma in addition to CT and other conventional staging modalities, because it can detect additional nodal and extranodal lymphomatous lesions, as well as bone marrow involvement even when bone marrow biopsy is negative.

Both HD and NHL exhibit marked FDG uptake, and FDG imaging is useful both for staging and monitoring therapy. During chemotherapy, FDG-PET imaging can identify the responders early in the course of treatment, allowing alterations in the chemotherapy regimen as indicated. FDG-PET can differentiate scarring from persistent or recurrent tumor in residual masses after the end of treatment and allows improved discrimination of rebound thymic hyperplasia from viable lymphoma. Because of the superior resolution of FDG images compared to ^{67}Gallium scintigraphy, FDG imaging, therefore, is replacing ^{67}Gallium for the evaluation of patients with lymphoma.

Staging Lymphoma

One of the most important factors influencing relapse-free and total survival of lymphoma patients, besides histology, is extent of disease, and accurate initial staging is essential for optimizing patient therapy and determining prognosis. Therapeutic implications for patients with HD and NHL emphasize the importance of initial staging accuracy—patients diagnosed with stage I or II HD and some subgroups of NHL may receive local external radiotherapy alone or in combination with chemotherapy, while those with stage III or IV disease are typically treated with chemotherapy.

Problems of Staging with Conventional Methods

Conventional modalities for staging lymphoma include physical examination; computed tomography (CT) of the chest, abdomen, and pelvis; and bone marrow biopsy. Although CT is the best imaging technique to provide detailed information about the relationship between organs and vascular structures, CT criteria for pathologic adenopathy are based on size alone, which is a major limitation. Benign lymph node enlargement may lead to over-staging, while small malignant lymph nodes may not be recognized, resulting in under-staging. In addition, CT has limited sensitivity for detection of the spleen, liver, and bone marrow involvement. Equivocal CT lesions are common and frequently require additional imaging or biopsy. This prolongs the staging workup and adds to patient expense and morbidity. Patients who are under-staged may receive inadequate therapy for their disease, jeopardizing the opportunity for remission or cure. Conversely, over-staged patients may receive unnecessarily aggressive or investigational therapy and be given an overtly grim prognosis.

Problems with Gallium Scintigraphy

Scintigraphy with ^{67}Gallium is often included in the initial staging of patients with lymphoma. However, it has many limitations. These include suboptimal photon energy leading to noisy images, variable uptake of Gallium by tumor, particularly low-grade NHL, limited detection of abdominal disease secondary to marked physiologic hepatic and colonic activity, and the potential for false-positive findings related to infectious or inflammatory processes. ^{67}Gallium scintigraphy is also inconvenient for the patient as multiple visits to the imaging facility on consecutive days are typically required. In addition, the value of ^{67}Gallium scintigraphy is not in the initial staging of patients with lymphoma but in the evaluation of the response to treatment and assessment of residual mass after therapy. A pretherapy scan is, however, necessary to confirm that the tumor is ^{67}Gallium avid before trusting follow-up ^{67}Gallium scans as an accurate measure of tumor response.

Staging of lymphomatous involvement of the bone marrow requires invasive bone marrow aspiration; their sensitivity for detecting bone marrow disease is limited by sampling error. Bone scintigraphy is often unreliable for the demonstration of skeletal involvement because of its low sensitivity. Although MRI appears to be the most sensitive imaging technique, whole-body MRI is not applicable as a screening technique and should be reserved for areas that are clinically suspect.

Advantages of PET Scanning in Lymphoma Management

Many of the above limitations of conventional staging modalities for lymphoma can be overcome with the use of FDG-PET. Unlike CT, functional imaging with FDG-PET imaging directly identifies increased metabolic activity in malignant tissue, and does not depend on anatomical distortion or enlargement for the determination of abnormalities. Compared with ^{67}Gallium, FDG is avidly trapped by virtually all lymphomas, although the degree of FDG uptake does seem to correlate with the histological grade of malignancy and proliferation rate. Although physiologic gastrointestinal activity does occur with FDG, the better quality of the images usually allows the differentiation of physiologic activity in the bowel from abdominal and pelvic lesions. With FDG-PET imaging, most of the skeleton is imaged during scanning, enabling noninvasive detection of focal bone marrow disease that may be missed through sampling error with standard iliac crest biopsy.

The degree of FDG uptake seems to be a prognostic factor. The cost-effectiveness of PET compared to conventional staging modalities has also been established.

Role of PET in Breast Cancer

Breast cancer is the most common cancer diagnosed among women in advanced countries and appears to be the same at least in urban India. Once diagnosed, appropriate surgery and adequate therapy can lead to decrease morbidity and improve quality of life. The assessment of axillary lymph nodes for cancer involvement is mandatory for planning and staging breast cancer for the appropriate treatment of patients.

Mammography, magnetic resonance imaging (MRI), ultrasound, computed tomography (CT), and bone scintigraphy play a significant role in breast cancer detection, assessment of treatment response, detection of recurrence, and assessment of complications. In recent times, scintimammography has been playing an important role in detection and follow-up.

Imaging with the positron-emitting isotope 18-F attached to glucose (FDG) and imaging with PET-CT have been a standard imaging procedure in breast cancer. Such PET-CT images are superior to PET or CT alone. The intensity of FDG uptake is related to the biological and histological characteristics with uptake being usually more marked in invasive ductal carcinoma than in invasive lobular carcinoma. Uptake in ductal carcinoma in situ (DCIS) is usually poor or absent, rendering it unsuitable for very early detection. Positive correlation between FDG uptake, tumor grade, and tumor proliferation index has been demonstrated (Fig. 6.4).

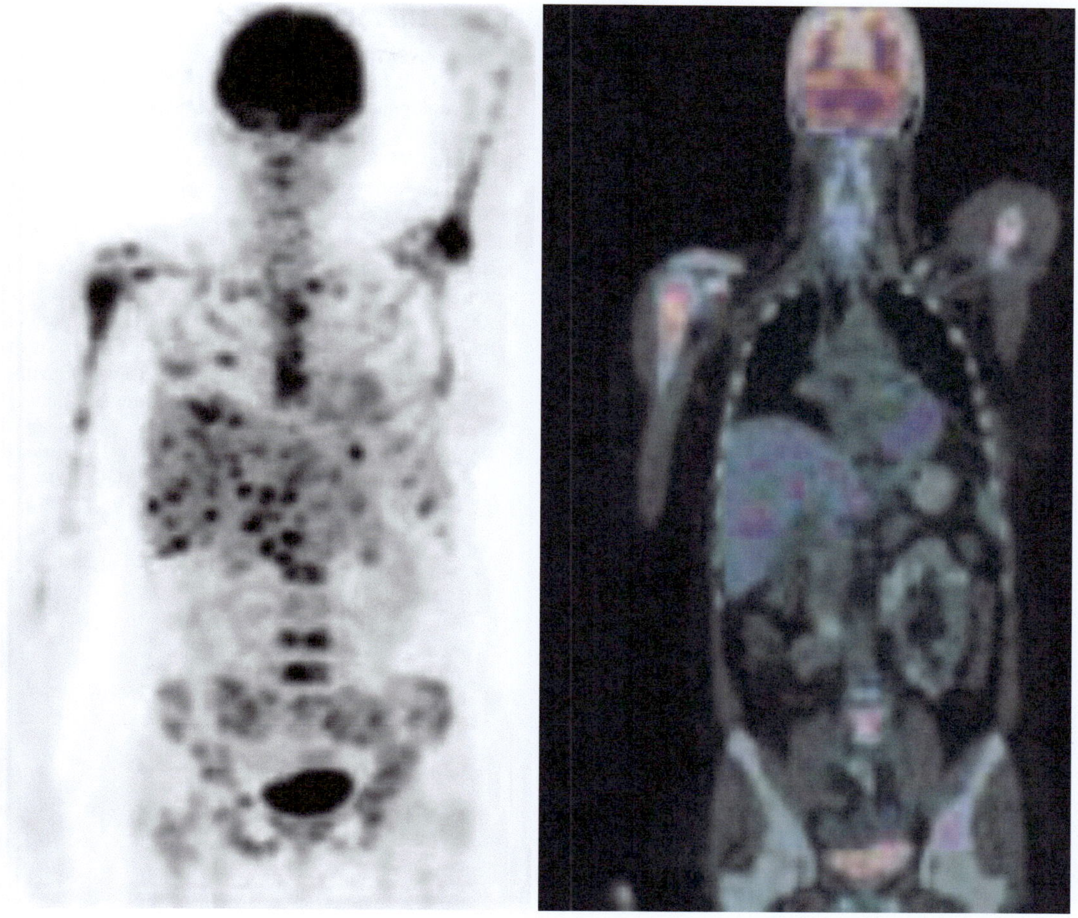

Fig. 6.4 Staging of breast cancer. FDG image (*left side*) and coronal fused images (*right side*) of 18 F-FDG PET-CT showed extensive metastases in the liver and skeleton

Role of PET in Neuroendocrine Tumors

Neuroendocrine tumors develop from cells originating from neural crest and maintain the capacity for amine precursor uptake and decarboxylation, a process essential for the production of monoamine transmitters. They can be located in the thyroid (C-cells), adrenal medulla, lung, skin (melanocytes), nervous system, gastrointestinal tract, and pancreas. These tumors secrete a variety of peptide hormones. Direct assessment of these peptide hormones or their metabolites can be used as tumor markers, both for diagnosis and monitoring treatment response.

The diagnosis of neuroendocrine tumors is mainly biochemical. Localization and therapy of these tumors pose a significant challenge. Conventional morphological imaging modalities such as ultrasound, CT, and MRI are the primary imaging modalities. However, in significant percentage of cases, the localization becomes difficult, and radionuclide images are indicted.

Conventional radiopharmaceuticals used for the diagnosis of neuroendocrine tumors are as follows:

In-111 octreotide (OctreoScan)

I-123 and I-131 metaiodobenzylguanidine (MIBG)

Tc-99 m DMSA (used mainly for medullary carcinoma of the thyroid)

In recent years, FDG-PET and coincidence imaging are gaining considerable importance and popularity in tumor imaging. Comparison of FDG images with OctreoScan in the assessment of neuroendocrine tumors has, however, shown that OctreoScan is more sensitive and specific. But, in case of medullary carcinomas of the thyroid, FDG-PET images have fared better than OctreoScan. The general recommendation is that OctreoScan be a primary imaging in tumors of neuroendocrine origin, and only when OctreoScan fails to localize the tumor or in undifferentiated neuroendocrine tumors, FDG-PET can be recommended.

In the absence of Octreotide Scan or FDG-PET facility, MIBG scans can be recommended. The advantage of MIBG is that it can be used for therapeutic purposes of such tumors.

For Further Reading

1. Esteves FP, Schuster DM, Halkar RK. Gastrointestinal tract malignancies and positron emission tomography: an overview. Semin Nucl Med. 2006;36:169–81.
2. Eubank WB, Mankoff DA. Evolving role of positron emission tomography in breast cancer imaging. Semin Nucl Med. 2005;35:84–99.
3. Fischer B, Lassen U, Mortensen J, Larsen S, Loft A, Bertelsen A, et al. Preoperative staging of lung cancer with PET-CT. N Engl J Med. 2009;361:32–9.
4. Kato H, Miyazaki T, Nakajima M, Takita J, Kimura H, Faried A, et al. The incremental effect of positron emission tomography on diagnostic accuracy in the initial staging of esophageal carcinoma. Cancer. 2005;103:148–56.
5. Rohren EM, Lowe VJ. Update in PET imaging of nonsmall cell lung cancer. Semin Nucl Med. 2004;34:134–53.

Applications of PET-CT in Neurological Disorders: An Overview

7

Madhvi Tripathy and C.S. Bal

Abbreviations

18F-MPPF	Methoxyphenyl-(*N*-2′-pyridinyl)-*p-fluoro-benzamidoethyipiperazine* that binds to the *serotonin-1A receptor*
AD	Alzheimer disease
AADC	Amino acid decarboxylase
CBS	Corticobasal syndrome
DAT	Dopamine transporter
DLBD	Diffuse Lewy body dementia
DOTATOC	DOTA-Tyr^3-octreotide (a somatostatin analog)
DNA	Deoxyribonucleic acid
FDOPA	*Fl*uorodopa min
F-18 FLT	32-Deoxy-32 fluorothymidine (marker for cellular proliferation)
FTD	Frontotemporal dementia
FMISO	Fluoromisonidazole (a hypoxia imaging agent)
F-18 FDDNP	[F-18]fluoroethyl) (methyl) amino] -2-naphthyl ethylidene malononitrile for studying binding patterns of amyloid plaque and tau tangles in Alzheimer disease
5-HT1A	A subtype of 5-hydroxytryptamine
GLUT	Glucose transporter
C-11 DTBZ	*D*ihydro *t*etra*b*ena*z*ine (used for the study of dopaminergic neuronal function)
C-11 CFT	*C*arbomethoxy-3-â-(4-*f*luorophenyl) *t*ropane (for the study of distribution of *dopamine* transporters in the *brain*)
C-11 MP4A	C-11-labeled N-methyl-4-piperidyl -acetate for measurements of regional acetylcholine esterase activity in degenerative dementia
C-11 PMP	Methylpiperidin-4-yl propionate for the study of brain AChE activity in patients with AD and PD
C-11 CPK	1-[2-Chlorophenyl]-*N*-methyl-*N*-[1-methyl-propyl]-3-isoquinoline carboxamide used for quantification of activated microglia
C-11 PiB	Carbon-11-labeled Pittsburgh compound B, a marker of cortical fibrillar amyloid-beta load
C-11 SCH	A selective dopamine D1 antagonist
5-HT2A	5-Hydroxytryptamine (neurotransmitter serotonin (5-HT))
IPD	Idiopathic Parkinson's disease
MCI	Mild cognitive impairment
MMSE	Mini-mental state examination

M. Tripathy, MD, DNB (✉)
Asst. Prof., Department of Nuclear Medicine and PET, All India Institute of Medical Sciences, Ansari Nagar, New Delhi, India
e-mail: madhavi.dave.97@gmail.com

C.S. Bal, MD, DSc
Prof. & Head, Department of Nuclear Medicine and PET, All India Institute of Medical Sciences, Ansari Nagar, New Delhi, India
e-mail: csbal@hotmail.com

B.K. Das (ed.), *Positron Emission Tomography: A Guide for Clinicians*,
DOI 10.1007/978-81-322-2098-5_7, © Springer India 2015

MSA	Multiple system atrophy
PCNSL	Primary CNS lymphoma
PSP	Progressive supranuclear palsy
SPM	Statistical parametric mapping
TLE	Temporal lobe epilepsy
VMAT	Vesicular monoamine transporter
VMAT	Volumetric modulated arc therapy

When positron emission tomography (PET) was introduced into medicine more than 30 years ago, the first organ of major interest was the brain. Presently, the applications of PET in neurology can be divided into the more commonly performed:

1. *Resting brain metabolic studies* using F-18 fluorodeoxyglucose (^{18} F- FDG) PET
2. *Neuroreceptor and brain activation studies* primarily used for research purposes

   The broad areas of neurology where PET studies are being used in clinical practice are as follows:
   - Epilepsy
   - Dementia
   - Movement disorder
   - Stroke
   - Paraneoplastic syndromes
   - Brain tumors
3. *Psychiatric applications*: Psychiatry studies mostly involve the neuroreceptor applications of PET, and are not in routine clinical practice.

The single most commonly utilized PET agent in clinical neurological imaging is ^{18}F-FDG, but the list of promising radioligands used for imaging various aspects of brain function and pathology is increasing. Table 7.1 presents an elaborate list of radioligands that are presently available for neurological PET imaging.

When interpreting ^{18}F-FDG PET (half-life of ^{18}F is 110 min) images, one should keep in mind that the imaged glucose uptake pattern reflects transport and trapping during a prolonged period of time (approximately 35–40 min), thus representing a summation of cellular metabolic processes during the uptake period. Therefore, it is not an ideal agent to

Table 7.1 List of PET radioligands used in neurological imaging

Parameter	PET radioligand	Assessment of cerebral activity
Functional activity		
	F-18 FDG	Glucose metabolism
	O-15 H2O	Blood flow
	C-11 methionine/choline	Amino acid metabolism
	F-18 choline/fluoroethyltyrosine	Amino acid metabolism
Neurotransmitters		
Dopamine	F-18 fluoroDOPA (FDOPA)	Dopamine synthesis
	C-11 raclopride	D2 receptors
	C-11 SCH 23390	D1 receptors
	C-11 DTBZ	VMAT
	C-11 CFT	Dopamine reuptake
Acetylcholine	C-11 PMP	Acetylcholinesterase activity
	C-11 MP4A	Butyrocholinesterase activity
	C-11 methyl-4-piperidinyl-N-butyrate	Nicotinic receptors
	C-11 nicotine	Nicotinic receptors
	F-18 fluoro-A-85380	Muscarinic receptors
	C-11 benztropine	Muscarinic receptors
Serotonin	C-11 WAY-100635	
	18 F-MPPF	5-HT1A receptors
	18 F-altanserin	5-HT2A receptors
Neuro-inflammation	C-11 CPK11195	Glial inflammation
Pathology in vivo	C-11 PiB	Amyloid
	F-18 florbetapir	Amyloid
	F-18 FDDNP	Tau proteins

measure rapid short-term activation neuronal processes, which are better detected by using ^{15}O-labeled water or functional MRI or SPECT perfusion studies.

Epilepsy

Approximately one-third of patients with epilepsy will be medically refractory and will continue having disabling seizures despite an adequate trial of appropriate antiepileptic medications. These patients are assessed for surgical candidacy. This evaluation involves use of clinical, neuropsychological, electrophysiological, and radiological data to determine whether seizure originates from a focal area of the brain—lesional epilepsy. Surgery may be considered the treatment of choice in lesional epilepsy. It has been observed that there is very low probability of long-term seizure freedom without surgery in pharmacoresistant patients. In those patients where the seizure focus is accurately identified and is located in noneloquent brain, complete excision results in good postoperative seizure control.

The principal aim of presurgical evaluation is the lateralization and localization of the epileptogenic zone, which can be achieved in 85 % patients; the remaining 15 % would require intracranial electrodes for this determination. There are no evidence-based guidelines to direct the course of investigations. However, epilepsy teams at surgical centers usually follow set protocols for workup of these patients. Most importantly, there has to be a concordance of interictal scalp EEG, ictal video EEG, and MRI for the epileptogenic focus before patient is taken up for surgery.

Temporal Lobe Epilepsy (TLE)

MRI is the first-line neuroimaging procedure for the localization and lateralization of the epileptogenic lesion, and hippocampal sclerosis, the most common pathological substrate of TLE, can be readily detected by MRI in most cases. ^{18}F-FDG PET is useful in cases with normal or inconclusive MRI (present in 53 %) and in patients with discordant MRI and electroclinical studies. $^{18F\text{-}FDG}$ PET, which was most frequently carried out after unsatisfactory MRI/EEG, influenced surgical decision in 71 % cases, and surgical candidacy was based on ^{18}F-FDG PET findings in 17 % cases. MRI-negative, PET-positive TLE is likely a distinct subgroup of TLE where the pathophysiology more often involves the temporal neocortex rather than being confined to mesial temporal structures. In the case of mesial temporal lobe epilepsy, the minority of patients with electroclinically well-localized temporal lobe seizures but no evidence of hippocampal sclerosis on MRI may be brought to surgery based on hypometabolism seen on ^{18}F-FDG PET in the temporal lobe, assuming it to be the origin of seizures.

Neocortical Epilepsy

^{18}F-FDG PET may detect 70–90 % of cortical dysplasias (Fig. 7.1), the most common pathological substrate of neocortical epilepsies. Interictal PET may be of assistance in nonlesional neocortical epilepsy cases for general localization and for guiding intracranial electrode placement. ^{18}F-FDG PET has been shown to have 71 % overall detection rate in patients with extratemporal epilepsy and normal MRI. Also, many pediatric epilepsy surgery centers routinely perform PET MRI fusion in all cases of epilepsy and have shown an incremental value of PET in localizing dual pathology, cortical dysplasia type I, and mild malformations of cortical development.

Usually, interictal ^{18}F-FDG PET studies are performed because of the duration of tracer uptake (extends over 35–40 min). However, rarely ictal ^{18}F-FDG PET may prove useful in a child with "epilepsia partialis continua" suspected to have "Rasmussen's encephalitis" and a normal MRI. The localization of the epileptogenic focus (which appears hypermetabolic) would justify the use of intravenous immunoglobulin in such patients (see Fig. 7.2).

A number of additional tracers like C-11 flumazenil and C-11 methyl tryptophan have been used for epilepsy imaging, but are presently not available in India.

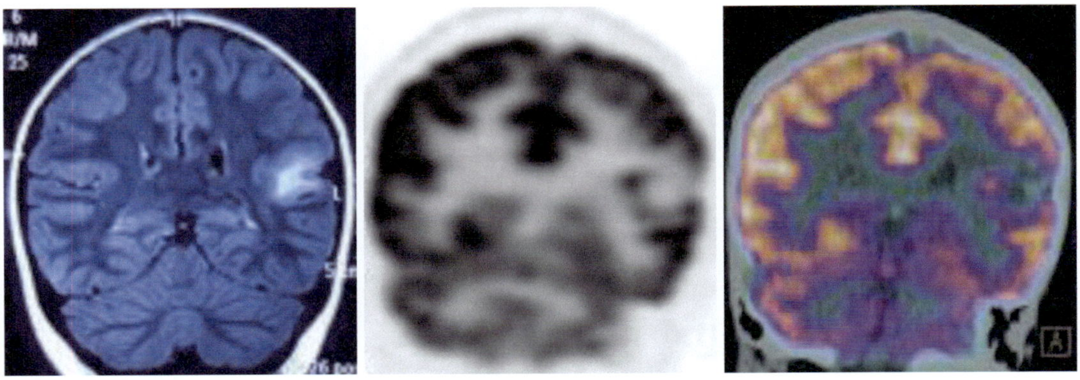

Fig. 7.1 An 11-year-old male child with drug refractory epilepsy. T1W MRI shows cortical dysplasia in the left parietal lobe. Plain ^{18} F-FDG PET and fused PET-CT images show hypometabolism corresponding to the lesion on MRI

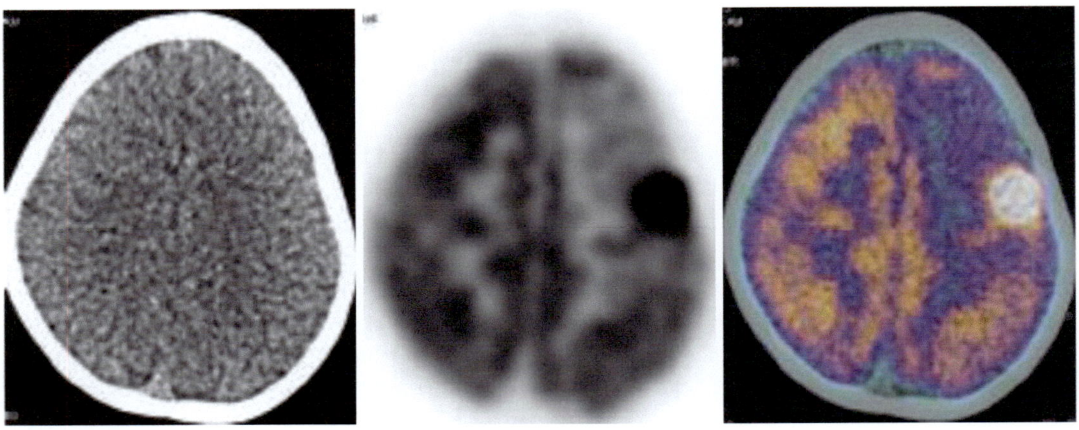

Fig. 7.2 A 3-month-old male child with epilepsia partialis continua and a normal MRI. Pediatric neurologist referred the child for ^{18} F-FDG PET and on the basis of focal hypermetabolism in the left posterior frontal region (which was again reviewed to be normal on MRI) gave the child intravenous immunoglobulin to which he responded

Dementia

Three distinct phases for Alzheimer's dementia (AD) have been described: presymptomatic, mild symptomatic but predementia, and dementia caused by AD. Large collections of data corroborating the power of brain imaging in helping to establish diagnoses and predict cognitive decline are emerging from multicenter international imaging efforts. Also, imaging can effectively be used for differential diagnosis of dementia subtypes.

^{18}F-FDG *PET:* Though the diagnosis of dementia type is primarily clinical, there may be considerable overlap of phenotypes in early cases. Hypometabolism on ^{18}F-FDG PET can be used as a downstream marker of neuronal injury. Specific patterns of hypometabolism on the $^{18F\text{-}FDG}$ PET study have been used for the differential diagnosis of neurodegenerative dementias. Hypometabolism involving the parietotemporal cortices (unilateral or bilateral), including the posterior cingulate and precuneal cortices, is the characteristic metabolic pattern for AD (Fig. 7.3). This is used to differentiate AD from frontotemporal dementia (FTD), which is characterized by hypometabolism in the frontal and temporal cortices (behavioral variant) (Fig. 7.4). ^{18}F-FDG PET has a high sensitivity of 97 and 86 % specificity in distinguishing AD from FTD. Occipital and visual cortices hypometabolism on ^{18}F-FDG PET is useful for differentiating

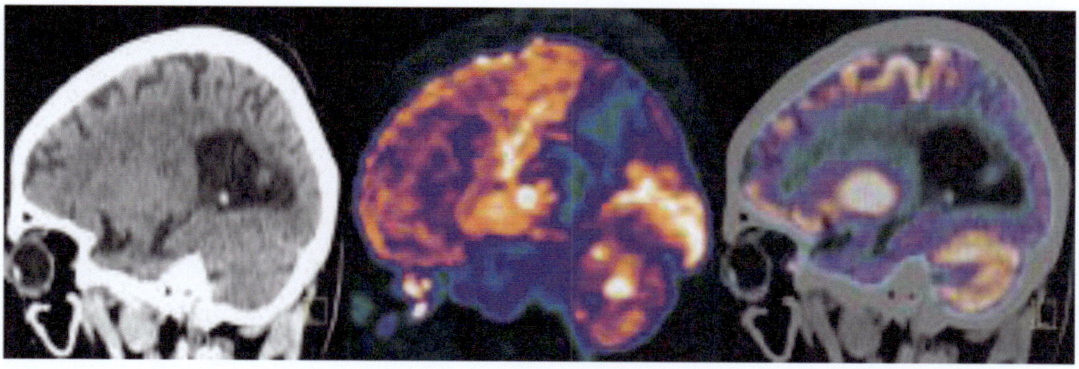

Fig. 7.3 Sagittal plain CT (*left*), sagittal MIP (*middle*), and sagittal fused ^{18} F-FDG PET-CT image (*right*) of a case of Alzheimer's dementia, typical metabolic pattern with parietotemporal hypometabolism and relatively preserved uptake in both visual and sensorimotor cortices, basal ganglia, thalami, and cerebellar hemispheres

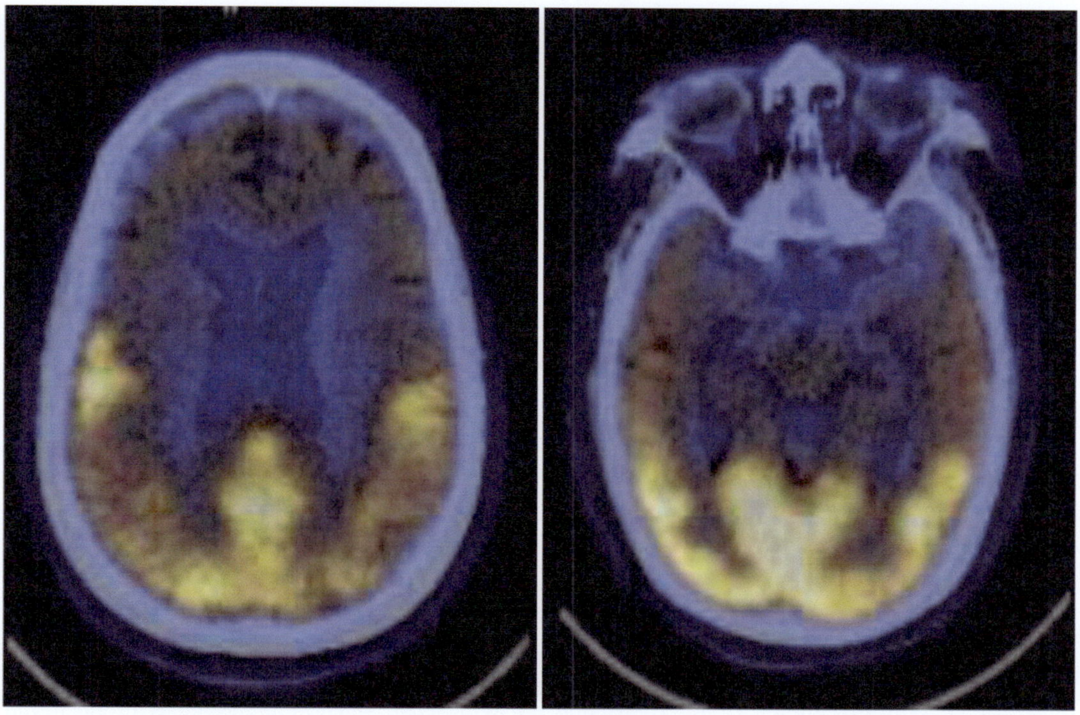

Fig. 7.4 Transaxial fused ^{18} F-FDG PET-CT image of a case of frontotemporal dementia (behavioral variant), classical metabolic pattern with frontal (*left*) and temporal (*right*) hypometabolism

diffuse Lewy body dementia (DLBD) from AD and FTD with nearly 80–87 % specificity and 83–90 % sensitivity in autopsy-validated studies. Further, a DAT agent can be used to differentiate AD from DLBD, while the basal ganglia would show normal DAT binding in AD; DAT binding would be reduced in DLBD.

^{18}F-FDG PET can be used to classify mild cognitive impairment (MCI) patients at risk for cognitive decline and progression to Alzheimer's based on the metabolic pattern of AD (Fig. 7.5). The pooled diagnostic accuracy of ^{18}F-FDG PET for differentiating AD from normal subjects was 93 %, in studies that used clinical assessment as

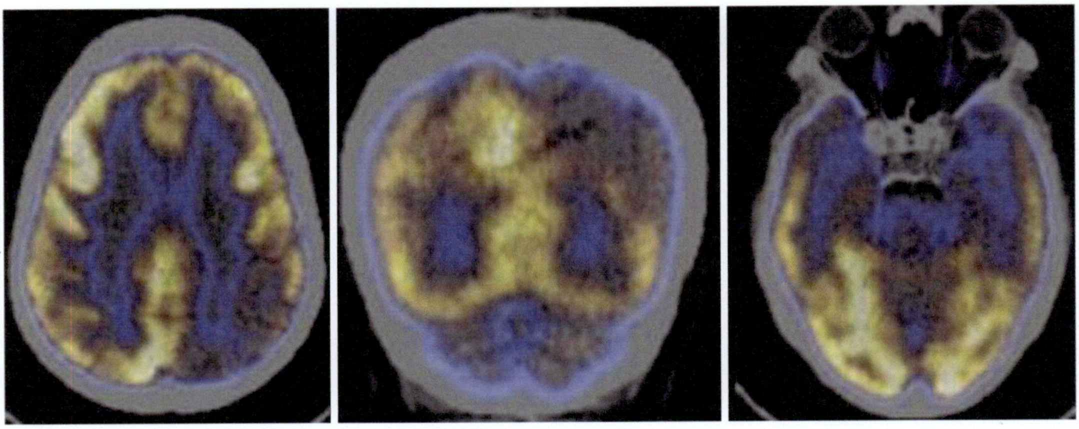

Fig. 7.5 Transaxial fused [18] F-FDG PET-CT image of a case of amnestic MCI, MMSE=26, showing left parietal (*left*), precuneus (*middle*), and temporal (*right*) hypometabolism suggesting high likelihood for progression to AD

the reference standard. When neuropathological confirmation was used as the reference standard, sensitivity of ^{18}F-FDG PET for AD diagnosis was 94 %, and specificity was 73 % for patients presenting with other causes of dementia.

Amyloid Imaging

Amyloid beta deposition is a relatively early event on the path to AD. The amyloid cascade hypothesis states that overproduction or inability to remove â-amyloid results in amyloid deposition and subsequently in neurofibrillary tangles, inflammation, cell death, and cognitive impairment. PET tracers have been developed that bind to amyloid in vivo. The oldest of these is C-11 Pittsburgh (C-11 PiB) compound, which is limited by the presence of C-11 label with a short half-life (20 min). Thus, presently, F-18-labeled tracers like F-18 florbetapir have gained entry into markets worldwide. Appropriate use criteria for amyloid imaging have been defined, but this agent still lacks availability in India. The specific conditions when the use of amyloid is expected to make a difference have been defined as:

Patients with persistent or progressive unexplained MCI.

In patients satisfying core clinical criteria for possible AD but having an atypical clinical presentation or clinical course or an etiologically mixed presentation.

Patients with progressive dementia and an atypically early age of onset. Amyloid imaging can also be used to differentiate FTD from AD where it is expected that FTD patients would not show cortical binding of the amyloid agent.

Presently, amyloid agent is not being marketed in India.

Movement Disorders

The diagnosis of idiopathic Parkinson's disease (IPD) is primarily clinical. However, a common dilemma for doctors assessing patients in movement disorder clinics, especially early in the disease course, is that the main features of IPD are shared, at least in part, by several other disorders, which can be broadly classified into two categories, one which would not be associated with a dopaminergic deficit, like essential tremors, psychogenic Parkinsonism, and drug-induced Parkinsonism, and the second which share nigrostriatal degeneration as a common pathological feature with IPD and are called atypical Parkinsonian syndromes of multiple system atrophy (MSA), progressive supranuclear palsy (PSP), and corticobasal syndrome (CBS). The prognosis and management of each of these conditions differ significantly from that of IPD. The ability to reach a correct diagnosis and distinguish between the abovementioned entities is of clinical importance. In addition to the development and implementation of diagnostic clinical assessments,

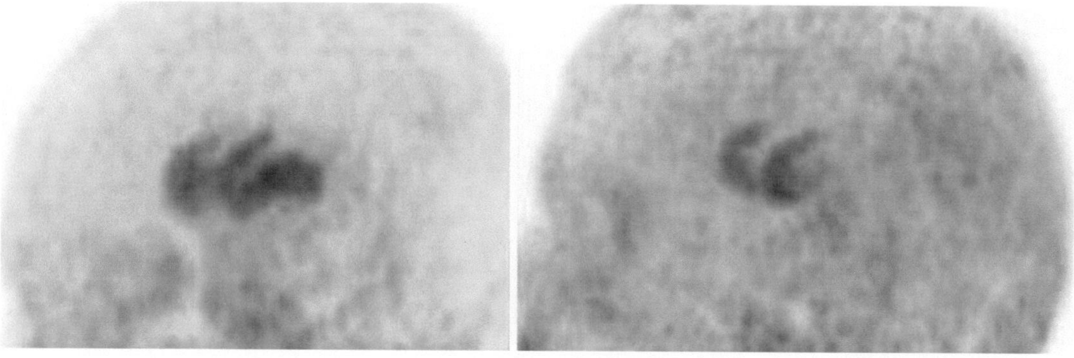

Fig. 7.6 Maximum intensity projection (MIP) F-18 FDOPA images. *Left*—is of a normal subject showing the "rabbit-like" appearance of the basal ganglia with ears, head, and body. MIP (*right* image) is of a case of idiopathic Parkinson's disease showing loss of the rabbit's body (caudal putamen)

there is a need for objective markers to aid the physician in the diagnosis and differential diagnosis of IPD and the atypical syndromes. Thus, the most developed area in terms of providing an objective assessment is neuroimaging. Importantly, it can serve as a biomarker of trait, state, and rate of progression for Parkinson's disease.

Functional imaging using PET has a two-pronged approach for differential diagnosis of Parkinsonism:

I. Evaluating changes in striatal pharmacology by targeting various sites of the dopaminergic pathway, which can provide objective evidence of the presence of dopaminergic deficit and thus provide in vivo evidence of Parkinson's pathology. *Currently, the gold standard imaging technique with the highest resolution and power to differentiate between normal and abnormal nigrostriatal innervation is PET.* These tracers can be used to a limited extent to distinguish IPD from atypical Parkinsonian syndromes also.

II. Metabolic imaging using ^{18}F-FDG, which can be used to distinguish IPD from atypical Parkinsonian syndromes using a qualitative or quantitative image-based approach.

(i) PET studies using presynaptic dopaminergic tracers have objectively demonstrated nigrostriatal nerve terminal loss in Parkinson's disease, even at very early or preclinical stages. Choice of tracer to use would depend on the stage of disease as during the very early or preclinical stages or in asymptomatic carriers of Parkin genes; a dopamine transporter (DAT) agent is preferable as DATs are downregulated quite early into the disease process, while amino acid decarboxylase (AADC) activity may be upregulated and vesicular monoamine transporter (VMAT) shows an intermediate effect. Classically, imaging reveals what is called a rabbit in the brain, with a body, head, and ears (Fig. 7.6-left), the head and body representing the rostral and caudal putamen, respectively, and the ears the caudate nuclei. Reductions are more severe in the posterior putamen and contralateral to the clinically most affected side; thus, the rabbit loses its body early, while the head and ears (caudate and rostral putamen uptake) are visualized till quite late into the disease process (Fig. 7.6-right). Nigrostriatal denervation is not specific for Parkinson's disease and is demonstrated in patients with atypical syndromes of MSA, PSP, and CBS. Imaging of the dopaminergic pathway is especially useful to differentiate dopamine deficiency state (Fig. 7.7) from non-dopaminergic deficient states like essential tremor and vascular, psychogenic, and drug-induced Parkinsonism, but may not be truly useful to differentiate IPD from the atypical subgroups. Dopaminergic imaging can also be used to assess graft viability following transplantation of embryonic dopaminergic tissue in vivo.

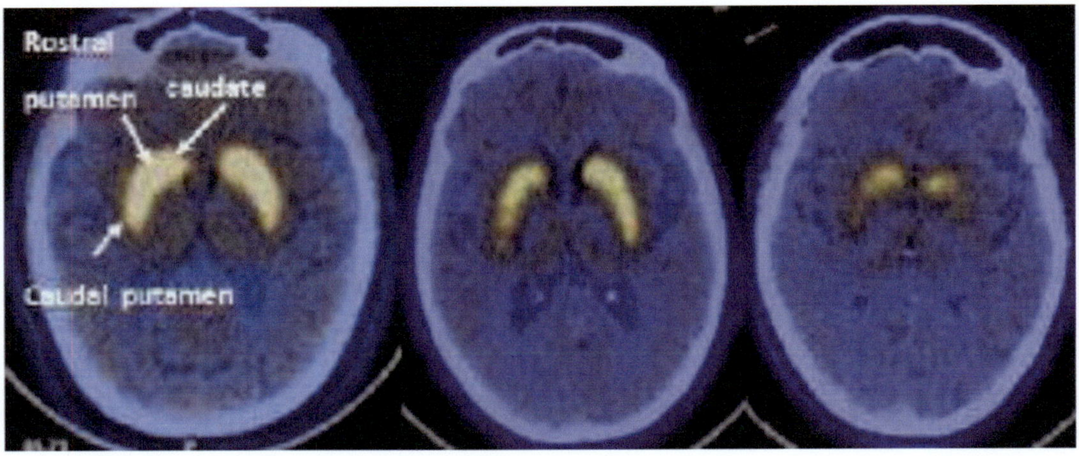

Fig. 7.7 Transaxial fused F-18 FDOPA PET-CT images at the striate level in a normal healthy volunteer (*left*) and in a patient with early IPD (*middle*) and in advanced IPD (*right*). In the patient with IPD (*middle*), there is an asymmetric loss of uptake of tracer, more pronounced in the caudal right putamen than in the caudate and rostral putamen. In advanced IPD (*right*), there is marked decrease in tracer uptake in both basal ganglia

(ii) Postsynaptic D2 receptor imaging, on the other hand, can be used to differentiate IPD from the atypical group as D2 receptor availability is normal or upregulated in untreated IPD (as a compensatory response to the decrease in presynaptic dopamine), whereas in the atypical groups there is a loss of D2 receptors (GABAergic spiny interneuron loss), which can be effectively imaged using tracers like C-11 raclopride and F-18 fallypride. In treated or long-standing IPD, D2 binding normalizes or is only mildly reduced. This is in contrast to MSA or PSP where a significant reduction in D2 receptor binding is seen, and thus a 100 % separation is achieved. However, D2 receptor imaging would not be useful to differentiate MSA from PSP or these two from CBGD.

Thus, depending on the question to be answered, the agent that could best be used for:

- Early identification of a dopaminergic deficit states—*DAT agent like F-18 FP-CIT*
- Establishing IPD, MSA, PSP, and CBS (dopaminergic deficiency)—*F-18 FP-CIT, F-18 FDOPA, and F-18 AV133*
- Differentiate IPD from MSA, PSP, or CBS—*D2 agent like F-18 fallypride*
- Differentiating MSA from PSP or CBS—*none of these would help (^{18}F-FDG PET is useful)*

(iii) Metabolic imaging with ^{18}F-FDG in Parkinsonism

In addition to changes in striatal pharmacology (receptor, transporter, or enzyme status), PET can be used to quantify resting changes in regional glucose utilization using ^{18}F-FDG. In early IPD, striatal metabolic activity is increased in the lentiform nucleus contralateral to the affected limb. A number of studies using ^{18}F-FDG have described characteristic patterns of glucose metabolism in patients with IPD (Fig. 7.8-right), MSA (Fig. 7.9), PSP (Fig. 7.10), and CBS (Fig. 7.11). Based on these patterns, visual analysis and computer-supported reading using statistical parametric mapping (SPM) have been useful for differential diagnosis of IPD from MSA and PSP (Table 7.2).

Others

While the initial studies of PET were in stroke with interesting applications, but as of date it has not found great utility in day-to-day neurology practice.

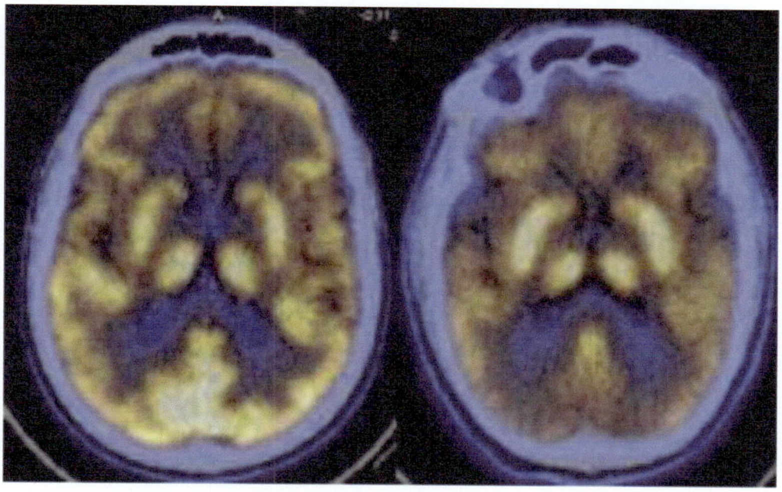

Fig. 7.8 Transaxial fused [18] F-FDG brain PET-CT image of a normal subject (*left*) and in idiopathic Parkinson's disease (*right*), hypermetabolism is noted in basal ganglia (caudally) with parieto-occipital hypometabolism; this is the typical metabolic pattern of IPD

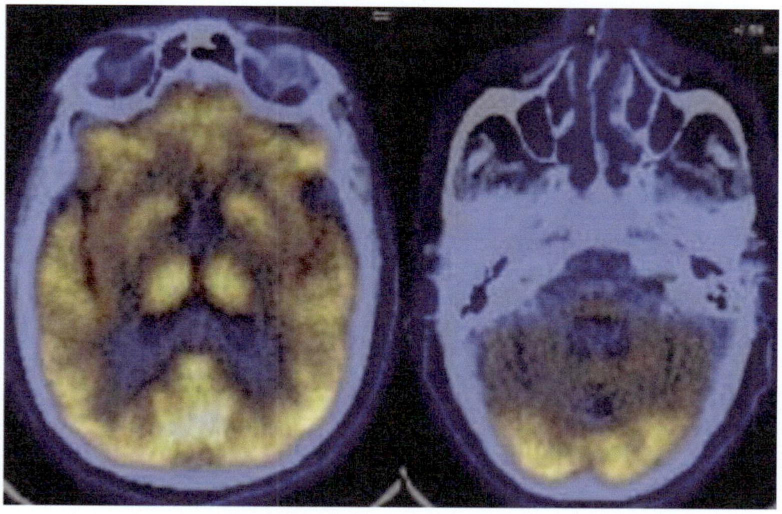

Fig. 7.9 Transaxial fused F-18-FDG PET-CT image (*left*) of a case of MSA-C; hypometabolism is noted in both basal ganglia posteriorly and in both cerebellar hemispheres (*right*)

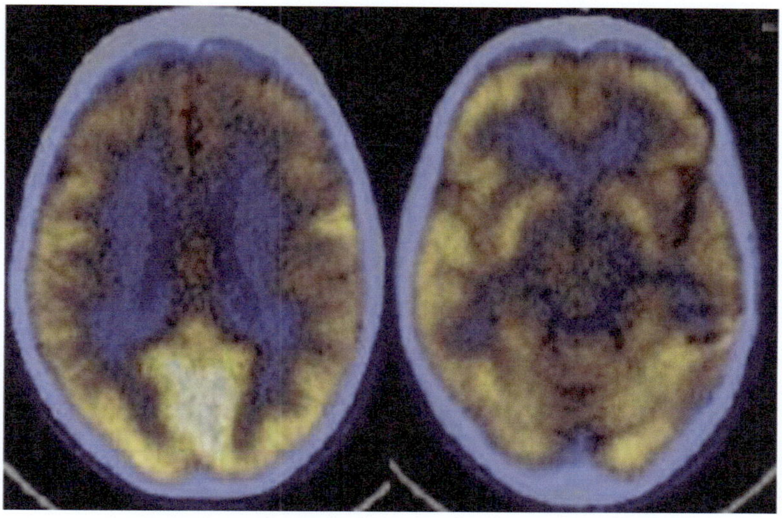

Fig. 7.10 Transaxial fused F-18-FDG PET-CT image of a case of PSP; hypometabolism is noted in both frontal and anterior cingulate cortices (*left*) and in both basal ganglia and midbrain (*right*)

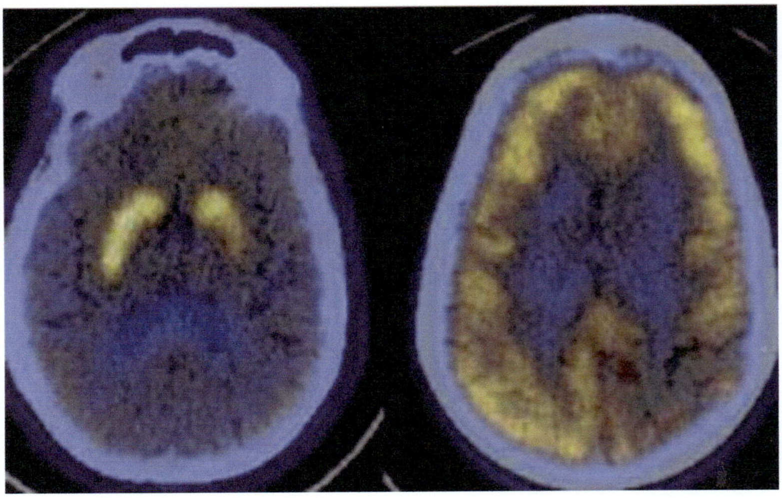

Fig. 7.11 Transaxial fused F-18-FDOPA image (*left*) showing decreased tracer uptake in caudal left putamen in a case of CBS symptomatic on the right side. Transaxial fused [18] F-FDG PET-CT image (*right*) of the same case of CBS; hypometabolism is noted in the left parietal cortices

Table 7.2 An overview of all functional markers that can be used for the diagnosis and differential diagnosis of Parkinsonian syndromes

Disease	Presynaptic markers (DAT, FDOPA, VMAT)	Postsynaptic markers (D2 receptor targets)	Metabolic patterns (F-18 FDG)
Idiopathic Parkinson's disease	Decreased asymmetrically with rostrocaudal gradient	Normal or upregulated	Pallidothalamic and pontine hypermetabolism, parieto-occipital hypometabolism
Multiple system atrophy	Decreased involving both caudate and putamen	Decreased	Hypometabolism in both basal ganglia marked posteriorly ± cerebellar hypometabolism
Progressive supranuclear palsy	Decreased involving both caudate and putamen	Decreased	Hypometabolism in both basal ganglia, midbrain, and anterior cingulated cortices
Corticobasal Syndrome	Decreased posteriorly opposite to clinically affected side	Decreased	Hypometabolism in the basal ganglia and parietal cortices opposite to the clinically affected side

Another interesting application of PET is in patients with paraneoplastic syndromes (PNS). PNS represent rare symptom complexes resulting from the ability of tumor cells to disrupt the homeostatic processes of various body systems. Suspecting and investigating PNS are crucial as up to 80 % of patients present with PNS before there is any other indication of malignancy. A PET scan and regular surveillance may reveal occult malignancies better than CT or MRI (Fig. 7.12). Neuromodulatory therapies and treatment of the underlying malignancy remain the best management options in the patients being evaluated for primary malignancies. ^{18}F-FDG PET is a useful whole-body screening modality for localizing the primary site in these patients.

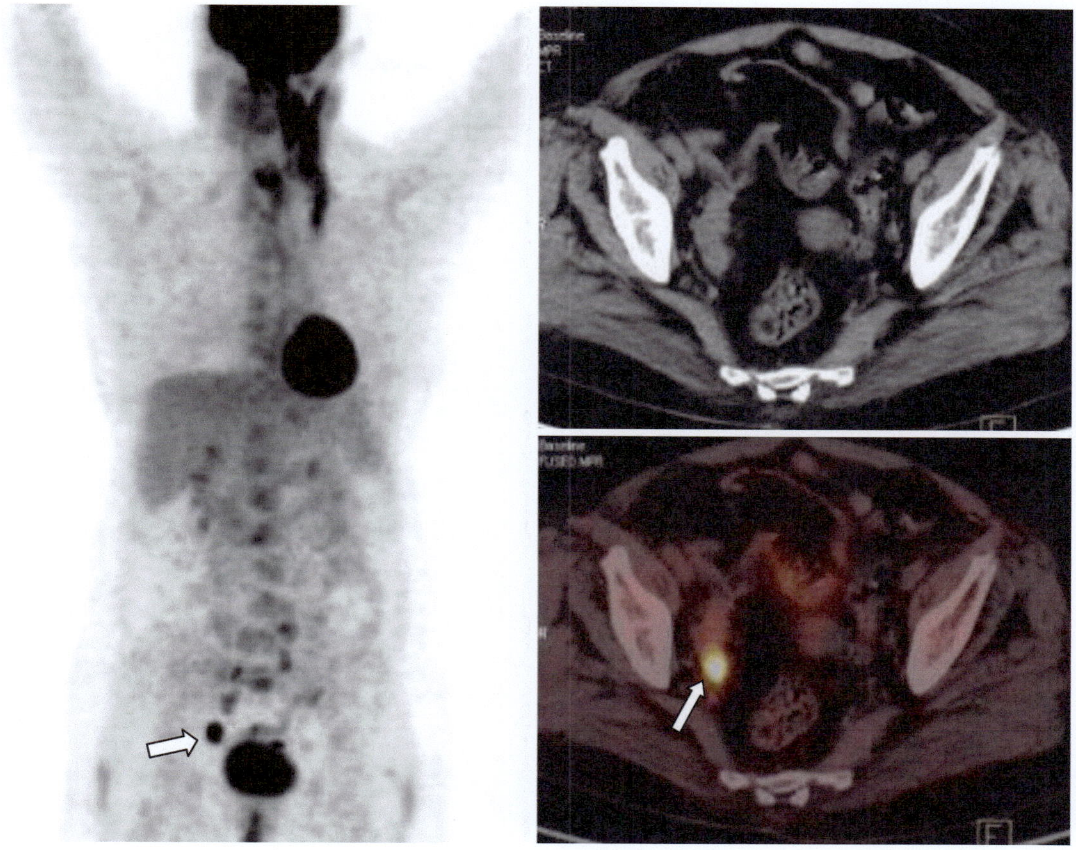

Fig. 7.12 A 66-year-old female presenting with acute onset of cerebellar signs. A PNS was suspected. The whole-body FDG PET-CT revealed a hypermetabolic soft tissue density mass in the right adnexal region which was confirmed to be an Ovarian neoplasm on biopsy (*arrows*)

For Further Reading

1. Berti V, Pupi A, Mosconi L. PET-CT in diagnosis of dementia. Ann N Y Acad Sci. 2011;1228:81–92.
2. Bohnen NI, Djang DS, Herholz K, et al. Effectiveness and safety of F-18 F-18 FDG PET in the evaluation of dementia: a review of recent literature. J Nucl Med. 2012;53:59–71.
3. Carne RP, O'Brien TJ, Kilpatrick CJ, et al. MRI-negative PET –positive temporal lobe epilepsy: a distinct surgically remediable syndrome. Brain. 2004;127:2276–85.
4. Foster NL, Heidebrink JL, Clark CM, et al. F-18 FDG-PET improves accuracy in distinguishing fronto-temporal dementia and Alzheimers disease. Brain. 2007;130:2616–35.
5. Juhasz C, Chugani HT. Positron Emission tomography: glucose metabolism in extratemporal lobe epilepsy. In: Chugani HT, editor. Neuroimaging in epilepsy. New York: Oxford University Press; 2011. p. 141–55.
6. Silverman DHS, Small GW, Chang CY, et al. Positron emission tomography in evaluation of dementia. JAMA. 2001;286:2120–7.
7. Uijl SG, Leijten FS, Arenda JB, et al. The added value of F-18 Fluorodeoxyglucose PET in screening for temporal lobe epilepsy surgery. Epilepsia. 2007;48(11): 2121–9.

Applications of PET in Management of Brain Tumors

8

Madhvi Tripathy and C.S. Bal

Brain tumors being structural lesions of the brain and their diagnosis and management are primarily dependent on computed tomography (CT) and magnetic resonance imaging (MRI), the latter being the standard method for evaluating both primary and metastatic tumors of the brain with excellent sensitivity. Positron emission tomography (PET) provides physiological information related to tumor metabolism, proliferation rate, aggressiveness, and invasiveness. It indirectly also provides information on the functional interaction of the tumor with the remaining brain. In fact, the first oncological application of PET was for the assessment of brain tumors. The state-of-the-art PET scanners available today provide excellent anato-metabolically coregistered images with the added utility of CT for attenuation correction of PET images. The future of fusion technology, which has recently been launched in global markets, is the integrated PET/MRI scanner, and the one important area which definitely stands to benefit from this technology is the brain.

A list of PET tracers, which can be used to image various aspects of tumor biology, is enlisted below:

Fluorine-18 (F-18)-labeled tracers

- F-18 fluorodeoxyglucose (FDG) (glucose metabolism)
- F-18-fluroethyltyrosine (FET) (amino acid metabolism)
- F-18-fluoro-dihydroxyphenylalanine (FDOPA) (amino acid metabolism)
- F-18-fluorocholine (cell membrane synthesis)
- F-18 fluorothymidine (FLT) (tumor proliferation rate)
- F-18 fluoromisonidazole (FMISO) (intratumoral hypoxia)

Carbon-11 (C-11)-labeled tracers

- C-11 methionine (amino acid metabolism)
- C-11 choline (cell membrane synthesis)

Nitrogen-13 NH3 (NH_4^+) (K^+ analog)

Gallium-68-labeled tracers

DOTA-D-phe1-Tyr3-octreotide (DOTATOC-somatostatin receptor expression)

Each of the abovementioned tracers has their own advantages and disadvantages based on the radiolabels used and the aspect of tumor biology targeted. The most widely used PET tracer for evaluating brain tumors is F-18 fluorodeoxyglucose (FDG), the workhorse of PET in oncology. The advantage of F-18-labeled tracers is the favorable half-life of 110 min, which allows distribution to centers away from the cyclotron

M. Tripathy, MD, DNB (✉)
Asst. Prof., Department of Nuclear Medicine and PET, All India Institute of Medical Sciences, Ansari Nagar, New Delhi, India
e-mail: madhavi.dave.97@gmail.com

C.S. Bal, MD, DSc
Prof. & Head, Department of Nuclear Medicine and PET, All India Institute of Medical Sciences, Ansari Nagar, New Delhi, India
e-mail: csbal@hotmail.com

B.K. Das (ed.), *Positron Emission Tomography: A Guide for Clinicians*,
DOI 10.1007/978-81-322-2098-5_8, © Springer India 2015

facility. The short half-life of C-11 (T1/2 = 20 min) and N-13 (T1/2 = 10 min) makes availability restricted to centers with an in-house cyclotron facility. Ga-68 (T1/2 = 68 min)-labeled tracers have the advantage of being generator produced, and therefore would not depend on an expensive cyclotron facility for availability. The advantages and disadvantage of each radiopharmaceutical would be elaborated as we discuss the clinical applications.

Applications of PET-CT in brain tumors can primarily be in the setting of *primary brain tumors* or *brain metastasis* and are enumerated below:

Primary Brain Tumors

- Grading and prognosis of tumors
- Evaluation of recurrence versus radiation necrosis
- PET-guided biopsy/treatment
- Monitoring therapy
- Role in meningiomas
- Differentiate between central nervous system (CNS) opportunistic infections and CNS lymphoma

Metastatic Brain Tumors

- Carcinoma unknown primary presenting with brain metastasis
- Detection of brain metastasis in oncology patients undergoing scanning

Grading of Primary Brain Tumors

Though MRI is the modality of choice for the primary evaluation of intracerebral masses, F-18 FDG PET can play a role for differentiating a low-grade from high-grade pathology, especially when MRI features are equivocal, thus indirectly helping in deciding the urgency of treatment. F-18 FDG is actively transported across the blood-brain barrier, and the basis for its high uptake in tumor tissues is the overexpression of GLUT 1 transporters on tumor cells with increased glucose utilization and high metabolic rates. A positive correlation between F-18 FDG uptake and the degree of malignancy has been demonstrated with the high-grade tumors (Gd III and IV gliomas) showing a higher glucose consumption than the low-grade gliomas (Gd I and II). Cerebral glucose metabolism adjacent to or neurally connected to tumor site is often depressed because of vasogenic edema and diaschisis. F-18 FDG uptake has also been shown to be useful to prognosticate patients with brain tumors; mean survival time of patients with tumors exhibiting high glucose uptake is shorter than for tumors showing low glucose uptake. The association of tumor F-18 FDG uptake with survival has also been correlated in various studies. The limitation of its use for grading brain lesions is that F-18 FDG uptake is relatively nonspecific and is seen to occur in inflammatory and granulomatous tissues also. So, characterization of the nature of pathology is difficult.

Evaluation of Recurrence Versus Radiation Necrosis

The risk of radiation necrosis increases with the increasing use of stereotactic radiosurgery and when chemotherapy is combined with radiation for treatment of high-grade gliomas. It is clinically challenging to evaluate disease status with the use of MRI in brain tumor patients who have been treated. Treatment-induced changes such as radiation necrosis appear as enhancing lesions on conventional MRI images and can be difficult to distinguish from similarly enhancing recurrent tumors. Differentiating radiation necrosis (RN) from viable tumor is important for optimizing patient care in management of cerebral gliomas. Studies of the accuracy of F-18-FDG PET for discrimination between viable tumor and necrosis have demonstrated a sensitivity range of 43–83 % with a specificity range of 40–100 %. Glucose is an obligatory energy substrate for the brain, which results in high glucose uptake in normal gray matter, and this makes evaluation of brain tumors difficult. This is especially so for the low-grade tumors (grade I and II gliomas) with F-18-FDG uptake similar to that in normal white matter, thus decreasing the sensitivity of detection of these

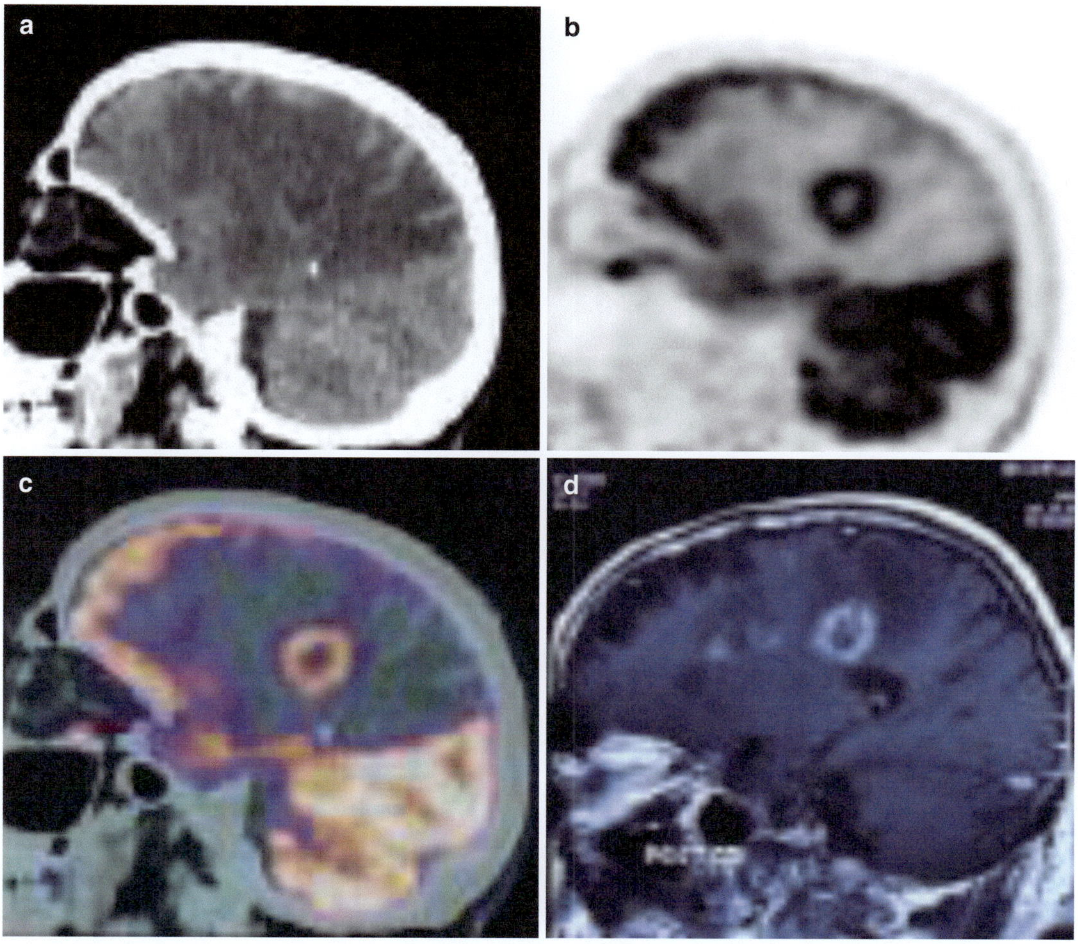

Fig. 8.1 Sagittal CT (**a**), plain F-18 FDG PET (**b**), fused F-18 FDG PET-CT (**c**), and gadolinium-enhanced MR images (**d**) of a 40-year-old male patient with recurrent Gd II glioma. MR findings were equivocal for recurrence or radiation necrosis. F-18-FDG was suggestive of viable recurrent tumor in the right frontoparietal lobe

lesions unless they are located in or have extension into the white matter regions (Fig. 8.1) or have uptake greater than the surrounding gray matter (high-grade tumors). Having the MRI images is useful in these cases and coregistration or correlation of uptake with contrast enhancement of these lesions on MRI is useful for interpretation. The specificity of F-18 FDG is more than MRI in evaluating these lesions. High F-18 FDG uptake in previously known low-grade tumor is diagnostic of anaplastic transformation and indicates a poorer prognosis. It is generally recommended that a gap of at least 6 weeks be given between radiation and the F-18 FDG PET study. Another advantage of F-18 FDG PET in this regard is that the entire craniospinal axis can be assessed, thus picking up leptomeningeal metastasis in primary brain tumors that are known to metastasize through CSF (Fig. 8.2).

Amino acid and amino acid analog PET tracers are an important class of tumor imaging agents developed for improving tumor detection ability. Studies have demonstrated the usefulness of C-11 methionine, F-18 FDOPA, and F-18 FET, especially in recurrence of low-grade gliomas. The mechanism of uptake of amino acid tracers is via transport by neutral amino acid transporters, which is upregulated in tumors. Thus, while F-18 FDG is useful for the detection of recurrence in high-grade gliomas (have uptake greater than gray matter uptake) (Fig. 8.3), amino acid tracers are useful for evaluating recurrence of both low- and high-grade gliomas. Amino acid tracers score over F-18-FDG in the evaluation of brain tumors because of

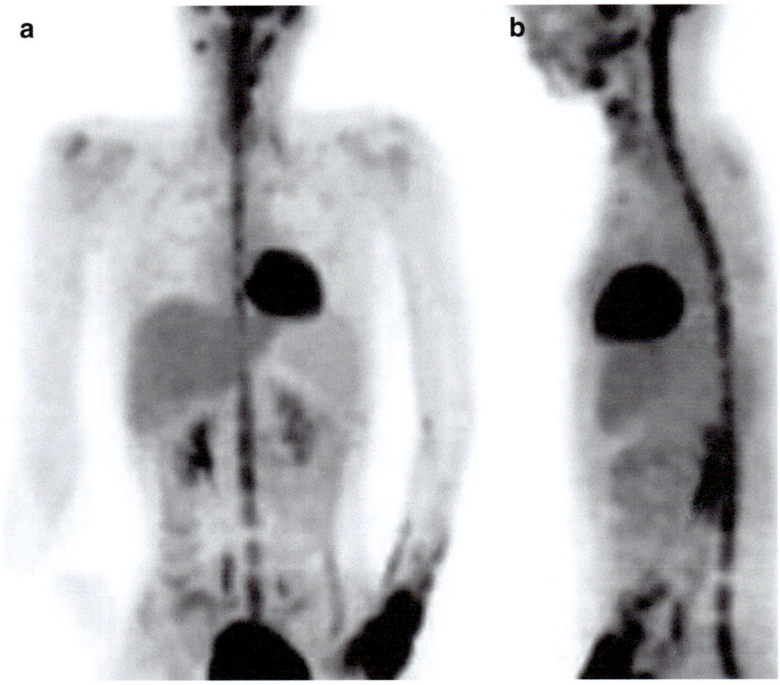

Fig. 8.2 A 20-year-old male with recurrent desmoplastic medulloblastoma showing leptomeningeal metastasis on FDG PET-CT coronal (**a**) and sagittal (**b**) maximum intensity projection (MIP) images

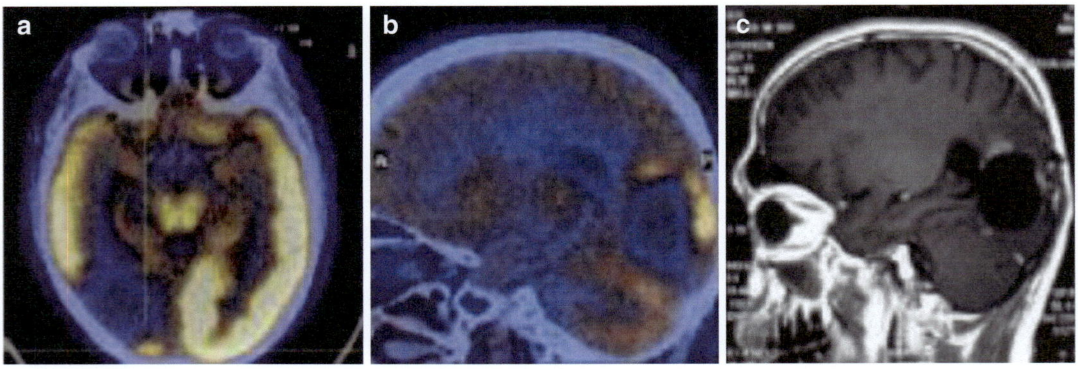

Fig. 8.3 Transaxial fused F-18 FDG PET-CT, (**a**) sagittal fused C-11 methionine PET-CT, (**b**) and gadolinium-enhanced MRI (**c**) images of a 58-year-old male with recurrent glioblastoma multiforme. F-18 FDG PET revealed recurrent viable tumor in the posterior aspect of the lesion in the right occipital lobe, while C-11 methionine showed viable tumor corresponding to enhancing lesion on MRI

high uptake in tumor tissue and low uptake in the normal gray matter cortex, thus giving good lesion to background ratios. Diagnostic accuracy of C-11 methionine (Fig. 8.4), F-18 FDOPA (Fig. 8.5) and F-18 FET has been shown to be superior to F-18 FDG in this regard. Amino acid uptake, on the other hand, may not be useful for prognostication as even low-grade gliomas like oligodendroglioma have high methionine uptake and F-18 FDG is the agent of choice for prognostication. Thus, a combination of F-18 FDG and amino acid PET can be performed for evaluating recurrence and prognosticating the patient. Recently, though, it has been shown that C-11 methionine can predict prognosis in gliomas and is better than F-18 FDG PET and MRI in predicting survival in low-grade gliomas. Another tracer that is useful in the recurrence of high-grade neoplasms is F-18 FLT (Fig. 8.6). F-18 FLT PET is a noninvasive marker of cellular proliferation. Its uptake correlates with thymidine

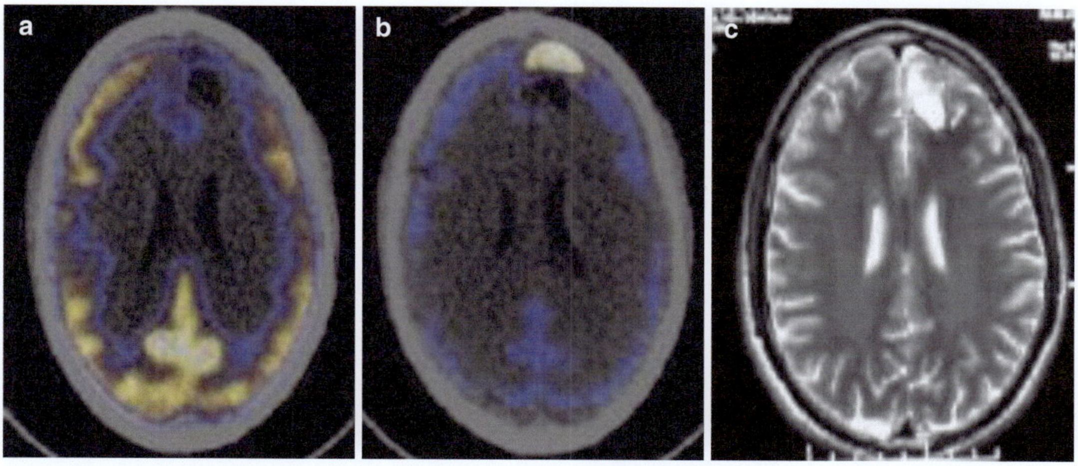

Fig. 8.4 Transaxial fused F-18 FDG PET-CT (**a**), transaxial fused C-11 methionine PET-CT, (**b**) and T2W MRI (**c**) images of a 38-year-old male with recurrent astrocytoma grade II in the left frontal lobe. F-18 FDG PET-CT was equivocal for lesion recurrence, while C-11 methionine showed definite localization on the anterior aspect of the resection cavity suggestive of viable tumor. MRI was also equivocal for tumor residue/recurrence

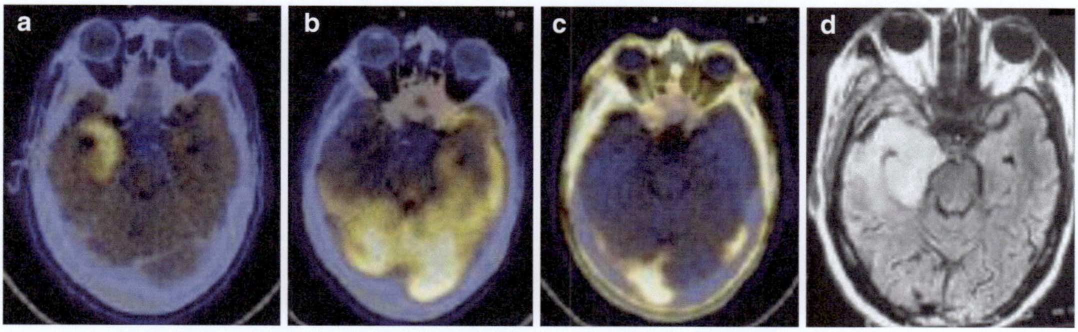

Fig. 8.5 Transaxial fused F-18 FDOPA PET-CT (**a**), F-18 FDG PET-CT (**b**), F-18 FLT PET-CT (**c**), and T2W MRI (**d**) images of a 56-year-old female, operated case of astrocytoma grade II with proven recurrence in the right temporal lobe, which was demonstrated on the FDOPA images, but not on FDG or FLT images

kinase-1 activity, which is an enzyme expressed during the salvage pathway of DNA synthesis in the cell cycle. Phosphorylation of F-18 FLT intracellularly by thymidine kinase-1 results in trapping of negatively charged F-18 FLT monophosphate. Uptake of F-18 FLT has been found to correlate with Ki-67, the immunohistochemistry marker of cellular proliferation.

PET-Guided Biopsy and Treatment

Accurate grading and diagnosis are especially important for directing the therapeutic approach and providing prognosis in patients with unresectable tumors. MRI-guided stereotactic biopsy may not always yield a valid diagnosis and tumor grading because some nonenhancing areas may be high grade. F-18 FDG PET-guided metabolic biopsies can help in obtaining specimens from metabolically most active areas of the lesion, thus improving the yield of brain biopsies. Amino acid tracers can also be used for this purpose, especially when low-grade tumors are being considered that may be hypo- or isometabolic on F-18 FDG. Studies combining the use of amino acid tracers with MRI have given good yields and identification of tumor tissue.

PET has been used to delineate tumor volumes for dose escalation radiation therapy in brain tumors. F-18 FDG uptake has been shown to be a

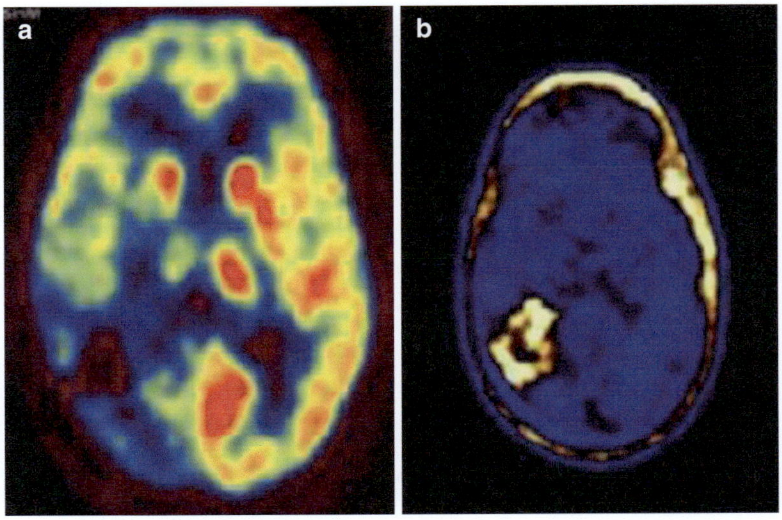

Fig. 8.6 Transaxial plain F-18 FDG (**a**) and F-18 FLT PET (**b**) images of a case of recurrent glioblastoma well delineated on the FLT image, but not with FDG

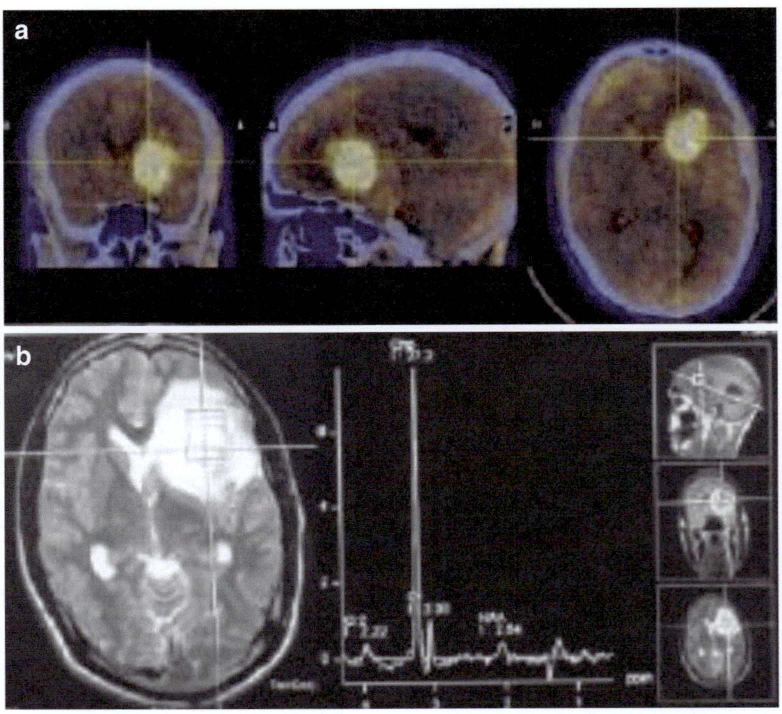

Fig. 8.7 Fused coronal, sagittal, and transaxial F-18 FMISO PET-CT images in a case of left frontal anaplastic astrocytoma (biopsy proven) showing F-18 FMISO uptake (**a**) in the lesion suggesting intratumoral hypoxia. MR spectroscopy showing choline peak (**b**)

significant parameter for predicting survival and time to tumor progression, but the results have not been consistent. C-11 methionine PET has also been used to delineate tumor volumes for fractional stereotactic radiotherapy with a survival advantage for patients in whom C-11 methionine scans were fused with CT or MRI. Tumor hypoxia is associated with tumor progression and resistance to radiotherapy in brain tumors also. F-18 FMISO is a nitroimidazole derivative that has been developed as a PET agent to image hypoxia on the basis that its metabolites are trapped exclusively in hypoxic cells. F-18 FMISO uptake is found in high-grade gliomas (Fig. 8.7), and a significant relationship has been found between F-18 FMISO uptake and expression of angiogenesis marker VEGF-R1. F-18 FMISO may have a role in directing and monitoring hypoxia therapy.

Monitoring Therapy

Individualized care of glioma patients is likely to benefit from imaging biomarkers as precocious predictors of therapeutic efficacy. F-18 FDG, F-18 FLT, and amino acid tracers like F-18 FDOPA and F-18 FET can be used to predict therapeutic response in patients with recurrent high-grade gliomas put on bevacizumab-irinotecan. They have been found useful for predicting failure of therapy, overall survival, and progression-free survival in these patients.

Ga-68 DOTATOC

Meningioma is the most frequent nonglial brain tumor. Human meningioma cells strongly express somatostatin receptors (SSTR), thus the somatostatin analog Ga-68 DOTATOC, which binds to somatostatin receptor subtype 2 (SSTR-2) and is useful for imaging meningiomas. Ga-68 DOTATOC PET-CT shows high uptake in meningiomas with good target-to-background ratios and can be used for distinction of meningiomas from other brain tumors. Together with FDG, Ga-68 DOTANOC PET-CT can be used for the grading of meningiomas. Grade II and III meningiomas show high FDG uptake. Ga-68 DOTANOC PET-CT is complementary to MRI for evaluating residual/recurrent meningiomas (Fig. 8.8). Ga-68 DOTATOC allows detection of additional lesions in patients with multiple meningiomas. It is also useful for improved target volume delineation when IMRT is used for treatment of residual/recurrent meningiomas.

C-11 methionine and F-18 tyrosine are other tracers that can be effectively used for meningioma evaluation.

Differentiate Between Central Nervous System (CNS) Opportunistic Infections and Primary CNS Lymphoma (PCNSL)

Discrimination between PCNSL and toxoplasmosis is sometimes difficult in an immunocompromised patient with acquired immunodeficiency syndrome on MRI. Cerebral lymphomas have a high cell density and high glucose metabolism, thus resulting in high FDG uptake (even higher than that in metastasis and gliomas). Thus, F-18 FDG PET-CT is useful for differentiating PCNSL with high uptake (Fig. 8.9) from toxoplasmosis with low uptake (Fig. 8.10). F-18 FDG PET is also useful to demonstrate a response to chemotherapy in lymphoma patients after initiation of therapy.

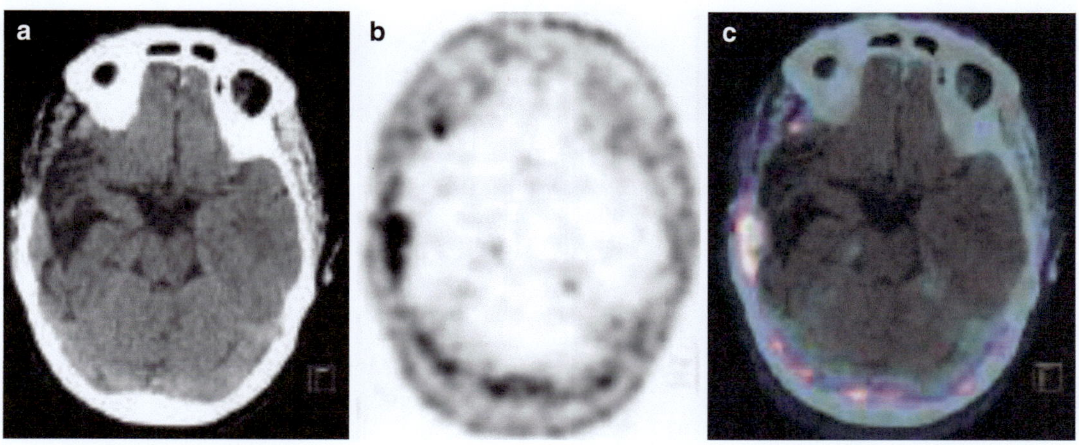

Fig. 8.8 Transaxial plain CT (**a**), plain Ga-68 DOTANOC PET (**b**), and fused Ga-68 DOTANOC PET-CT (**c**) images of a 30-year-old male operated thrice for right sphenoid wing meningioma still showing an extra-axial hyperdense lesion with DOTANOC accumulation suggestive of residual meningioma

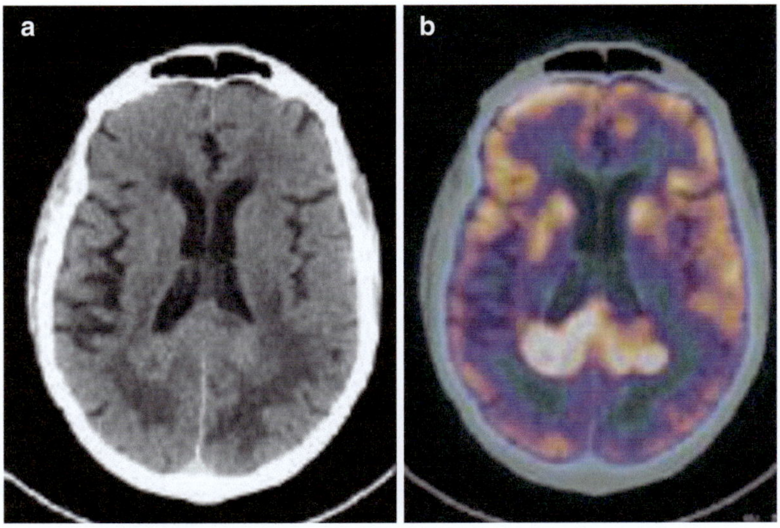

Fig. 8.9 Transaxial plain CT (**a**) and fused F-18 FDG PET-CT (**b**) images of a 51-year-old male treated for testicular lymphoma now presenting with a contrast-enhancing lesion on MRI showing intense FDG accumulation in a mass involving the corpus callosum posteriorly on either side of midline (mimicking butterfly glioma) with SUVmax 16.8 suggesting a PCNSL

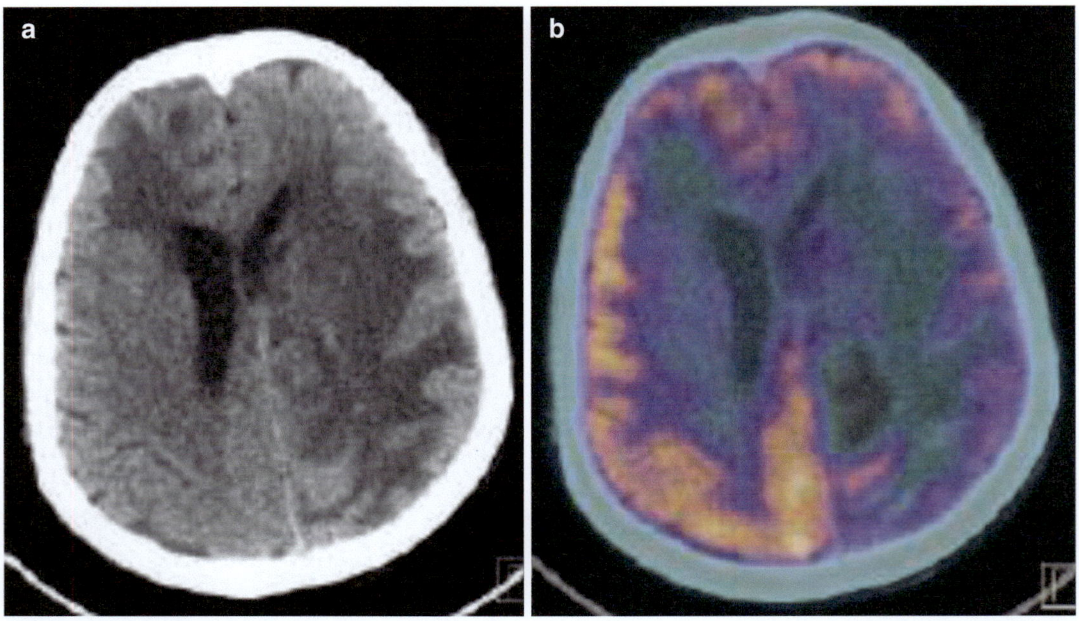

Fig. 8.10 Transaxial CT (**a**) and fused F-18 FDG PET-CT images (**b**) of a 42-year-old renal transplant recipient presenting with intracerebral lesions, lymphoma, and toxoplasmosis. The hypo- to isometabolic lesions in the left parietal region favor the possibility of toxoplasmosis. Toxoplasma serology was positive, and the patient showed improvement on amphotericin

Metastatic Brain Tumor

CNS metastasis is a common complication of systemic tumors. The most common malignancies metastasizing to the brain are lung and breast tumors and melanomas. Though MRI is the standard modality for assessing brain metastasis, F-18 FDG PET-CT plays an important role in detecting asymptomatic CNS metastasis that may exist in patients with other metastatic site undergoing F-18 FDG PET-CT for evaluation of disease. It can also play an important role in assessing response to therapy in these patients. However, brain metastasis can be hypometabolic and may not be

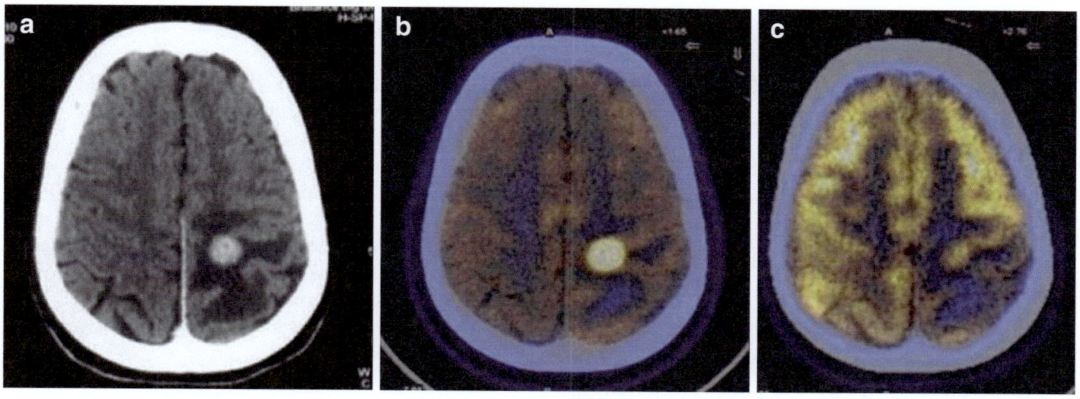

Fig. 8.11 Transaxial contrast-enhanced CT (**a**), fused C-11 methionine PET-CT (**b**), and F-18 FDG PET-CT (**c**) images reveal a round contrast-enhancing lesion in left parietal cortex showing avid methionine accumulation but isometabolic on FDG PET

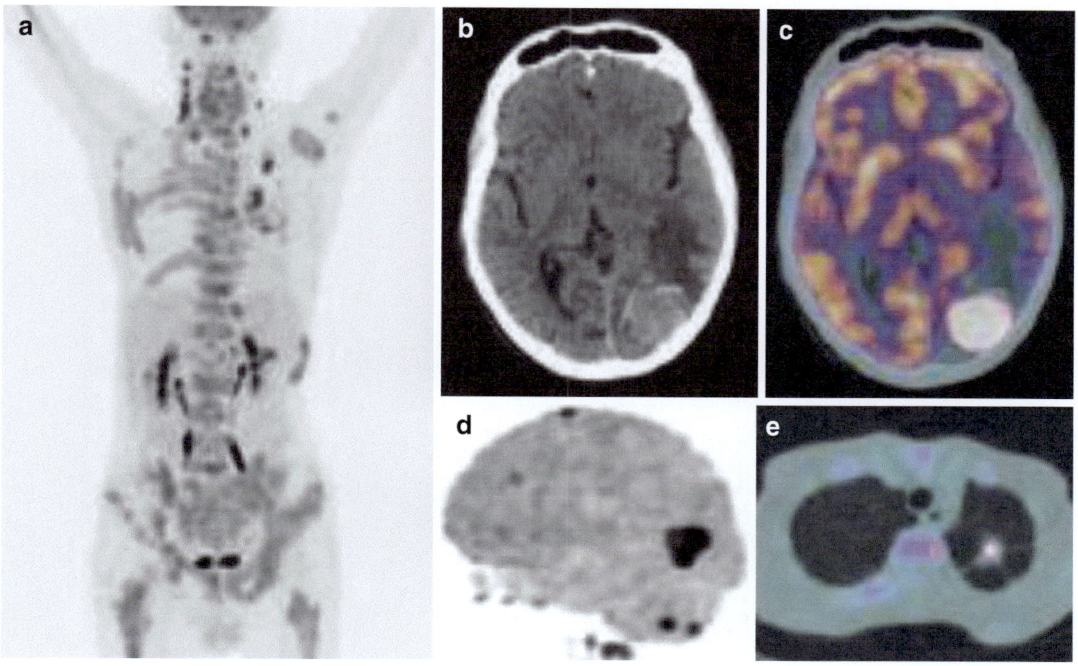

Fig. 8.12 A 50-year-old male presenting with multiple enhancing lesions in the brain on MRI suspected to have intracranial metastasis was sent for FDG PET-CT to localize the primary site. MIP PET (**a**) whole-body image showing extensive skeletal metastasis. Transaxial plain CT images (**b**) of the head showing a hyperdense lesion in the left parietal lobe, transaxial fused PET-CT images showing abnormal tracer accumulation in the lesion (**c**). MIP PET (**d**) image of the brain showing multiple hypermetabolic lesions suggestive of metastasis. The transaxial fused PET-CT image (**e**) of the chest showing a speculated soft tissue mass in the left lung upper lobe, which was the likely primary site

adequately assessed with F-18 FDG; in these patients, amino acid tracers like C-11 methionine (Fig. 8.11) can play an important role, especially when leptomeningeal metastasis is also present.

Another presentation in this regard is a patient presenting with single- or multiple-ring enhancing lesions on MRI, which are suspicious for metastasis from a primary site. F-18 FDG PET-CT is a one-stop modality for assessing the entire body for a primary site, which is most commonly the lung in these patients (Fig. 8.12).

Concluding Remarks

F-18 FDG PET-CT is extremely useful for prognostication of patients with brain tumors.

Amino acid tracers may improve diagnostic accuracy for detection of recurrence and for differentiating radiation necrosis from recurrence in both high- and low-grade gliomas in comparison to F-18 FDG.

High F-18 FDG uptake in a previously low-grade glioma is indicative of conversion to high grade.

Ga-68 DOTATOC PET-CT is complementary to MRI for evaluating recurrent and multiple meningiomas.

F-18 FDG PET-CT is an excellent one-stop shop for evaluating patients presenting with brain metastasis for a primary site.

For Further Reading

1. Herholz K, Langen KJ, Sceipers C, et al. Brain tumors. Semin Nucl Med. 2012;42:356–70.
2. Popperl G, Gotz C, Rachinger W, et al. Value of F-18 fluoroethyltyrosine PET for the diagnosis of recurrent glioma. Eur J Nucl Med Mol Imaging. 2004;31: 464–1470.
3. Singhal T, Narayanan TK, Jacobs MP, Bal C, Mantil JC. C-11 Methionine PET for grading and prognostication in gliomas: a comparison study with F-18 FDG PET and contrast enhancement on MRI. J Nucl Med. 2012;53:1709–15.
4. Tripathi M, Sharma R, Varshney R, et al. Comparison of F-18 FDG and C-11 methionine PET-CT for the evaluation of recurrent primary brain tumors. Clin Nucl Med. 2012;37:158–63.

The Role of PET–CT in Management of Nasopharyngeal Carcinoma

9

Vandana Dhingra, Alok Pawaskar, and Sandip Basu

Anatomy of the Nasopharynx

The *pharynx* is a part of the digestive tube, which is placed behind the nasal cavities, mouth, and larynx. It is a musculomembranous tube, somewhat conical in form, broad upward, and narrow downward. Its extent is from the inferior surface of the skull upto the cricoid cartilage (sixth cervical vertebra). Pharynx is about 12.5 cm in length and is subdivided into three parts: (a) Nasopharynx, (b) oropharynx, and (c) laryngopharynx from upward down.

Radiological Anatomy of the Nasopharyngeal Region

The nasopharynx begins anteriorly at the nasal choanae and measures about 4 cm in height and 2 cm in width. Its roof slopes beneath the sphenoid bone and clivus. It is closely related to the foramina of the central skull base, thus accounting for the frequent neurological involvement in nasopharyngeal carcinomas. The floor of the nasopharynx is formed by the hard and soft palates, thereby transitioning into the oropharynx. The auditory tube anteriorly and levator palatine muscle posteriorly pass through the sinus of Morgagni. Further, posteriorly, the pharyngeal mucosa herniates out to form the lateral recess, the fossa of Rosenmuller (site of origin of >50 % nasopharyngeal carcinomas). These structures produce the characteristic appearance of the lateral nasopharyngeal wall on axial CT and MRI scans, with bilateral paired recesses separated by ridge of levator palatii muscle (red arrow) (Figs. 9.1 and 9.2). On coronal scans, the Fossa of Rosenmuller appears superior to the cartilaginous opening of the auditory tube, torus tubarius.

Nasopharyngeal Carcinoma

Incidence and Etiology

Nasopharyngeal carcinoma (NPC) is a tumor arising from the epithelial cells that cover the surface and line the nasopharynx. NPC was first described as a separate entity by Regaud and Schmincke in 1921.

Though rare among most parts of the world, NPCs are common in the certain regions.

V. Dhingra, MD (✉)
Consultant, Radiation Medicine Centre, Bhabha Atomic Research Centre, Tata Memorial Hospital Annexe, Mumbai, India
e-mail: modisbanu@yahoo.com

A. Pawaskar, DRM, DNB
Consultant, Department of Nuclear Medicine, Radiation Medicine Centre, Bhabha Atomic Research Centre, Tata Memorial Hospital Annexe, Mumbai, India
e-mail: alokpawaskar@rediffmail.com

S. Basu, MD
Professor, Radiation Medicine Centre, Bhabha Atomic Research Centre, Tata Memorial Hospital Annexe, Parel, Mumbai, India
e-mail: drsanb@yahoo.com

B.K. Das (ed.), *Positron Emission Tomography: A Guide for Clinicians*,
DOI 10.1007/978-81-322-2098-5_9,

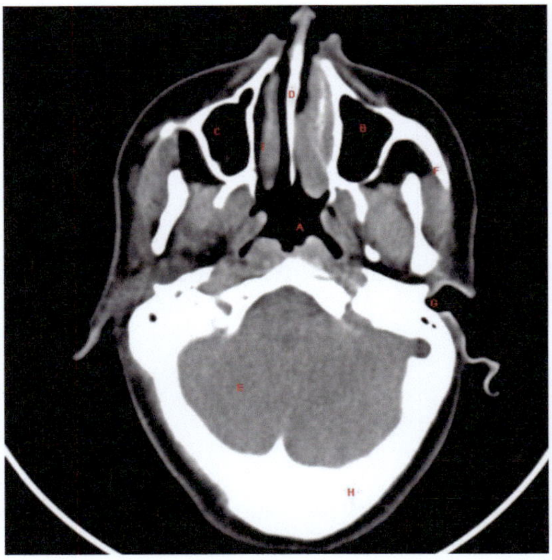

Fig. 9.1 Radiological anatomy of the nasopharynx: axial CT scan. (*A*) Nasopharynx, (*B*) Left maxillary sinus, (*C*) Right maxillary sinus, (*D*) Nasal septum, (*E*) Cerebellum, (*F*) Zygomatic arch, (*G*) External acoustic meatus, (*H*) Occipital bone, (*I*) Inferior nasal turbinate

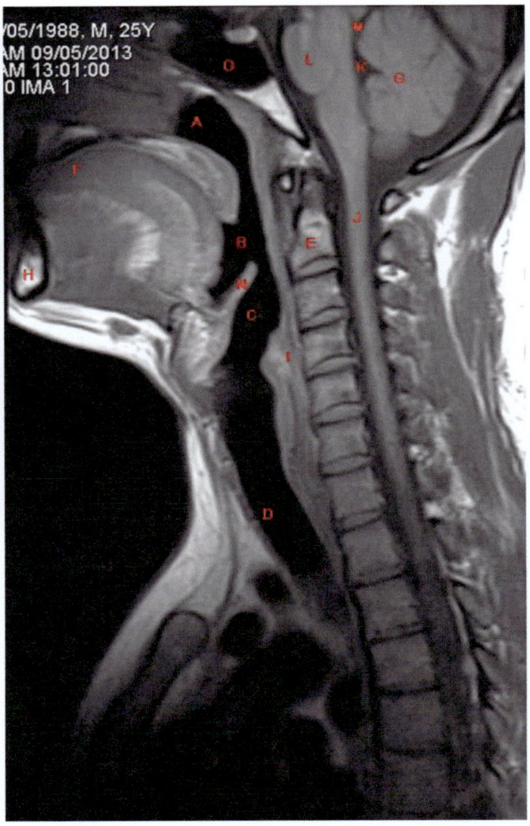

Fig. 9.2 *Radiological anatomy of nasopharyngeal region: sagittal view of MRI.* (*A*) Nasopharynx, (*B*) Oropharynx, (*C*) Hypopharynx, (*D*) Trachea, (*E*) Second cervical vertebra, (*F*) Tongue, (*G*) Cerebellum, (*H*) Body of mandible, (*I*) Prevertebral fascia, (*J*) Spinal cord, (*K*) Fourth ventricle, (*L*) Pons, (*M*) Acqueduct of Silvius, (*N*) Epiglotis, (*O*) Sphenoidal sinus

Particularly high incidences are seen in the South China and Southeast Asian regions. A very common cancer found in Odisha. Chinese emigrants continue to have a high incidence of the disease, but the rate of NPC among ethnic Chinese born in North America is considerably lower than those born in China. Certain other regions like the Maghrebian Arabs in North America, and Eskimos in the Arctic also show higher incidences . This epidemiologic evidence implies that both environmental factors and genetic susceptibility play roles in the development of NPC. The environmental factors that have been implicated include (a) exposure to nitrosamines in salted and pickled foods, (b) certain human leukocyte antigen subtypes, as they have various genetic polymorphisms, and (c) Epstein–Barr Virus (*EBV) infection* (recent data suggests that almost all NPC tumors, regardless of their histological subtype, have comorbid EBV infections, especially type 2 and 3 tumors, which is strong evidence for EBV as the etiology of NPC; this close association with EBV is what makes NPC unique from other head and neck cancers).

The most widely used classification currently is the WHO classification where there are three types of NPCs:

(i) Type 1: Squamous cell carcinoma, (ii) Type 2: Non-keratinizing carcinoma, and (iii) Type 3: Undifferentiated carcinoma.

Clinical Presentation

Origin of Primary Tumor

NPC usually originates in the lateral wall of the nasopharynx, commonest site being the fossa of Rosenmuller.

Local Extent

The NPCs extending from the nasopharynx onto lateral wall and/or posterosuperiorly to the base of the skull or the palate, nasal cavity, or oropharynx.

Depending on its extent, symptoms related to the primary tumor include trismus, pain, otitis media, nasal regurgitation due to paresis of the soft palate, hearing loss, and cranial nerve palsies. Larger growths may produce nasal obstruction or bleeding and a "nasal twang".

Metastatic Spread

NPCs typically metastasize to cervical lymph nodes. Being difficult to detect by simple clinical means, it is one of the commonest head and neck tumor responsible for cervical lymphadenopathy of unknown primary. Cervical lymphadenopathy is the initial presentation in many patients, and the diagnosis of NPC is often made by lymph node biopsy.

Distant metastases may occur in bone, lung, mediastinum, and, more rarely, the liver.

Metastatic spread may result in bone pain or organ dysfunction. Rarely, a paraneoplastic syndrome of osteoarthropathy may occur with widespread disease.

Histopathological Classification

Three subtypes of NPC are recognized in the World Health Organization (WHO) Classification:

- Type 1: squamous cell carcinoma, typically found in the older adult population
- Type 2: non-keratinizing carcinoma
- Type 3: undifferentiated carcinoma

Most cases in childhood and adolescence are type 3, with a few type 2 cases. Type 2 and 3 are associated with elevated EBV titers, but type 1 is not. The Cologne modification of the WHO scheme by Krueger and Wustrow includes the degree of lymphoid infiltration. Types 2 and 3 may be accompanied by an inflammatory infiltrate of lymphocytes, plasma cells, and eosinophils, which are abundant, giving rise to the term lymphoepithelioma.

Diagnostic Methods

It is challenging to diagnose NPCs due to nonspecific symptoms, which may mimic ear infection as well as its difficult location for clinical assessment. However, evaluation under following broad headings can be done for patients presenting with suspicion of NPC.

1. Clinical Examination: For evaluation of the size and location of cervical lymph nodes.
2. Indirect nasopharyngoscopy: To assess the primary tumor.
3. Neurological examination of cranial nerves.
4. Regional Imaging with Anatomical Imaging: Computed Tomography (CT)/Magnetic Resonance Imaging (MRI) scan of the head and neck to below clavicles to assess base of skull erosion.
5. Staging/Metastatic workup:
   (a) Chest radiography (anteroposterior and lateral views) to see if NPC has spread to the lungs.
   (b) Baseline whole body evaluation with PET–CT
   (c) Bone scintigraphy by Tc 99 diphosphonate to look for metastatic skeletal spread.
6. Full blood count.
7. Baseline Urea, electrolyte, creatinine, liver function, Ca, PO_4, alkaline phosphate.
8. Prognostication: EBV viral capsid antigen and EBV DNA.
9. Pretreatment Histopathology: Biopsy of either the lymph nodes or primary tumor for histological examination.

PET–CT Imaging in Nasopharyngeal Carcinomas

The role of PET, especially PET–CT imaging, has been well established in the field of oncology. Nasopharyngeal carcinomas pose an inherent challenge for PET–CT imaging due to its proximity to the skull and due to the fact that the most widely used agent F-18 FDG, which is currently the workhorse of PET–CT imaging, demonstrates high physiological tracer uptake in the adjacent brain and choroid plexus and thus pose problems in the detection of extent of the primary tumor. In this aspect, the significance of simultaneous CT increases; and now MR along with PET, opens up a new option, which needs to be fully explored.

Patterns of Physiological Uptake of FDG in the Region of Nasopharynx

The importance of recognition of the physiological patterns is important for the evaluation of every region with FDG imaging as the variations in physiological patterns may be subtle and may prevent false reporting of images.

In the nasopharyngeal region, physiological uptake of following key anatomical points are to be considered:

- Soft palate may show intense FDG uptake physiologically
- The inferior concha generally shows low uptake
- Parotid glands show variable intensity of uptake and upto 14 % may show intense, non-specific FDG uptake
- Palatine and lingual tonsils also commonly show intense FDG uptake in >70 % of scans [average SUV[1] >3.00]
- The tongue has no or mild accumulation of FDG

Role of PET Imaging with F-18 FDG in Nasopharyngeal Carcinomas

We will consider the following major areas of NPC and highlight potential indications of F-18 FDG-PET in each of the following domains:

A. Initial diagnosis and assessment [with focus on extent] of primary tumor
B. Nodal disease involvement assessment
C. Distant metastatic workup
D. Evaluation for Recurrence of tumors post-radiotherapy
E. Treatment Planning (Radiotherapy) of nasopharyngeal carcinomas

[1] Standardized uptake value, SUV, (also referred to as the dose uptake ratio, DUR) is a widely used, simple PET quantifier, calculated as a ratio of tissue radioactivity concentration (e.g., in units kBq/ml) at time T, C_{PET} (T) and injected dose (e.g., in units MBq) at the time of injection divided by body weight (e.g., in units kg).

$SUV_{bw} = C_{PET}$ (T)/(Injected dose/Patient's weight)

Initial Diagnosis and Assessment of Primary Tumour

When imaging patients for suspected or biopsy proven NPC, attention should be made to patterns of spread of the tumor, as well as features that assist in staging. Staging of the primary tumor is performed according to the American Joint Committee on Cancer (AJCC) TNM classification. T staging is based on the relationship of the primary tumor to anatomical structures (Tables 9.1 and 9.2).

Evaluation of Primary and Locoregional Involvement

The key question is whether to prefer CT/MRI/PET or PET–CT in most cases. For the initial assessment of primary tumor and in order to plan treatment (radiotherapy), MRI remains the investigation of choice as the details of primary tumor extent as well as intricate details of intracranial extent can be given by an MRI. By imaging, the primary tumor is assessed for the local extent and the presence of invasion in the nasopharynx, nasal cavity, oropharynx, parapharyngeal region, skull base, paranasal sinus, brain, orbit, infratemporal fossa, and hypopharynx(s).

In NPC patients, MRI has been shown to be superior to PET–CT for the assessment of locoregional invasion and retropharyngeal nodal metastasis.

Stage T2b (parapharyngeal involvement) can be detected best by MRI; this carries prognostic significance and chemotherapy needs to be added in these cases.

Overall, PET–CT is more accurate than PEt alone or CT alone for the depiction of NPC. Fused PET–CT is a valuable imaging tool in patients for staging diagnosis of NPC.

Usually, FDG uptake is increased, with an overall SUV of >2.5 taken as cutoff for delineation between benign and malignant disease, although there is overlap between them due to presence of physiological structures with varying FDG uptake. Sometimes, primary NPCs may have poor FDG uptake or FDG uptake may be obscured due to presence of lymph nodes in vicinity. On the other hand, false-negative result

Table 9.1 TNM staging of nasopharyngeal carcinomas as per AJCC (2002)

T staging	N staging	M staging
Tis in situ tumor	Nx (nodes cannot be assessed)	Mx: metastatic disease cannot be, or has not been, assessed
T1 tumor confined to the nasopharynx	N0 (no regional nodal metastases)	M0: (no metastasis)
T2 tumor extends to soft tissues	N1 (unilateral nodes <6 cm in greatest dimension)	M1: (distant metastasis)
T2a (extends to oropharynx or nasal cavity but no parapharyngeal extension)	N2 (bilateral nodes <6 cm in greatest dimension)	
T2b (parapharyngeal extension beyond pharyngobasilar fascia)	N3a (nodes >6 cm in dimension)	
T3 (tumor involves bone or paranasal sinuses)	N3b (nodal metastasis in supraclavicular fossa)	

Table 9.2 Final tumor stage using the AJCC staging system for NPC

Stage 0: Tis N0 M0
Stage I: T1 N0 M0
Stage II A: T2a N0 M0
Stage II B: T2b N0 M0 Or T1/T2a/T2b N1 M0
Stage III: T1/T2a/T2b N2 M0 Or T3 N0/N1/N2 M0
Stage IV A: T4 N0/N1/N2 M0
Stage IV B: Any T N3 M0
Stage IV C: Any T Any N M1

of CT, for example thickened mucosa showing intense FDG uptake would clinch the presence of disease. In such circumstances, fused images are of advantage as both CT and PET would both compensate for false positive or negative results of the other.

Usually, false-positive results of FDG PET are seen in infections and inflammation of adjacent structures like tonsils, salivary glands and muscles, uptake in reactive non-neoplastic lymph nodes and, recent surgery, noninfectious inflammation and granulation at the surgical site. Muscle uptake can mimic tumor in the cervical region, and muscle relaxant may be given in cases where doubtful cervical uptake is observed.

The major advantage of CT is to help correctly differentiate the physiological uptake that would otherwise be mistaken for tumor on PEt alone. Fused images with PET–CT allow for direct correlation between FDG metabolic uptake and anatomic structures, thus reducing false-positive results. It is important to evaluate the patient's clinical history and physical findings to distinguish between benign and malignant FDG uptake.

False-negative results: with PET–CT may occur under the following conditions:

- Malignancy present in structures with a physiologically elevated metabolism: like salivary glands, tonsils, soft palate, etc.
- Tumor size below the resolution of the current PET–CT scanner (>15 mm); there are reports that this can be partly overcome by using a dedicated head and neck PET–CT scanning device.
- The tumor is low in cell density or low metabolic rate of FDG. A delayed view between 2 and 3 h post injection; prolonged emission time may be helpful in detecting these tumors.

Key Learning Points

The value of PET–CT as an initial imaging tool is less because of limited ability to detect extent of local disease but more so for confirmation of extent of primary in gray zones of CT images. MRI is still the recommended tool for primary tumor delineation.

Nodal Disease Involvement

NPC is one of the few head and neck malignancies prone to neck nodes or distant metastases, regardless of the primary tumor size (Fig. 9.3).

AD King et al. in their study assessed the neck for the presence of any node from the retropharyngeal region down to the supraclavicular fossa. Nodes that were identified were then assessed for malignancy (Table 9.3).

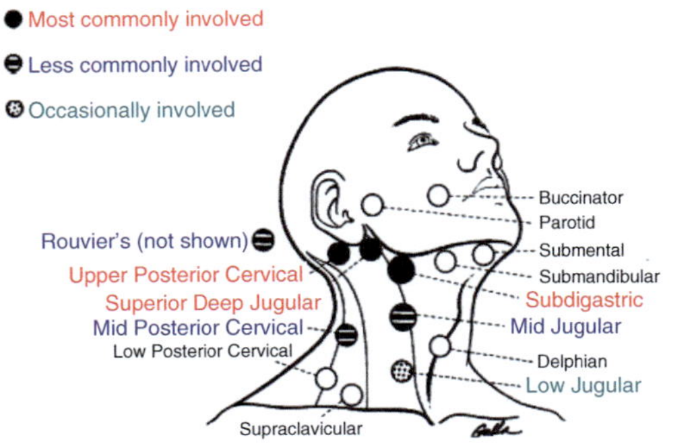

Fig. 9.3 Lymph node regions involved in nasopharyngeal carcinoma: a risk-wise depiction

Table 9.3 Criteria for nodal metastasis in nasopharyngeal carcinoma

Anatomical imaging criteria (CT/MRI)	Metabolic criteria (18 F-FDG-PET)
Shortest axial diameter of >11 mm in the jugulodigastric region, >5 mm in the retropharyngeal region and >10 mm in all other regions of the neck	Intense uptake with SUV_{max} >2.5 (except for necrotic lymph nodes)
A group of three or more nodes that were borderline in size	
Nodes with necrosis or extracapsular spread irrespective of size	

There is no consensus regarding the best imaging approach for staging head and neck cancer. Variable results have been seen for all modalities like PET/PET-CT, MRI, USG (with FNAC) favoring all in varied approaches in this regard.

Most of the recent studies have suggested PET–CT to be the investigation of choice for detection of nodal disease in head and neck cancers. Detection of nodal disease has major prognostic significance; it is important as it increases the risk of local recurrence and is associated with a higher risk of metastatic disease, affecting management.

Studies suggest that nodal metastases are seen in 60–90 % of cases, implying that N0 disease is seen in only 10–40 % of cases. Usually, nodal disease has been seen to occur to the retropharyngeal nodes (RPN), then to Levels II, III, and IV. It has also been seen that RPN has been bypassed in few cases, which suggests an alternate pathway of spread to level II directly (Fig. 9.4).

Rarely, skip metastasis also have been noted in supraclavicular fossa and thoracic and abdominal nodes. Nodal involvement has more predilection for bilateral involvement than other squamous cell carcinomas; this is of significance while assessing imaging results.

Criteria for nodal metastasis on FDG PET: It has been shown that the median SUV_{max}* for positive nodes was 6.8 (mean, 9.1; range, 1.3–24.0), and the median SUV_{max} for negative nodes was 2.0 (mean, 2.8; range, 0.6–6.5) **SUV* on the highest image pixel in the tumor regions (SUV_{max}).

Unknown Primary with Neck Lymphadenopathy

FDG-PET has been shown to reveal unknown primary tumors in more patients with metastatic disease to the lymph nodes in the neck who had no detectable primary tumor during clinical examination than CT/MRI. Of these, nasopharyngeal carcinomas will be an important entity due to their routinely difficult location to access location and nonspecific presenting complaints.

Key Learning Points

PET–CT is the modality of choice for accurate nodal staging of nasopharyngeal carcinomas.

Provides valuable information for localizing primary tumors in patients with neck nodal metastases from an unknown primary.

Metastatic Workup of NPC

18 F-FDG PET–CT has the obvious advantage of being able to assess the whole body for metastases in one examination and, at the same time, assess the primary tumor and cervical nodes. For patients with NPC, mortality may be up to 90 % within a year when distant metastases are found before treatment is initiated. Therefore, the presence of distant metastatic lesions greatly influences prognosis and treatment strategy.

18 F-FDG PET has been shown to be more sensitive than conventional imaging, and approximately 12 % of patients harbor distant metastases,

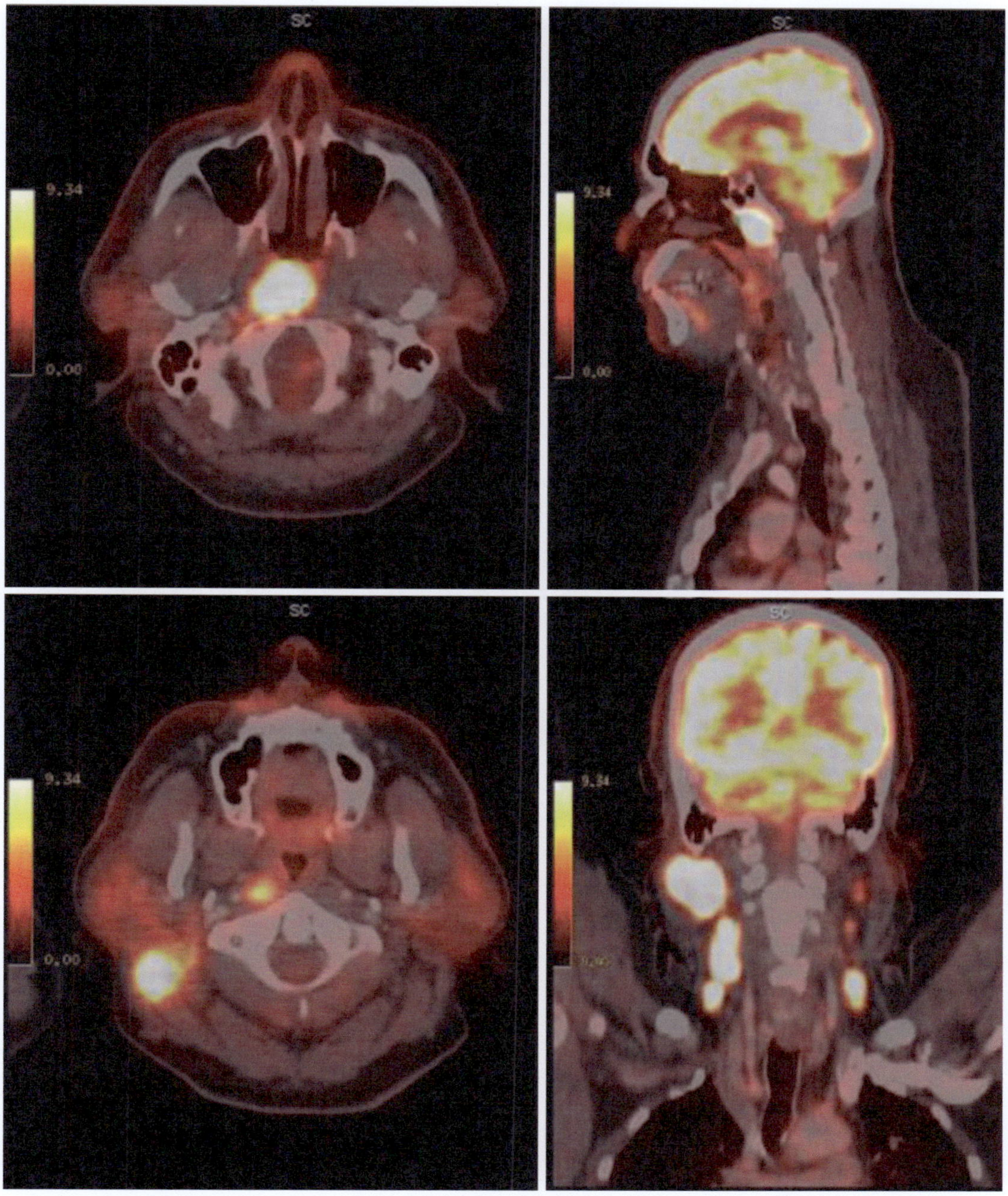

Fig. 9.4 A 62-year-old male was diagnosed as case of nasopharyngeal carcinoma. Whole body PET–CT showed increased FDG uptake in nasopharyngeal region with bilateral level II, III, IV nodes

which may be missed on conventional imaging studies (which include chest radiography, liver sonography, and whole body bone scanning). FDG-PET has been found to be more sensitive than whole body bone scanning for detection of skeletal metastasis.

Recurrence of Tumors Post-Radiotherapy

CT and MRI have limitations to differentiate residual or recurrent tumor from tumor necrosis and tissue fibrosis after radiotherapy. F-18 FDG PET–CT has been found most effective in

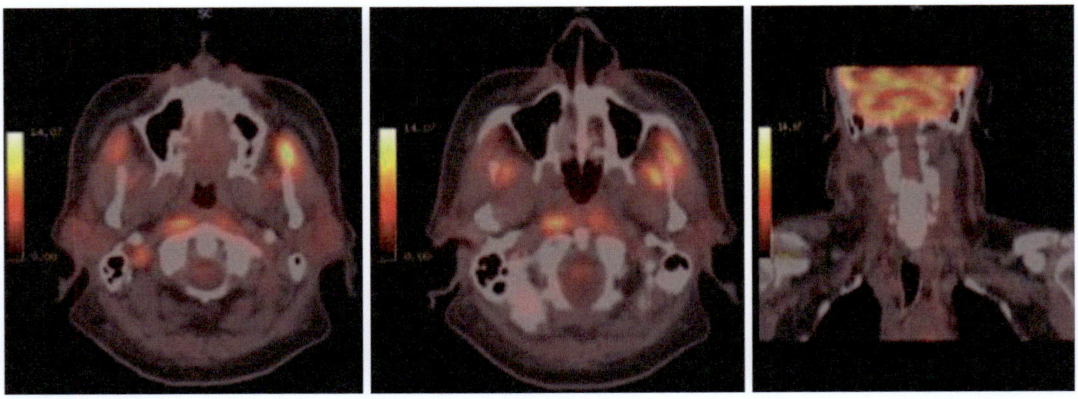

Fig. 9.5 A 45-year-old male was diagnosed as case of nasopharyngeal carcinoma. Initial whole body PET-CT had shown increased FDG uptake in nasopharyngeal region with bilateral level II, III, IV nodes (not shown in the figure). His post therapy scan after completion of concurrent chemo-radiotherapy showed no significant metabolically active disease in the nasopharynx as well as nodes suggesting good response to therapy

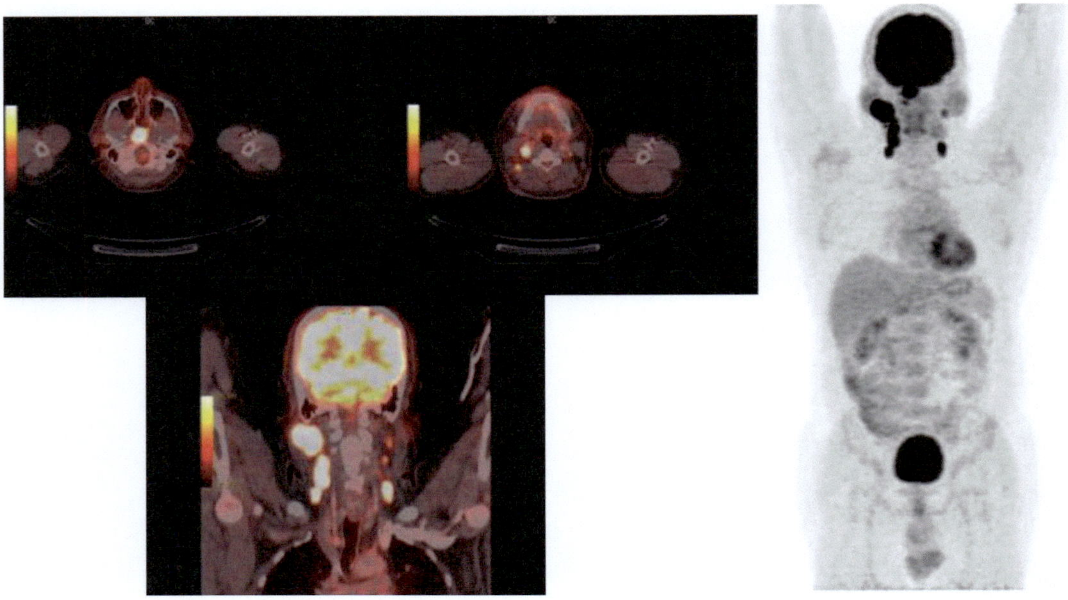

Fig. 9.6 A 45-year-old male, K/C/O nasopharyngeal carcinoma with bilateral nodal metastases. He underwent concurrent chemo-radiotherapy. He was referred for evaluation of response after completion of treatment. Whole body FDG PET-CT with contrast was done. Contrast CT showed no residual disease in the nasopharyngeal region as well as no significant lymphadenopathy. However, there was asymmetrically increased FDG uptake seen in the right nasopharyngeal region as well as mild FDG uptake in the right parapharyngeal node. Delayed PET-CT of head and neck region showed increase in FDG uptake over time in right nasopharyngeal region as well as right parapharyngeal node suggesting residual disease. This would have been treated as complete response if not for FDG PET study

detection of recurrence post therapy (Figs. 9.5 and 9.6). Radiation-induced inflammatory processes may lead to a transient increase of 18 F-FDG uptake and result in false-positive 18 F-FDG PET readings. False-negative 18 F-FDG PET results may occur as the metabolic mechanism of the residual tumor tissue may be temporarily inhibited after radiation.

In light of these potential pitfalls as well as prohibitive costs of a PET–CT scan, a study at 4 months post-therapy is an ideal time to perform the study.

Key Learning Points

PET–CT has a higher sensitivity for detection of distant metastasis, including skeletal lesions in comparison to conventional imaging and bone scintigraphy, respectively

Detection of recurrence needs to be evaluated carefully and should be well timed approximately 4 months post therapy

Planning Management (Radiotherapy) of Nasopharyngeal Carcinomas

A very radiosensitive tumor, radiotherapy with or without chemotherapy is the mainstay of management of NPCs. With the advent of IMRT and IGRT, advanced planning of fields with maximal sparing of local normal tissue is the intent, especially in head and neck cancers. NPCs are surrounded on all sides by vital structures and need more sparing of surrounding normal tissues than other head and neck cancers. However, due to the notoriety of NPCs for local recurrence and occult metastasis, the radiation oncologist often includes larger areas in the fields during primary therapy.

Many studies have been taken up to look for the advantage of PET-based radiotherapy of head and neck and a few on its impact on NPCs.

Potential Areas of Investigation of FDG-PET in Management of NPC

1. *FDG-PET can act as biomarker for response of radiotherapy :* Many parameters like baseline SUV, SUV-max, TLG, etc. have been used to predict outcome of treatment, overall consensus is lacking though. 18 F-FDG-PET is a more sensitive method for assessing early response to treatment than conventional anatomical imaging.
2. *Planning of radiation field*: Estimation of biological tumor volume (BTV) with F-18-FDG PET may improve overall outcome of radiotherapy in the sense of inclusion of diseased areas through PET and helping to reduce dose to physiological structures, especially in context to NPC, which is surrounded by crucial structures all around.

   This has been done by various methods, all of which are still under evaluation.

   - *Visual interpretation*, which is highly operator-dependent and susceptible to window-level settings and interpretation differences.
   - *Isocontouring* based on a fixed standardized uptake value (SUV).
   - *Fixed threshold of maximum tumor signal intensity* (40 or 50 %).
   - *Variable threshold based on adaptive signal-to-background ratio (SBR).*
   - *Iterative background-subtracted relative-threshold* using watershed transformation and hierarchical cluster analysis in which the optimal relative-threshold depends on the lesion size not the SBR.

Clearly, 18 F-FDG-PET (BTV) can alter the conventionally defined GTV and clinical target volume (CTV). Though there is vast literature in this area, streamlined guidelines for incorporation of FDG PET-CT in planning of radiation are yet to be defined.

Timing for detection of recurrence post radiotherapy/chemotherapy: An important aspect of including PET–CT in treatment planning would include response assessment and detection of recurrence. There is no consensus on timing of FDG scan in patients post treatment. The optimum time for post-treatment assessment seems to be between 1 and 4 months as per current data. Same is true for post chemotherapy detection of recurrence too. FDG-PET favorably predicts loco-regional control and improved survival in those with a negative scan (at a median of 9 weeks post-CRT).

Key Learning Points

PET–CT is coming to the forefront in management of nasopharyngeal carcinomas

Assessment of response to chemo or radiotherapy and planning of radiotherapy are the two thrust areas

PET–CT in planning of Radiation fields is still in its early phase of development

Future Trends

PET-CT in Nasopharyngeal Carcinoma: Tracers Beyond FDG

1. *Hypoxia imaging of NPCs* with Nitroimidazole PET Tracers: PET-CT imaging can provide a spatial map of the intra-tumoral distribution of hypoxia b*efore and during radio*therapy using 18 F-fiuoromisonidazole (18 F-FMISO), a nitroimidazole PET tracer. This can have impact on predicting outcome of conventional therapies, in newer treatments with chemotherapeutic agents and in planning of radiation therapy .
2. *Cell Proliferation Imaging* with Fluorothymidine: 3′ Deoxy-3′18 F-fiuorothymidine (18 F-FLT) is a tracer that reflects the activity of thymidine kinase 1, a key enzyme in DNA synthesis and is taken up by dividing tumor cells. This can help predict therapy response during radiotherapy.
3. *^{11}C-choline PET-CT for T staging in Nasopharyngeal Carcinoma:* Higher physiologic uptake of ^{18}F-FDG in normal brain, which makes intracalvarial lesions or lesions near the skull base in NPC difficult to detect on 18 F-FDG PET-CT. 11C-choline uptake in the normal brain has been shown to be quite low, suggesting that 11C-choline PET-CT may be helpful for T staging of NPC, especially for the detection and delineation of intracranial invasion.

For Further Reading

1. Bussink J, et al. PET-CT for response assessment and treatment adaptation in head and neck cancer. Lancet Oncol. 2010;11:661–9.
2. Daisne JF, et al. Tumor volume in pharyngolaryngeal squamous cell carcinoma: comparison at CT, MR imaging, and FDG PETand validation with surgical specimen. Radiology. 2004;233:93–100.
3. Heron DE, et al. Hybrid PET-CT simulation for radiation treatment planning in head and neck cancers: a brief technical report. Int J Radiat Oncol Biol Phys. 2004;60:1419–24.
4. King AD, et al. The impact of 18 F-FDG PET-CT on assessment of nasopharyngeal carcinoma at diagnosis. Br J Radiol. 2008;81:291–8.
5. Ng SH, et al. Staging of untreated nasopharyngeal carcinoma with PET-CT: comparison with conventional imaging work-up. Eur J Nucl Med Mol Imaging. 2009;36(1):12–22.
6. Wu HB. Preliminary study of 11C-choline PET-CT for T staging of locally advanced nasopharyngeal carcinoma: comparison with 18 F-FDG PET-CT. J Nucl Med. 2011;52:341–6. doi:10.2967/jnumed.110.081190.

Role of PET in Coronary Artery Disease

10

P.K. Pradhan and Gowri Sankar

Introduction

The use of positron-emitting radionuclides, especially ^{18}F labeled fluorodeoxy glucose (^{18}F-FDG), has increased recently due to its role in oncology. This has paved the way to look for other uses for these agents. The use of positron emission tomography (PET) as a noninvasive tool for imaging the heart in coronary artery disease to look for perfusion defects and viability has already been established. This has helped us to know the extent of myocardial blood flow, the metabolic changes occurring in the myocardium, and changes in the cardiac autonomic innervations in various pathological conditions.

Why PET Rather Than SPECT?

Myocardial perfusion imaging to look for stress-induced ischemia and viable myocardium using gamma rays-emitting radionuclides and single photon emission and computed tomography (SPECT) technique has been in use for long time. The most commonly used radiopharmaceutical has been sestamibi or tetrofosmin labeled with 99mTc.

Positron-emitting radionuclides undergo beta (+) decay, resulting in the emission of a positron (mass similar to that of an electron but oppositely charged), which then rapidly combines with a nearby electron undergoing annihilation. This results in the emission of two 511-keV photons, which travel in opposite directions. The basic principle of PET lies in the coincidence detection of these photons in a ring scanner.

The spatial resolution of reconstructed clinical PET images is currently in the range of 4–7 mm, and it is superior to conventional nuclear imaging techniques like SPECT. With the usage of the combined PET–CT imaging system, this can be further improved to be in the range of 1–2 mm.

PET also has high temporal resolution, which allows for creation of dynamic imaging sequences. PET is also quantitative in nature, which is made possible because of the readily available correction algorithms for photon attenuation, scatter, and random events. This quantitative measure of PET can be used in predicting the absolute myocardial blood flow or glucose use.

Most of the advanced PET–CT scanners are now equipped with multislice CT, allowing for measurement of coronary calcium and/or CT coronary angiography in addition to PET imaging procedures.

P.K. Pradhan, MD, DNB (✉)
Professor, Department of Nuclear Medicine and PET, Sanjay Gandhi Postgraduate Institute of Medical Sciences, Lucknow, India
e-mail: pkp9918@gmail.com

G. Sankar, MD
Asst.Prof., Department of Nuclear Medicine, Sanjay Gandhi Postgraduate Institute of Medical Sciences, Lucknow, India
e-mail: apgs.saga@gmail.com

B.K. Das (ed.), *Positron Emission Tomography: A Guide for Clinicians*,
DOI 10.1007/978-81-322-2098-5_10, © Springer India 2015

Table 10.1 Positron-emitting radionuclides used for myocardial perfusion studies

Tracer	Half life	Positron range (mm)	Mechanism of uptake	FDA approved	First pass extraction (%)
^{82}Rb	76 s	2.6	Na/K ATPase	Yes	65
$^{13}NH_3$	10 min	0.7	Diffusion/metabolic trapping[a]	Yes	80
$H_2{}^{15}O$	2 min	–	Diffusion (freely diffusible/ not trapped)	No	100
^{18}F-FBnTP[b]	110 min	0.2	Mitochondrial binding	No	NA
^{18}F-flupiridaz[b]	110 min	0.2	Mitochondrial binding	No	NA

Rb rubidium, *NH_3* ammonia, *H_2O* water, *^{18}F-FBnTP* F-18-fluorobenzyl triphenyl phosphonium, *^{18}F-flupiridaz* ^{18}F-BMS747158, *FDA* food and drug administration, *NA* not applicable

With the usage of contrast-enhanced CT angiography, electrocardiogram-gated image acquisition for complementary functional analysis and the addition of respiratory gating for creation of motion-frozen images to reduce breathing artifacts, PET–CT imaging of the cardiovascular system is considered superior to other noninvasive imaging studies of the heart. Also, comparative studies with ^{82}Rb PET and ^{99m}Tc-sestamibi show that PET has greater diagnostic accuracy when compared with SPECT (89 % vs. 79 %).

Myocardial Perfusion Study Using PET

PET Perfusion Tracers

The most commonly used PET tracers used for myocardial perfusion imaging are listed in Table 10.1, some of which are discussed in detail.

$^{13}NH_3$ (Ammonia) has a high first-pass extraction of 80 % and it requires energy for myocardial retention. The images are of high quality and resolution, and uptake is linear over a wide range of myocardial blood flow except at very high flow rates. $^{13}NH_3$ is cyclotron produced and it has a half-life of about 10 min. Hence, the major disadvantage in using $^{13}NH_3$ for imaging is that it needs an on-site cyclotron for its production and usage.

^{82}Rb (Rubidium) is a potassium analogue and it enters the myocardial cell via Na/K-ATPase in an energy requiring process. Its first-pass extraction rate is 65 % and it decreases in a nonlinear manner with increasing blood flow. This effect is more pronounced when compared with ammonia, but it's still superior when compared with ^{99m}Tc-labeled SPECT tracers. It has an ultra short half-life of about 76 s. The ultra short half-life and high positron range in the tissue leads to a minimal deterioration in the image resolution and quality. A major advantage of ^{82}Rb over $^{13}NH_3$ is that it is a generator ($^{82}Sr/^{82}Rb$ generator) product without the need for an on-site cyclotron.

$H_2{}^{15}O$ is superior to ^{82}Rb and $^{13}NH_3$ because it is metabolically inert and it is freely diffusible across cell membrane. It has a 100 % extraction rate but the major disadvantage is that the tracer is not accumulated in myocardium and instead reaches equilibrium between extra- and intravascular compartments. Images of regional myocardial perfusion are not readily obtained and blood pool subtraction is usually needed.

Since the routinely used PET tracers have very short half-life, research for tracers with longer half-life is still undergoing. ^{18}F tagged with FDG is largely being used in PET for oncological imaging and follow up. Since ^{18}F has a half-life of 110 min, trials are undergoing to tag molecules with this agent and use for cardiac imaging. Two of these tracers are ^{18}F-flupiridaz and ^{18}F-fluorobenzyl triphenyl phosphonium, which have entered human clinical trials. Details about these new tracers are discussed in Table 10.1.

Protocol and Interpretation of PET Perfusion Studies

PET acquisition for perfusion studies is similar to that for any other PET imaging techniques. The usual steps in image acquisition are described as shown in Fig. 10.1. These include proper patient

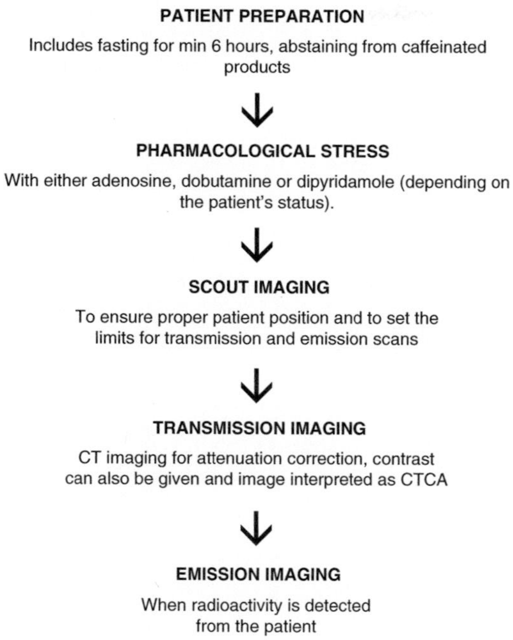

Fig. 10.1 Image acquisition protocol for pet perfusion study

preparation like fasting for a minimum of 6 h and complete abstinence from caffeinated products for a minimum of 24 h. Usually, pharmacological stress is preferred with adenosine, if there is no contraindication for its use. If adenosine is used, then drugs interacting with it like theophylline should also be stopped for at least 48 h before the study. Rest study is usually done with sublingual nitrate after measuring blood pressure.

Scout scanning is also done to ensure proper patient position and to set the limits for the subsequent scans. Transmission scan is done next along with ECG gating and/or respiratory gating for attenuation correction. Contrast can also be given at this step so that detailed visualization of the vessels can also be done. This is followed by emission scan when the radioactivity is detected.

Result

Comparison of stress and rest images may reveal:

(a) Normal study—where there is uniform tracer distribution in the myocardium in both stress and rest images.
(b) Reversible defect—where there is a perfusion defect in the stress images, which gets normalized in the rest images, also known as stress-induced ischemia.
(c) Fixed defect—where there is a perfusion defect in both stress and rest images. These fixed defects can contain nonviable myocardium or hibernating myocardium (has perfusion abnormalities but is still viable). To differentiate the former from the latter, a viability study using ^{18}F-FDG is usually required.

Uses of PET Myocardial Perfusion Studies

For Detection of Coronary Artery Disease

As mentioned earlier, PET perfusion study will be able to detect reversible or fixed defects of the myocardium. In a 2005 review, eight studies that compared PET perfusion imaging with that of the gold standard coronary angiography were done. This review included nearly 800 patients, were summarized, and a mean sensitivity of 93 % and specificity of 92 % for detection of significant CAD were observed. A more recent review, reporting a weighted sensitivity of 90 % and specificity of 89 % from nine studies, including 877 patients, scanned mostly with 82 Rb PET, confirmed these results. For detection of myocardial ischemia, myocardial perfusion PET is considered to have superior diagnostic accuracy when compared with the more widely available and more frequently used SPECT technique.

For Prognosticating the Patients

PET perfusion study also plays a greater role in prognosticating the patients. In a study involving 367 patients who were divided into three groups (normal, mild, moderate to severe), according to the difference in the stress perfusion abnormalities, had annual rates of adverse cardiac events of 0.4, 2.3, and 7 %, respectively. In this study, the

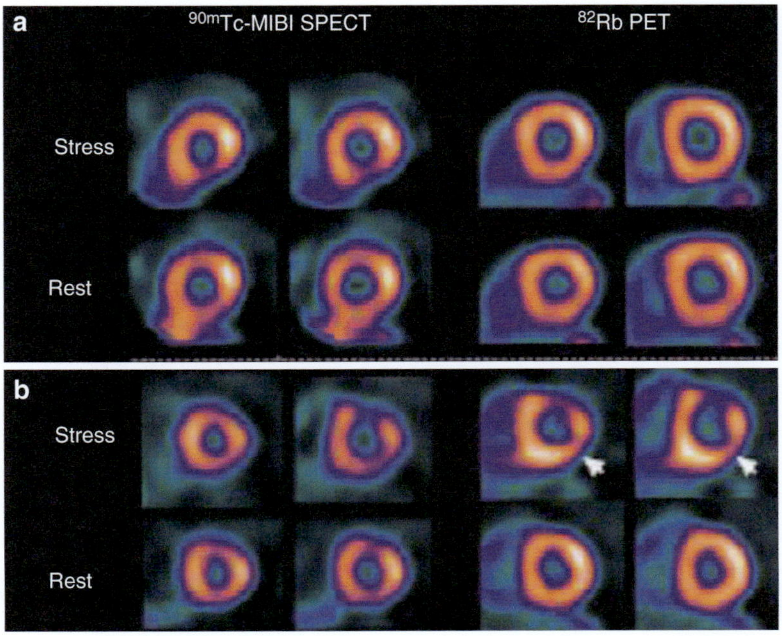

Fig. 10.2 Advantage of PET over SPECT MPI: Image *A* shows a fixed defect in the inferior wall on SPECT, which is absent on PET, suggesting the presence of a SPECT attenuation artifact. This demonstrates the improved specificity of PET. Image *B* shows stress-induced perfusion defects in the anterior and inferolateral walls (*arrows*), which are more clearly shown in PET rather than SPECT, thus demonstrating the improved sensitivity with PET

PET data was the strongest predictors of total cardiac events. In obese patients, who are at higher risk and in whom PET imaging is preferred, the annual total event rate was 11 % with an abnormal scan and 1.5 % with a normal scan. In another study done with dipyridamole ^{82}Rb PET, 685 patients were scanned and followed up. It was found that the annual mortality rate for a normal scan was 0.9 % and for an abnormal scan was 4.3 %. Multivariate analysis showed that the PET results had an independent and incremental prognostic value in these patients.

For Absolute Flow Quantification

Coronary flow reserve (CFR) is the ratio of peak MBF during near maximal pharmacologically induced vasodilatation to resting MBF. CFR is an index of the functional significance of a coronary stenosis. This can be quantified by PET–CT through compartmental modeling of multiframe dynamic acquisitions. Several studies have suggested the prognostic value of quantitative PET measurements of MBF and CFR for progression toward clinically overt CAD and in idiopathic and hypertropic cardiomyopathies leading to emphasis on microcirculation and endothelial function (Fig. 10.2).

Myocardial Viability Using PET

Cardiac viability study is done with ^{18}F-FDG (fluro deoxy glucose) PET tracer. This tracer has a half-life of 110 min, positron range of 0.2 mm and is FDA approved. FDG enters the viable cells via glucose transporters and then enters the anaerobic glycolysis cycle. It is metabolized by hexokinase into FDG-6-phosphate, which cannot be further metabolized and thus is trapped in the cell.

The basic principle is that even though there are perfusion abnormalities, if the cells are viable, they may pick up FDG. This might help in identifying hibernating or stunned myocardium, which is seen as tracer-deficient areas in perfusion studies but picks up FDG during viability study, thus producing a perfusion–viability mismatch. Immediate intervention, if done, may lead to improvement of these regions. Hence, FDG viability study is considered the gold standard for myocardial metabolic assessment (Fig. 10.3) (Table 10.2).

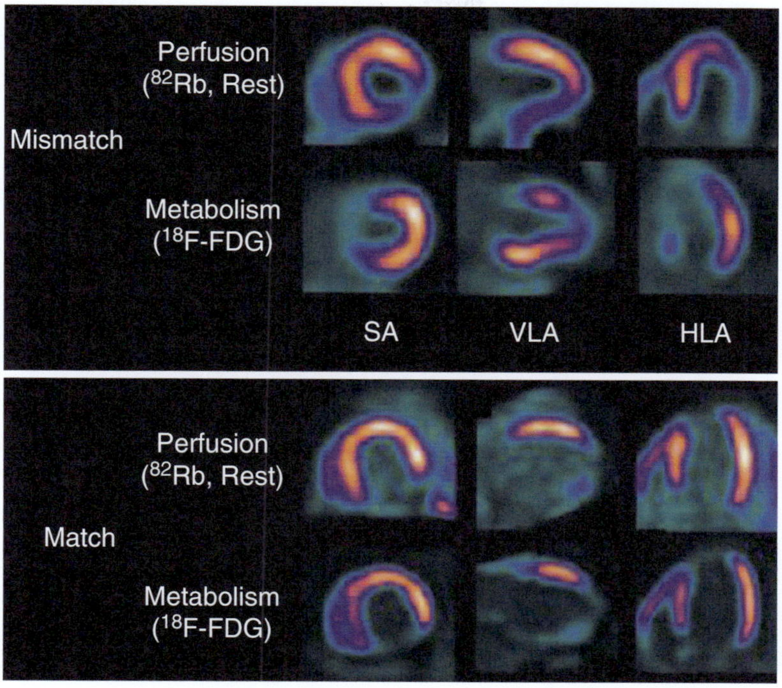

Fig. 10.3 A mismatch with reduced rest perfusion (measured by ^{82}Rb) and preserved metabolism (measured by ^{18}F-FDG) is shown in the inferolateral wall, indicating ischemically compromised but viable "hibernating" myocardium (*upper panel*). A matched perfusion/metabolism defect is shown in the inferior wall, indicating nonviable scar (*lower panel*)

Table 10.2 Various interpretations of perfusion and metabolic cardiac imaging

Perfusion study findings	Viability study findings	Final impression
Normal	Normal	Normal
Abnormal	Normal	Mismatch (hibernating myocardium)
Abnormal	Abnormal	Match (nonviable)
Normal	Abnormal	Reverse mismatch (altered myocardial glucose metabolism)

Concluding Remarks

Cardiac PET is a quantitative, noninvasive imaging technique that is being increasingly used in the clinical arena. Despite high single test costs, it shows overall cost-effectiveness as it can be used as an "one stop shop" for diagnosing myocardial perfusion abnormalities and viability study. With the inclusion of CT in PET and novel tracers being produced, the role of PET in cardiac imaging will still improve in the near future.

For Further Reading

1. Bateman TM, Heller GV, McGhie AI, et al. Diagnostic accuracy of rest/stress ECG-gated Rb-82 myocardial perfusion PET: comparison with ECG-gated Tc-99 m sestamibi SPECT. J Nucl Cardiol. 2006;13:24–33. PubMed: 16464714.
2. Di Carli MF, Hachamovitch R. New technology for noninvasive evaluation of coronary artery disease. Circulation. 2007;115:1464–80.
3. Lortie M, Beanlands RS, Yoshinaga K, Klein R, Dasilva JN, de Kemp RA. Quantification of myocardial blood flow with 82Rb dynamic PET imaging. Eur J Nucl Med Mol Imaging. 2007;34:1765–74.
4. Machac J. Cardiac positron emission tomography imaging. Semin Nucl Med. 2005;35:17–36.

11 Application of PET in Cancer of the Breast

Birendra Kishore Das

Breast cancer is the most common cancer in all countries of the world. One out of every 25 women will suffer from this cancer, which is potentially incurable. In India, breast cancer is the first cause of cancer death in women in urban areas. Early diagnosis and start of definitive therapy can save many lives. There are several methods available for diagnosis of breast cancer, the most important of them are as follows:

X-ray mammography (conventional and digital)
Scintimammography
Ultrasonography
Magnetic Resonance Imaging (MRI)
Computes Tomography (CT)
Positron Emission Tomography (PET)
FNAC and Biopsy

In this chapter, we are discussing specially about the modality of PET and its contribution in the diagnosis and management of breast cancer.

Detection of Breast Cancer with FDG PET

PET technology has been widely used in the management of breast cancer. Imaging modalities like computed tomography (CT) and magnetic resonance imaging (MRI) rely on detecting anatomic changes for the diagnosis, staging, and follow-up of cancer patients. However, positron emission tomography (PET) has the ability to demonstrate abnormal metabolic activity, and 18 F-2-deoxy-D-glucose (FDG) PET provides important tumor-related qualitative and quantitative metabolic information that may be critical for the diagnosis and follow-up. Moreover, the combination of PET and computed tomography (PET–CT) allows the functional PET and anatomical CT images to be acquired under identical conditions and then they are rapidly co-registered. This combined system has advantages over CT alone as functional information is added to morphological data, and this combined system has advantages over PEt alone because pathological areas of tracer uptake are better localized and the image acquisition time is reduced. Moreover, the limited specificity of PET that's due to the increased glucose metabolic activities of benign tumors and inflammatory tissues (such as those of tuberculosis) can be partially overcome by PET–CT.

FDG-PET has high sensitivity and specificity for detection of malignant lesions in general, but breast cancer detection requires the ability to demonstrate non-palpable, small (<1.0 cm) invasive and in situ malignancies. These requirements are beyond the capability of current whole body FDG-PET, and thus FDG-PET is not used for primary breast cancer detection. The ability of PET to detect breast cancer depends on the tumor's size and histology. The sensitivity of PET has been reported to be 68 % for small (<2 cm)

B.K. Das, MD, ANM, FAR, FAMS
Director and Chief of Nuclear Medicine,
Utkal Institute of Medical Sciences,
Jaydev Vihar, Bhubaneswar, India
e-mail: bkishoredas@yahoo.com

B.K. Das (ed.), *Positron Emission Tomography: A Guide for Clinicians*,
DOI 10.1007/978-81-322-2098-5_11, © Springer India 2015

tumors and 92 % for larger (2–5 cm) tumors, and its reported overall accuracy for detecting in situ carcinomas is low (sensitivity: 2–25 %). The major limitation of PET or PET–CT for breast imaging is its poor detection rate for small breast carcinomas and noninvasive breast cancers. However, PET–CT has a role to play in a selected group of patients, such as those with dense breasts or with implants, for determining tumor multiplicity, for localizing the primary tumor in those patients with metastases of a breast origin when the mammography is indeterminate, and for those patients whom biopsy is not a desirable option. PET–CT has a potential advantage over PET for evaluating small lesions in which the uptake may be low due to the partial volume effect of PET and areas of mild hyper glycolytic activity can be reliably assigned to normal or abnormal anatomical structures.

Detection of Axillary Involvement

Detection of axillary lymph node involvement noninvasively is a diagnostic challenge. Early studies of demonstrating abnormal FDG-PET uptake in metastatic axillary lymph nodes of breast cancer patients led to prospective multicenter trial to evaluate the ability to stage the axilla with FDG-PET before surgery. The PET results were compared with those of pathologic analysis of axillary nodes. In one well-designed study, overall FDG-PET was 61 % sensitive and 80 % specific for axillary metastases, with a positive predictive value of 62 % and a negative predictive value of 79 %. However, several other studies have repeatedly demonstrated a high specificity for axillary nodal metastases, particularly in patients at high risk for nodal disease. Although FDG-PET is clearly not a substitute for histologic sampling, there may be a clinical role for preoperative FDG-PET in certain patient populations, such as advanced axillary disease, plexopathy, and symptomatic metastases. In patients with symptomatic advanced disease, FDG-PET can help accurately determine the extent of disease and distinguish radiation plexopathy from recurrence. FDG-PET can also be helpful in assessing breast cancer spread to regional nodal sites outside the axilla, particularly the internal mammary (IM) chain. These nodal basins are not routinely sampled, given their relative inaccessibility and their uncertain clinical significance and treatment. However, FDG uptake in intermammary nodes has been anecdotally reported and FDG uptake has been demonstrated in up to 25 % of patients and that such uptake is predictive of both the likelihood and pattern of treatment failure (Fig. 11.1).

Experience of Staging Breast Cancer with FDG-PET and PET–CT

Although FDG-PET and PET–CT are not currently recommended for routine staging of breast cancer, there are several scenarios where FDG-PET is often helpful. FDG-PET can provide additional information in staging or restaging cases when results of conventional imaging are equivocal or conflicting. FDG-PET may be particularly helpful in restaging cases of locally recurrent disease if aggressive local therapy is being considered, since it may reveal unsuspected mediastinal or distant metastatic disease, the presence of which would change clinical management. The sensitivity of FDG-PET is superior to that of CT in detecting nodal disease of the inter- mammary and mediastinal nodal basins, which are common sites of involvement in patients with advanced or recurrent disease. The additional information provided by PET aids clinical decision making in this complex group of patients. Although FDG-PET is more sensitive than conventional imaging in the detection of metastatic or recurrent disease, the impact of increased sensitivity on patient care and outcome has not been conclusively demonstrated. In any case, the improved sensitivity and accuracy of FDG-PET compared with those of CT in restaging cases of advanced breast cancer could justify in obtaining a FDG-PET for this scenario. It has been demonstrated that FDG-PEt alters therapy options in up to 44 % of patients with suspected locoregional recurrence, by

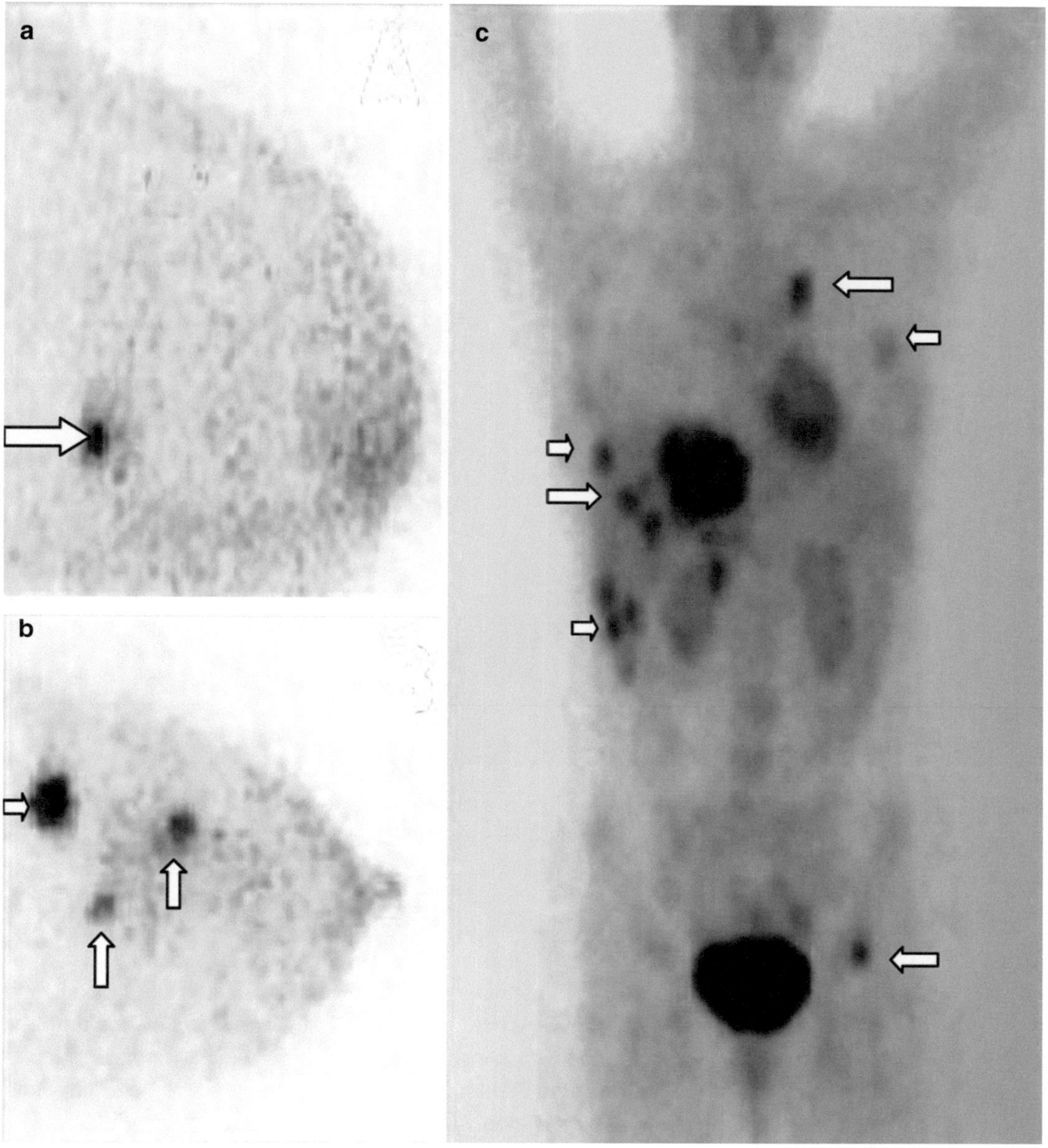

Fig. 11.1 PET study in breast cancer. (**a**) Small cancer near the thoracal wall of rt. Breast. (**b**) Multi-focal breast cancer. (**c**) Metastatic breast cancer with lymph node, lung, liver, and bone metastases (*arrows*)

demonstrating more widespread disease than CT and avoiding local surgical procedures for patients with metastatic disease. Combined PET–CT is likely to add even more clinically important information in this setting, since initial reports have demonstrated that FDG PET-CT adds incremental diagnostic confidence to PET results in more than 50 % of patients and that integrated PET–CT allows more accurate restaging of breast cancer than PEt alone. *FDG PET-CT may be useful for evaluating asymptomatic treated breast cancer patients with rising levels of tumor markers without clinical symptoms. In this clinical scenario, FDG-PEt allows more accurate diagnosis of metastatic disease compared with conventional imaging.*

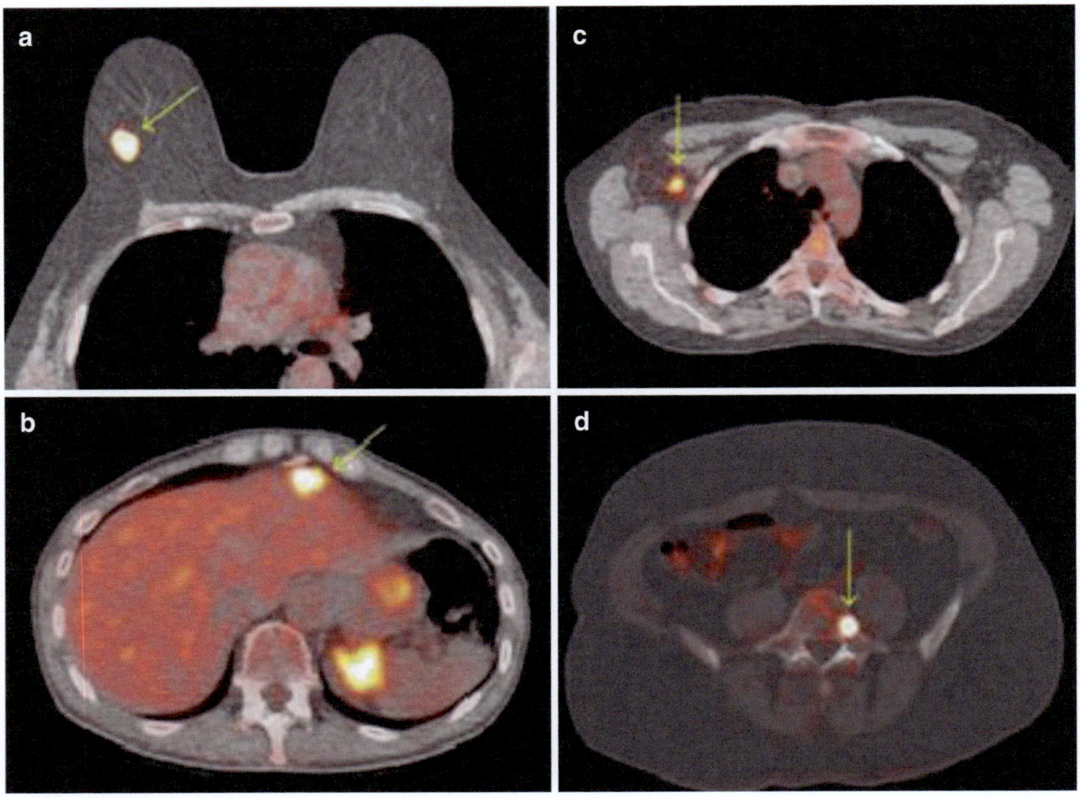

Fig. 11.2 Fused whole body PET–CT images depicting FDG uptake in the primary tumor (**a**), an axillary lymph node (**b**), a lymph node in the internal mammary chain, a solitary metastatic lesion in the left lobe of the liver and the liver (**c**), and the fifth lumbar vertebra (**d**)

Use of PET in Diagnosis of Distant Metastasis

Distant metastases from breast cancer are frequently found in the lungs, liver, and bones. One advantage of whole body PET imaging over conventional imaging modalities such as chest films, bone scanning, and abdominal ultrasound is its ability to detect metastasis at different sites and organs during a single examination. *It has been found that whole body PET imaging had high diagnostic accuracy for patients with suspected recurrent or metastatic breast carcinoma.* Although studies have proven its accuracy in detection of the primary tumor and axillary staging, its most important current clinical application is in detection and defining the extent of recurrent or metastatic breast cancer and for monitoring response to therapy. PET is complementary to conventional methods of staging in that it provides better sensitivity in detecting nodal and lytic bone metastases. However, it should not be considered a substitute for conventional staging studies, including computed tomography and bone scintigraphy (Fig. 11.2).

Detection of Skeletal Metastases

Skeletal metastases are the most common site of distant disease in breast cancer, accounting for 90 % of all metastatic lesions as well as representing the most common site of initial metastatic involvement. The role of FDG-PET and PET–CT for detection and evaluation of skeletal metastases is still controversial. Breast cancer is one of several malignancies that can result in bone metastases that are either osteolytic or osteoblastic.

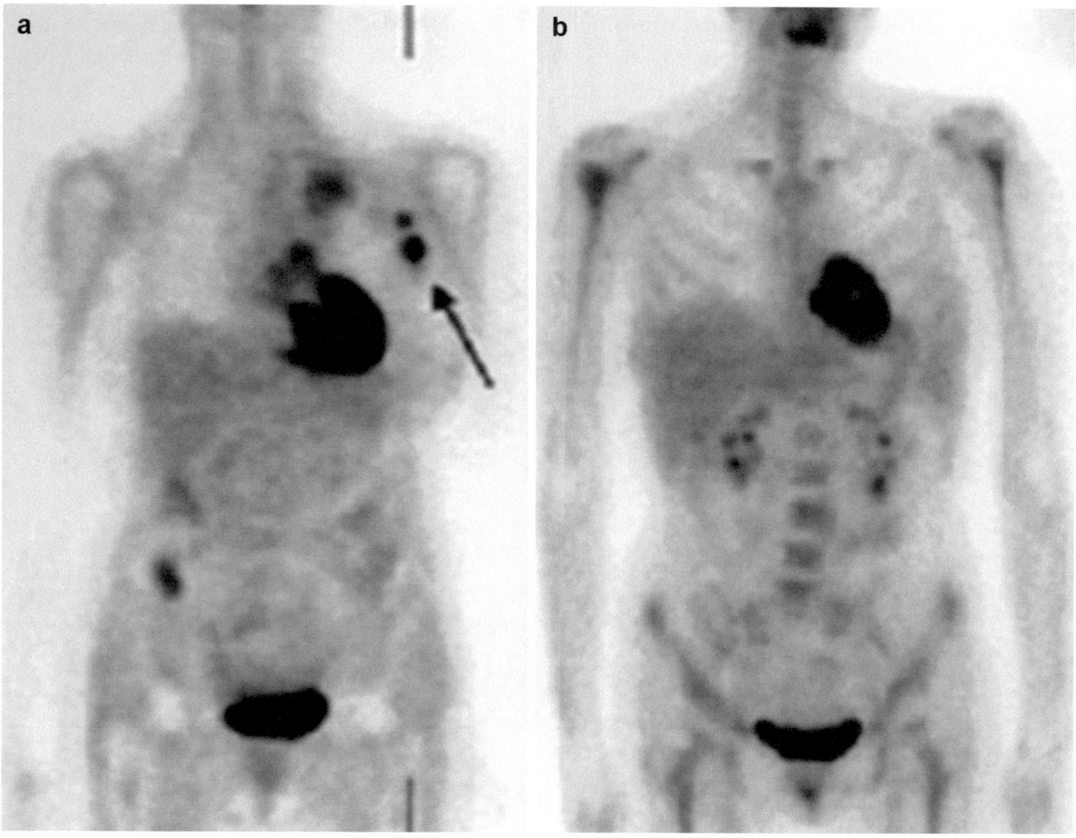

Fig. 11.3 PET scan of breast cancer showing multiple distant mets (**a**) and remission after 6 months of chemotherapy (**b**)

Although most lesions are mixed, with some combination of lytic and blastic components, some lesions are purely lytic or blastic, and these lesions can pose difficulties for imaging. Bone metastases have also presented a particularly vexing problem for measuring response. While bone scanning, MR imaging, and CT are effective in detecting bone metastases, it can be difficult to discern changes in response to therapy with these modalities. The bone scan can even worsen or "flare" in response to successful therapy. Recent studies have suggested that serial FDG-PET can be helpful in measuring bone metastasis response and that changes in FDG uptake correlate with clinical response and changes in breast cancer tumor markers. Recent data show that these changes are predictive of time to progression and the likelihood of a skeletal event. Further study is needed to evaluate the utility and accuracy of PET in this role. The combination of FDG and fluoride PET, to measure both sclerotic and lytic lesion response, may be helpful in this application. In the future, PET agents like fluorine-18 fluoride PET may offer improved bone metastasis detection compared to FDG and bone scintigraphy.

Treatment Monitoring

To downstage primary tumors prior to surgery and to abolish occult distant metastases, neoadjuvant chemotherapy is now being increasingly used to manage patients with large and locally advanced breast cancer. Moreover, several studies have demonstrated that patients with unresponsive tumor may achieve improved survival by administering alternative chemotherapy and/or prolonged courses of chemotherapy. Considering the side effects of chemotherapy, there is a need to quickly identify the non-responding patients early in their treatment. At present, the anatomical imaging modalities are commonly used to evaluate the response to treatment by evaluating the changes of the tumor's size. Nevertheless, sequential measurement of tumor size frequently does not allow the determination of early response. The effect of PET for evaluating the response to treatment has already been demonstrated for different types of neoplasm, including breast cancer. It has been reported that after a single cycle of chemotherapy, PET predicted the complete pathological response with a sensitivity of 90 % and specificity of 74 %, and by using a decrease of FDG uptake at the threshold of <55 %, as compared with the baseline scan, all the responders were correctly identified after the first treatment course (100 % sensitivity and 85 % specificity) (Fig. 11.3).

Neo-adjuvant (preoperative) systemic chemotherapy has become standard treatment for patients with locally advanced breast cancer (LABC), which is defined as primary breast cancer exceeding 5 cm, fixed axillary lymph nodes, or skin or chest wall invasion. But neoadjuvant systemic therapy (NST) is also being increasingly used in patients with non-metastatic operable breast cancer. Although NST has not been shown to improve survival compared to similar adjuvant therapy, it does improve surgical options and provides prognostic information. Studies have demonstrated that the extent of residual breast and axillary disease after treatment is prognostic for both disease-free survival and overall survival. Patients demonstrating complete pathologic response, defined as no residual invasive tumor at histopathological analysis, have improved long-term outcome compared to patients without complete pathologic response. One of the primary aims of NST is, therefore, to assess the response of the primary tumor to the treatment regimen.

Serial FDG-PET imaging has been widely studied as a method for assessing tumor response to NST, by using comparison to histopathological assessment of response from the post surgery specimen as the standard of reference. Multiple studies have evaluated serial FDG-PET imaging performed at different time points after initiation of neoadjuvant therapy and have demonstrated that a serial decrease in tumor FDG uptake, measured as SUV or the metabolic rate of FDG, is an indicator of response and FDG-PET performed early or at mid therapy is predictive of complete microscopic response and may serve as a surrogate marker for response. *Changes in FDG metabolism often precede morphologic changes in tumor and therefore PET can demonstrate response sooner than conventional imaging techniques*. Further, FDG-PET is likely to be most helpful as an early marker for resistance to therapy. FDG-PET imaging performed after completion of therapy allows confirmation of gross residual disease but does not allow exclusion of residual microscopic malignancy.

Metastatic breast cancer is often responsive to systemic therapy, and although cure is rarely achieved, with appropriate therapy, patients often have prolonged survival and preserved quality of life. FDG-PET can be particularly useful in this setting to evaluate the response of metastatic breast cancer to systemic therapy, since conventional imaging is often equivocal in this setting.

Detection of Recurrence of the Disease

Detecting early recurrence has an important survival benefit because it prompts clinical consideration for administering different therapies. However, it is difficult to differentiate true recurrence from post-surgical sequelae and radiation sequelae with using just the conventional imaging modalities. Locoregional recurrences predominately affect the breast, skin, the axillary and supraclavicular nodes, and the chest wall.

It has been found that the sensitivity and specificity of PET for detecting recurrence are 84 and 78 %, respectively, whereas the sensitivities and specificities of the conventional imaging modalities may be 63 and 61 %, respectively. PET is considered to be highly effective for evaluating patients with suspected recurrent breast cancer, and it surpasses the other conventional imaging modalities in terms of whole body evaluation.

Role in Prediction of Prognosis

Prospective studies designed for preoperative evaluation of prognosis more accurately than conventional TNM staging using FDG-PET suggests that FDG-PET is useful in the preoperative evaluation of prognosis in breast cancer patients with more accuracy than conventional TNM staging.

Patients with high SUVmax values show significantly ($P=0.011$) poorer prognosis than patients with low SUVmax values (5-year disease-free survival). It is expected that the indication of neoadjuvant chemotherapy can be decided more precisely by the preoperative evaluation of patient prognosis with FDG-PET due to a possible elimination of overtreatment for those who have good prognosis and, thus, need not to be treated with chemotherapy.

Follow-Up with PET

F-18 FDG PET has demonstrated significant advantage in the diagnosis of metastases in patients with breast carcinoma, compared with conventional imaging protocols. On a lesion base, significantly more lymph node and less bone metastases can be detected by using F-18 FDG PET compared with conventional imaging, including bone scintigraphy. In patients with clinical suspicion but negative tumor marker profile, too, F-18 FDG PET seems to be a reliable imaging tool for detection of tumor recurrence or metastases.

In the follow-up of breast carcinoma, evaluating tumor recurrence or detecting distant metastases presents a diagnostic challenge to the clinician. On one hand, scarred tissue after breast and axillary surgery and external radiation leads to many inconclusive results when using ultrasonography (US) or mammography (MG) in case of suspected local recurrence. On the other hand, tomographic procedures reveal many small lesions, for example, lymph nodes, which cannot be characterized precisely. Because of the heterogeneity of histology and tumor marker expression, there are several patients with negative marker profiles but clinical evidence of tumor recurrence or metastases. To cover all kinds of tumor relapse, several morphologic imaging methods such as common XR, US, bone scan, and tomographic procedures are combined to restage breast cancer patients during follow-up. Several papers have dealt with role of FDG-PET in the follow-up of breast cancer with very promising results and recommend this modality for this purpose.

Recent Development

Dedicated breast positron emission mammography (PEM) units have been developed to overcome the limitations of whole body PET and to provide a positron-emitting imaging platform capable of detection and depiction of primary breast carcinoma. In general, it consists of two sets of rotating planar detector heads, generating 3D reconstructed images with a FOV of $15 \times 15 \times 15$ mm^3 and a resolution of 1.84–2.04 mm. Patients are scanned in prone position after insertion of a single breast into an opening in the table. The scanner acquires fully tomographic images of the breast by rotating two PET detectors, a CT detector, and an X-ray tube in the coronal plane around a single breast. First clinical evaluation showed promising results. The advantages of such dedicated systems include improved geometric sensitivity, higher spatial resolution, shorter imaging time, and reduced attenuation compared with whole body PET systems. They also have a small physical footprint, which makes their use in a breast imaging facility feasible and allows correlation of the results with those of conventional breast imaging as well as PEM-guided biopsy.

Although, at present, mostly FDG and occasionally fluoride PET are used in clinical practice, it is likely that in future more and more PET–CT will be used due to widespread availability of the hybrid equipment and patients will be benefited. New tracers other than FDG, for example, ^{18}F fluorestradiol, to image estrogen receptor expression, to image a much wider range of tumor biologic features will be available.

Concluding Remarks

Breast cancer is the most frequent type of cancer in women all over the world.

FDG-PET and PET–CT have been shown to be most helpful in staging recurrent or metastatic breast cancer and in evaluating the response of locally advanced or metastatic breast cancer to treatment. FDG-PET demonstrates definite advantages in the diagnosis of metastases compared with conventional imaging on a whole body patient base. On a lesion base, significantly more lymph node and bone metastases can be detected by using FDG-PET compared with conventional imaging, including bone scintigraphy. In patients with clinical suspicion but negative tumor marker profile, FDG-PET seems to be a reliable imaging tool for detection of tumor recurrence or metastases. Considering the high predictive value of FDG-PET, tumor stage and therapeutic strategy may need to be reconsidered in several patients. Emerging data support the use of FDG PET-CT in advanced axillary disease and evaluation of regional nodal spread in LABC (Locally Advanced Breast Cancer).

For Further Reading

1. Aukema TS, Rutgers EJT, Vogel WV, et al. The role of FDG PET-CT in patients with locoregional breast cancer recurrence: a comparison to conventional imaging techniques. Eur J Surg Oncol. 2010;36(4):387–92.
2. Bowen SL, Wu Y, Chaudhari AJ, et al. Initial characterization of a dedicated breast PET-CT scanner during human imaging. J Nucl Med. 2009;50(9):1401–8.
3. Eo JS, Chun IK, Paeng JC, et al. Imaging sensitivity of dedicated positron emission mammography in relation to tumor size. Breast. 2012;21(1):66–71.
4. Groheux D, Giacchetti S, Moretti JL, et al. Correlation of high ^{18}F-FDG uptake to clinical, pathological and biological prognostic factors in breast cancer. Eur J Nucl Med Mol Imaging. 2011;38(3):426–35.
5. Groheux D, Espié M, Giacchetti S, Hindié E. Performance of FDG PET-CT in the clinical management of breast cancer. Radiology. 2013;266(2):388–405.
6. Sang Kyu Yang, Nariya Cho, Woo Kyung Moon. The role of PET-CT for evaluating breast cancer. Korean J Radiol. 2007;8(5):429–37.
7. Tatsumi M, Cohade C, Mourtzikos KA, Fishman EK, Wahl RL. Initial experience with FDG-PET-CT in the evaluation of breast cancer. Eur J Nucl Med Mol Imaging. 2006;33:254–62.

PET–CT Imaging of Lung Cancer: The Current Status and Future Potentials

12

Sunita Sonavane and Sandip Basu

Introduction

Lung cancer is the principal cause of cancer death worldwide; overall, 5-year survival of patients even with limited disease and resectable tumor is 35–40 %, making lung cancer one of the most important public health problems. Majority of patients die within 2 years of diagnosis. At the time of diagnosis, frequently, the disease is found to have spread beyond the primary site. Early localized lung cancer is treated with surgery with or without adjuvant chemotherapy; patients with positive ipsilateral mediastinal lymphadenopathy are treated with neoadjuvant chemotherapy followed by surgery, while more advanced cases are treated with palliative chemotherapy with or without radiotherapy.

S. Sonavane, DRM, DNB (✉)
Radiation Medicine Centre, Bhabha Atomic Research Centre, Tata Memorial Hospital Annexe, Parel, Mumbai, India
e-mail: drsunitatarsarya@gmail.com

S. Basu, MD
Professor, Radiation Medicine Centre, Bhabha Atomic Research Centre, Tata Memorial Hospital Annexe, Mumbai, India
e-mail: drsanb@yahoo.com

Clinical Problems

Tumor Diagnosis in Solitary Pulmonary Nodule (SPN)

Morphological images can detect a nodule or mass but cannot determine its benign or malignant nature in a conclusive fashion. Nonspecificity of anatomic imaging management depends upon invasive diagnostic procedures, which are associated with complications and have morbidity of 1–10 % depending on technique, the patient's medical status, and experience of the operator.

Pretreatment Staging in a Diagnosed Case of Lung Carcinoma

Defining the extent of disease, both locoregional and at distant sites, is key to choosing the appropriate therapeutic option. Anatomical imaging demonstrates substantial discordance in the tumor stage as determined preoperatively with conventional techniques and at surgery. Hence, FDG PET/CT has been investigated for the sensitive detection of tumor extent and disease spread to reduce futile surgery, without sacrificing specificity of diagnosis (Figs. 12.1, 12.2, 12.3, 12.4, and 12.5).

B.K. Das (ed.), *Positron Emission Tomography: A Guide for Clinicians*,
DOI 10.1007/978-81-322-2098-5_12, © Springer India 2015

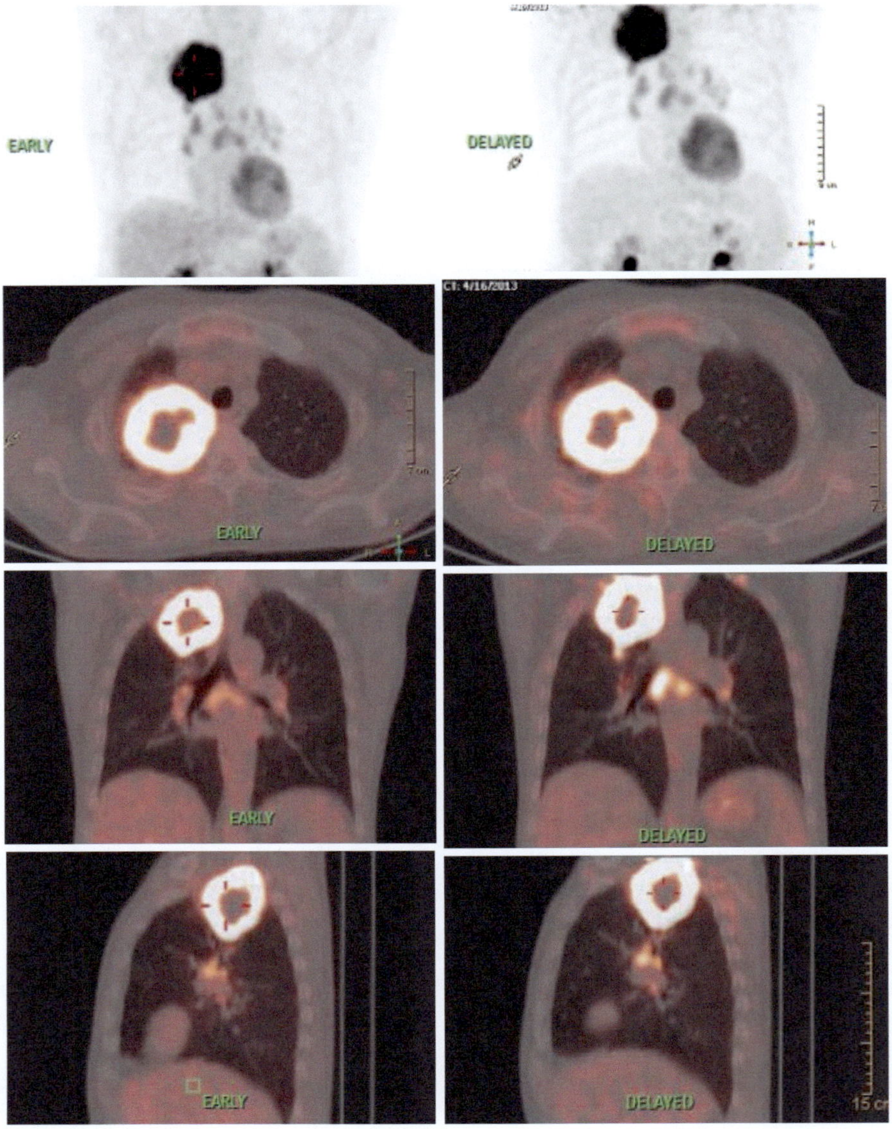

Fig. 12.1 Case record: A 55-year-old male biopsy-proven adenocarcinoma, diagnosed of left upper lobe mass with no locoregional lymph nodes by conventional contrast-enhanced CT scan, sent for whole-body PET–CT scan for staging. PET–CT revealed FDG-avid large cavitating pulmonary mass with irregular speculated margins in the upper lobe of the right lung, measuring 7.1 × 6.8 × 6.8 cm (SUVmax 12.5 g/ml). FDG-avid multiple subcentimeter bilateral hilar, right peribronchial, and subcarinal nodes (SUVmax 4.8 g/ml)

Message

FDG PET/CT when compared to the baseline diagnostic contrast-enhanced CT scan revealed contralateral hilar nodes, which changed the staging from operable to inoperable.

Message

PET–CT when compared to the baseline diagnostic contrast-enhanced CT scan upstaged the patient to stage IV and revealed additional hypermetabolic left hilar and subcarinal nodes and

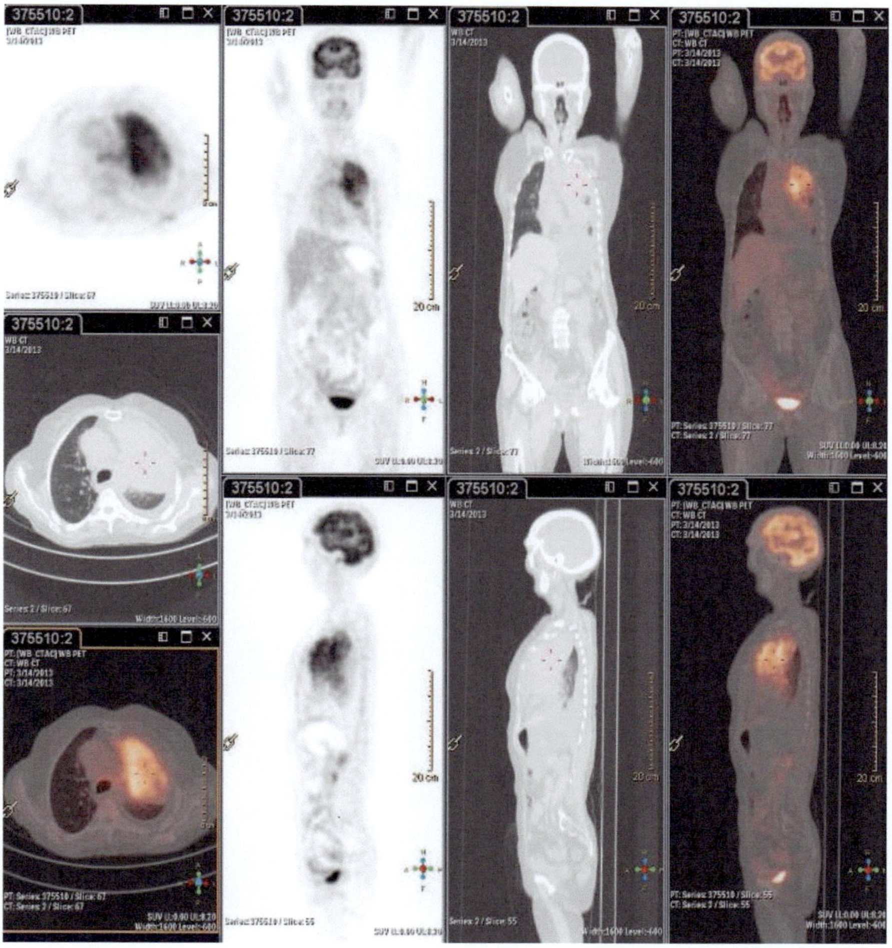

Fig. 12.2 Case record: A 55-year-old male biopsy-proven adenocarcinoma, diagnosed of well-defined soft tissue mass lesion 11.9 × 5.9 cm in the left lobe extending from the apex to the mediastinum abutting the adjoining mediastinum with no locoregional nodes by conventional contrast-enhanced CT scan, sent for whole-body PET–CT scan for staging. PET–CT lung window revealed marked heterogeneous FDG uptake in the large pulmonary mass involving almost the entire left lung (SUVmax 7.93 g/ml) with evidence of areas of necrosis, abutting the adjoining mediastinum arch of aorta with left hilar, subcarinal nodes (SUVmax 6.5 g/ml) and multiple thoracodorsal vertebrae, multiple ribs, right scapula (SUVmax 5.36 g/ml), right ischium, and sacrum with lytic component

multiple bone metastases. CECT chest was unable to detect the metastases to the right scapula, multiple ribs, and dorsal vertebrae.

Message
PET–CT-based disease upstaging is further demonstrated in this case, where disease is upstaged to stage IV. The asymmetry of FDG uptake in the vocal cord is also noteworthy, an indirect feature of recurrent laryngeal nerve palsy secondary to involvement.

Message
FDG PET/CT-based disease characterization of tumor heterogeneity is an area of major interest in recent times, which can highlight tumor biology

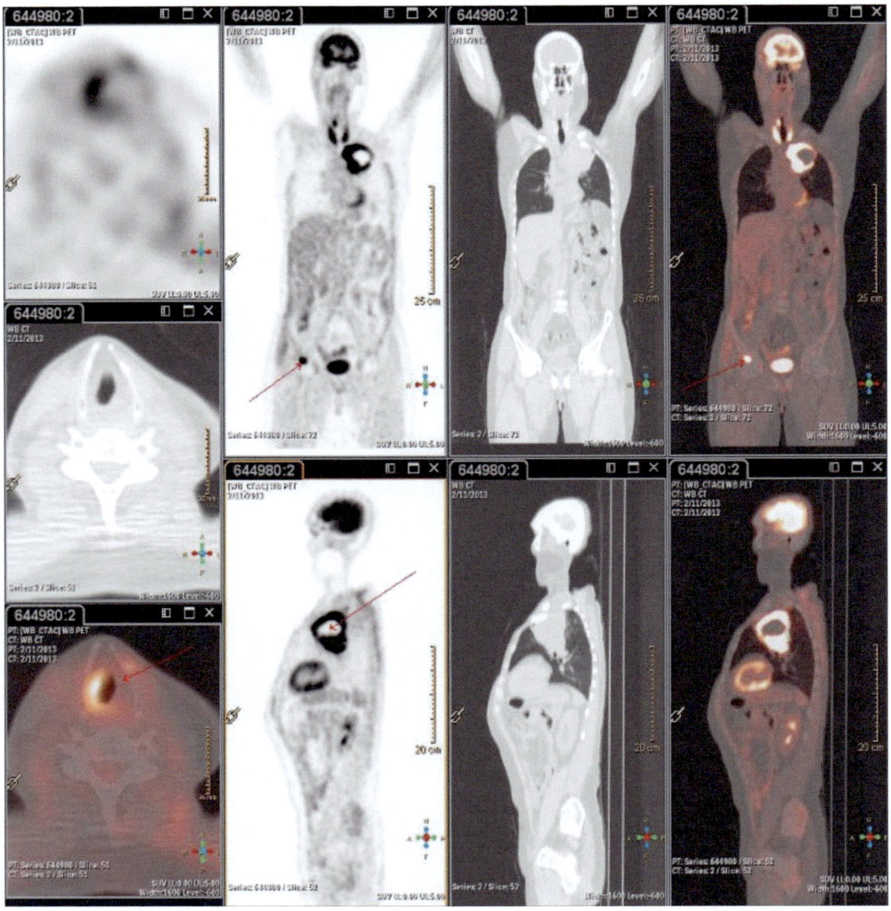

Fig. 12.3 Case record: A 50-year-old male presented with change in voice and left-sided chest pain; chest X-ray revealed a large mass in the upper and middle lobe of the left lung, thus was referred for PET–CT scan for staging, which revealed FDG-avid large cavitating pulmonary mass with irregular margins in the upper and middle lobe of the left lung, measuring 9.1 × 7.8 × 8.8 cm, SUV max 16.54 g/ml. FDG-avid multiple lytic and sclerotic skeletal metastases involving the skull, D3, D9, L2 vertebrae, right acetabulum, and sacrum (SUVmax 9.84 g/ml). Diffuse FDG uptake was noted in both lobes of the thyroid gland consistent with thyroiditis. No FDG uptake was noted in the left vocal cord as a result of left recurrent laryngeal nerve palsy from a left lung tumor, whereas benign intense FDG uptake was noted in the right vocal cord secondary to the compensatory hypertrophy

with various novel tracers too. This has implications for selecting patients for appropriate treatment options.

Message

This case highlights that PET–CT can clearly demarcate pleural involvement and the single lytic skeletal lesion that upstaged the disease. PET–CT aided in evaluating the abnormal FDG uptake in the right lobe of the thyroid gland, in which poorly differentiated thyroid carcinoma (synchronous second primary) with papillary thyroid carcinoma background which was proven by biopsy.

Assessment of Treatment Efficacy

Posttreatment detection of residual or recurrent tumor is an important clinical question, where functional imaging with FDG-PET has potential advantages (Fig. 12.6). Structural changes produced by radiotherapy and/or surgery further reduce the accuracy of anatomical imaging modalities.

Identification of metabolic abnormalities of lung cancer was the first non-CNS oncologic application of FDG-PET to be reported. Differentiating malignant from nonmalignant nodules was the first oncologic indication for FDG-PET to be established, leading the way for

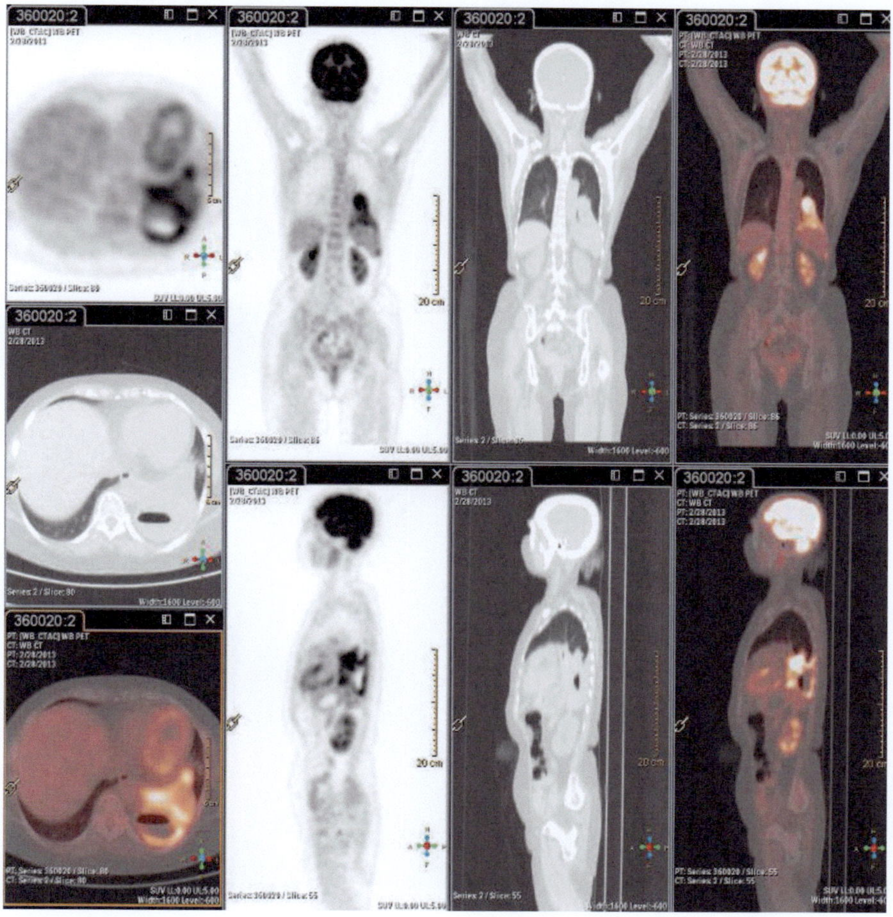

Fig. 12.4 Case record: A 65-year-old female presented with dry cough and breathlessness, diagnosed by CT chest as harboring dense pulmonary soft tissue of size approximately 3.5 × 2.5 cm in the left lower lobe and a single 1.7 × 1.6 cm left paraesophageal lymph node. FDG PET–CT revealed heterogeneous FDG concentrating lesion noted in the lower lobe of the left lung with area of central necrosis (SUVmax 9.59)

later advances in staging, therapy assessment, and follow-up, so the use of PET is now generally accepted for the management of lung cancer.

The most common staging system used is the TNM (tumor, node, metastases); this will be familiar to most readers.

Case Record

A 52-year-old male patient, biopsy-proven poorly differentiated squamous cell carcinoma, stage IIIb at presentation. Baseline CECT revealed heterogeneously enhancing spiculated mass lesion in the anterior segment of the right upper lobe causing cutoff of the upper lobe bronchus with areas of consolidation and volume loss (T2N2M0); patient received neoadjuvant chemotherapy and 3DCRT 60 Gy/30 # in 45 days with 5 # of weekly cisplatin and was referred for residual disease evaluation after 6 weeks of completion of EBRT. FDG PET/CT scan (Fig. 12.6a, b) revealed FDG-avid residual disease in the right upper lobe in paratracheal region with central photopenia (SUVmax 4.36) with CT evidence of patchy fibrosis surrounding the lesion. FDG-avid residual right tracheobronchial, pretracheal, subcarinal, and left hilar nodes are evident (Fig. 12.6b).

Message

This case highlights the role of FDG PET/CT in the evaluation of posttreatment residual disease assessment when CECT is inconclusive.

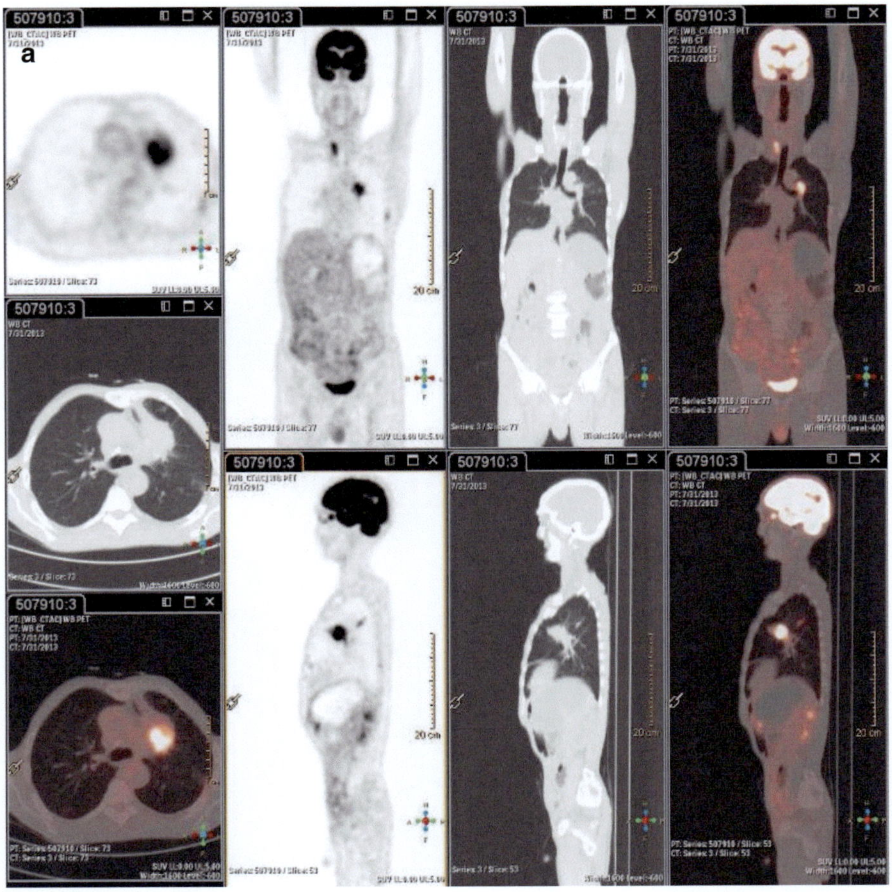

Fig. 12.5 (**a**–**c**) Case record: A 69-year-old male patient of biopsy-proven adenocarcinoma of the left lung. CT chest revealed mildly enhancing soft tissue mass with speculated margins (4.6 × 3.4 cm) involving perihilar region of anterior segment of the left upper lobe and left hilum abutting the mediastinum medially and pushing the left pulmonary artery, partial obliteration of the left upper lobe anterior segment bronchus and partial volume loss. Staging PET–CT (**a**–**c**) revealed FDG-avid soft tissue mass with spiculated margins (4.6 × 3.4 cm) involving perihilar region of anterior segment of the left upper lobe and left hilum (SUVmax 11.14). Low-grade FDG-avid subcentimeter parenchymal nodule along the fissure in the left lung upper lobe (SUVmax 3.88), and subpleural nodule in the left lung upper lobe (SUVmax 3.36) is seen; FDG uptake noted in the multiple foci in the left pleura, FDG-avid lytic lesion in the right ilium (SUVmax 6.61). FDG uptake in the right lobe of thyroid gland was also observed

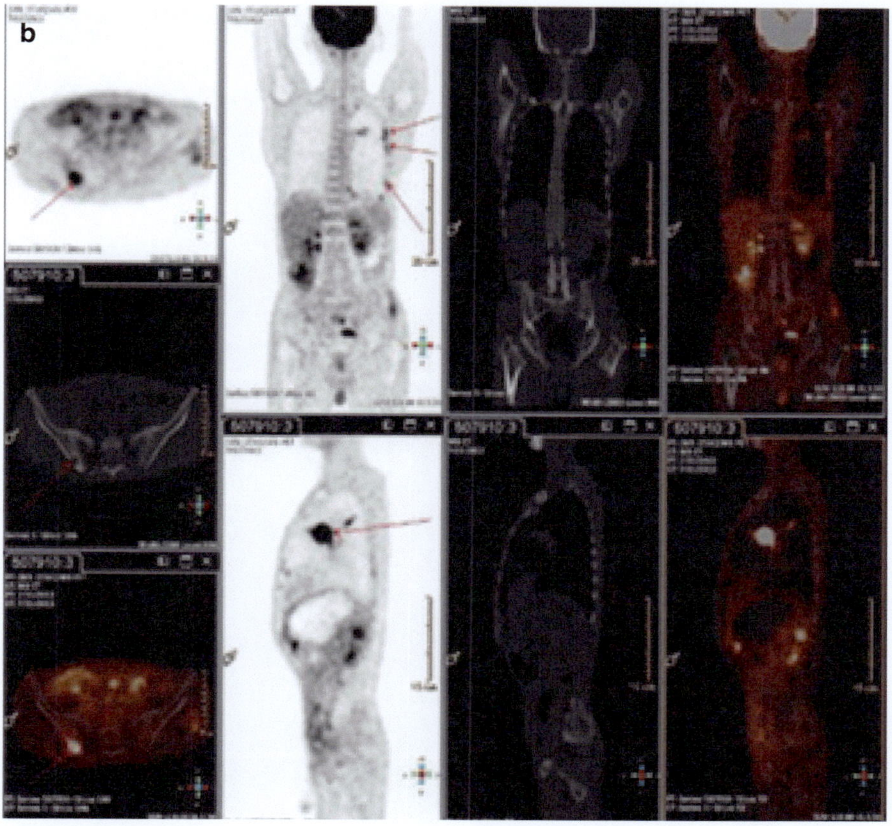

Fig. 12.5 (continued)

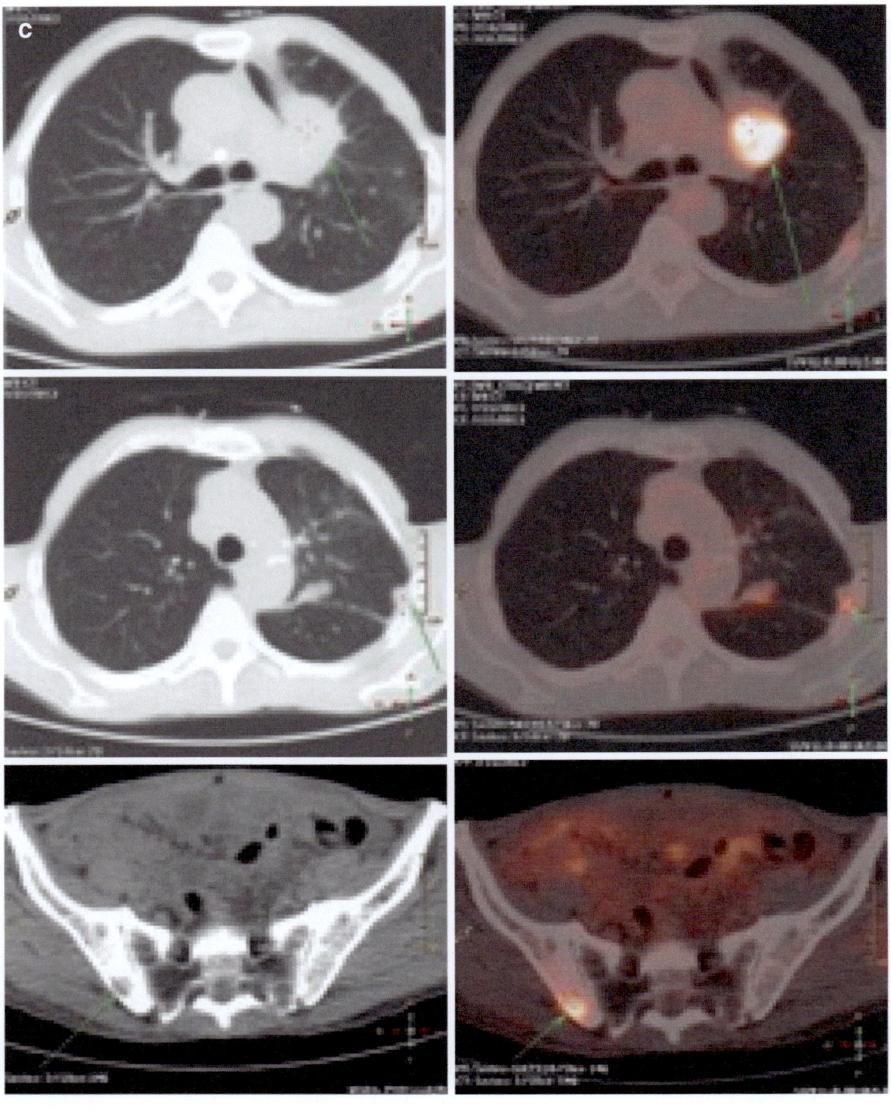

Fig. 12.5 (continued)

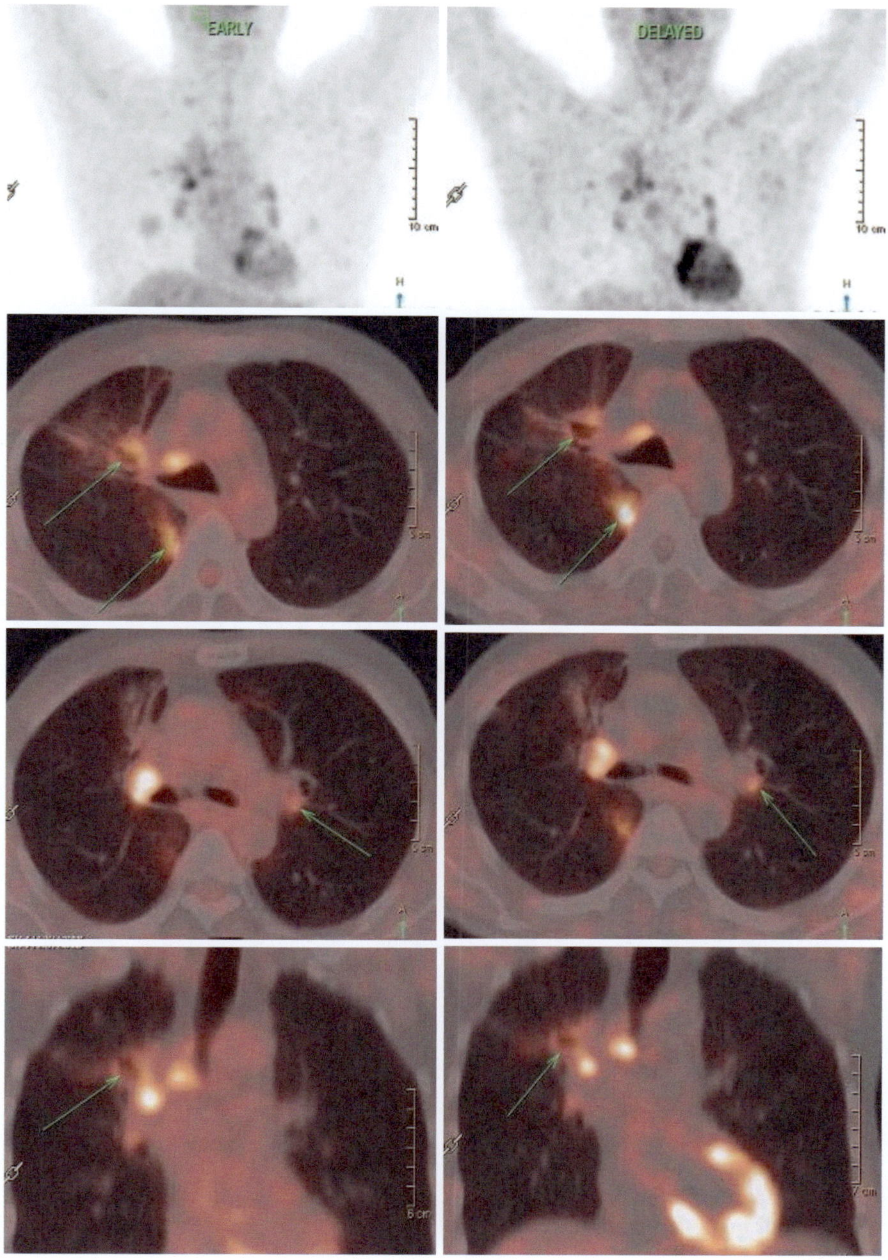

Fig. 12.6 (**a**, **b**) Case record: a 52-year-old male patient, biopsy-proven poorly differentiated squamous cell carcinoma, stage IIIb at presentation. Baseline CECT revealed heterogeneously enhancing spiculated mass lesion in the anterior segment of the right upper lobe causing cutoff of the upper lobe bronchus with areas of consolidation and volume loss (T2N2M0); patient received neoadjuvant chemotherapy and 3DCRT 60 Gy/30 # in 45 days with 5 # of weekly cisplatin and was referred for residual disease evaluation after 6 weeks of completion of EBRT. FDG PET/CT scan (**a**, **b**) revealed FDG-avid residual disease in the right upper lobe in the paratracheal region with central photopenia (SUVmax 4.36) with CT evidence of patchy fibrosis surrounding the lesion. FDG-avid residual right tracheobronchial, pretracheal, subcarinal, and left hilar nodes are evident (**b**)

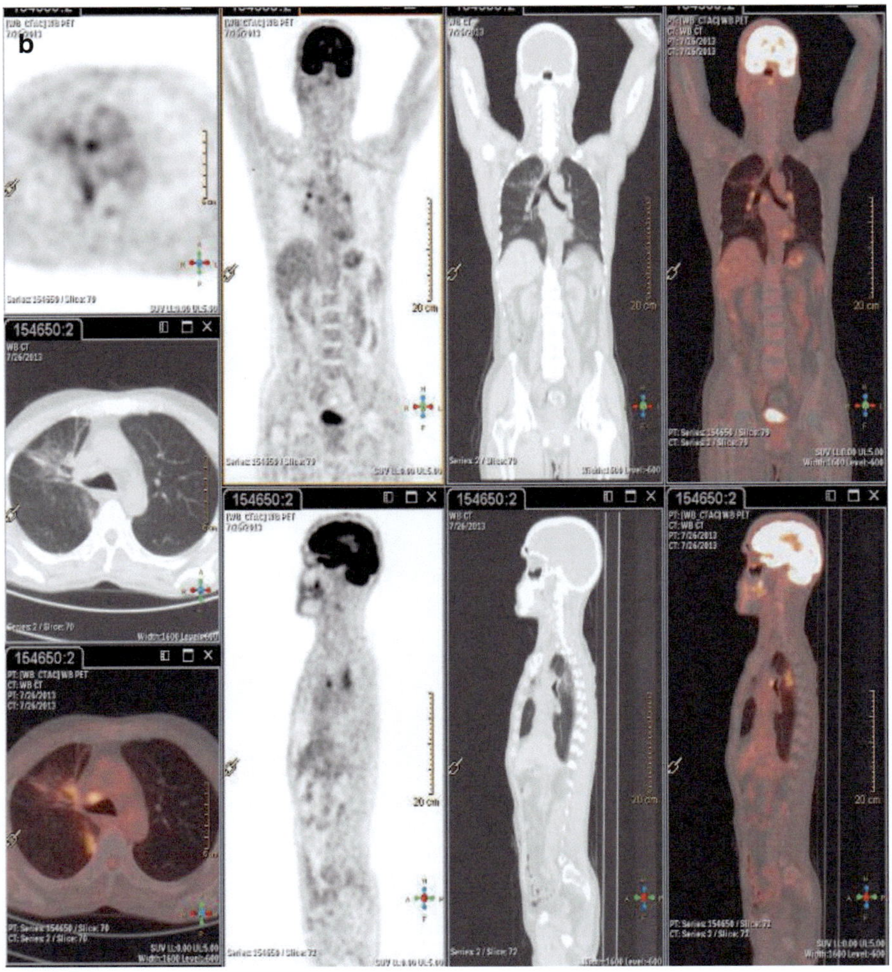

Fig. 12.6 (continued)

Classification and Evaluation of Lung Cancer

Histological Classification of Non-small-Cell Lung Cancers

The new World Health Organization/International Association for the Study of Lung Cancer histologic classification of non-small-cell lung cancers.

1. *Squamous cell carcinoma*
   - Papillary
   - Clear cell
   - Small cell
   - Basaloid
2. *Adenocarcinoma*
   - Acinar
   - Papillary
   - Bronchioloalveolar carcinoma

- Non-mucinous
- Mucinous
- Mixed mucinous and non-mucinous or indeterminate cell type
- Solid adenocarcinoma with mucin
- Adenocarcinoma with mixed subtypes variants
- Well-differentiated fetal adenocarcinoma
- Mucinous ("colloid") adenocarcinoma
- Mucinous cystadenocarcinoma
- Signet ring adenocarcinoma
- Clear cell adenocarcinoma

3. *Large cell carcinoma variants*
   - Large cell neuroendocrine carcinoma
   - Combined large cell neuroendocrine carcinoma
   - Basaloid carcinoma
   - Lymphoepithelioma-like carcinoma
   - Clear cell carcinoma
   - Large cell carcinoma with rhabdoid phenotype
4. *Adenosquamous carcinoma*
5. *Carcinomas with pleomorphic, sarcomatoid, or sarcomatous elements*
   - Carcinomas with spindle and/or giant cells
   - Spindle cell carcinoma
   - Giant cell carcinoma
   - Carcinosarcoma
   - Pulmonary blastoma
6. *Carcinoid tumor*
   - Typical carcinoid
   - Atypical carcinoid
7. *Carcinomas of salivary gland type*
   - Mucoepidermoid carcinoma
   - Adenoid cystic carcinoma
   - Others
8. *Unclassified carcinoma*

TNM Classification and Stage Grouping

Primary tumor (T)

TX	Primary tumor cannot be assessed, or tumor proven by the presence of malignant cells in sputum or bronchial washings but not visualized by imaging or bronchoscopy
T0	No evidence of primary tumor
Tis	Carcinoma in situ
T1	Tumor 3 cm or less in greatest dimension, surrounded by lung or visceral pleura, without bronchoscopic evidence of invasion more proximal than the lobar bronchus[a], (i.e., not in the main bronchus)

T2 Tumor with any of the following features of size or extent

More than 3 cm in greatest dimension

Involves main bronchus, 2 cm or more distal to the carina

Invades the visceral pleura

Associated with atelectasis or obstructive pneumonitis that extends to the hilar region but does not involve the entire lung

T3 Tumor of any size that directly invades any of the following

Chest wall (including superior sulcus tumors), diaphragm, mediastinal pleura, parietal pericardium; or tumor in the main bronchus less than 2 cm distal to the carina, but without involvement of the carina; or associated atelectasis or obstructive pneumonitis of the entire lung

T4 Tumor of any size that invades any of the following

Mediastinum, heart, great vessels, trachea, esophagus, vertebral body, carina; or separate tumor nodules in the same lobe; or tumor with malignant pleural effusion[b]

When investigations and clinical judgment dictate that the effusion is not related to the tumor, the effusion should be excluded as a staging element and the patient should be staged T1, T2, or T3

[a]*Note*: The uncommon superficial tumor of any size with its invasive component limited to the bronchial wall, which may extend proximal to the main bronchus, is also classified T1

[b]*Note*: Most pleural effusions associated with lung cancer are due to tumor. However, there are a few patients in whom multiple cytopathologic examinations of pleural fluid are negative for tumor. In these cases, fluid is non-bloody and is not an exudate. Such patients may be further evaluated by videothoracoscopy (VATS) and direct pleural biopsies

Regional lymph nodes (N)	
NX	Regional lymph nodes cannot be assessed
N0	No regional lymph node metastasis
N1	Metastasis to ipsilateral peribronchial and/or ipsilateral hilar lymph nodes and intrapulmonary nodes including involvement by direct extension of the primary tumor
N2	Metastasis to ipsilateral mediastinal and/or subcarinal lymph nodes(s)
N3	Metastasis to contralateral mediastinal, contralateral hilar, ipsilateral or contralateral scalene, or supraclavicular lymph nodes(s)

Distant metastasis (M)	
MX	Distant metastasis cannot be assessed
M0	No distant metastasis
M1	Distant metastasis present

Note: M1 includes separate tumor nodule(s) in a different lobe (ipsilateral or contralateral)

Stage Grouping

Occult carcinoma TX N0 M0	
Stage 0	Tis N0 M0
Stage IA	T1 N0 M0
Stage IB	T2 N0 M0
Stage IIA	T1 N1 M0
Stage IIB	T2 N1 M0
	T3 N0 M0
Stage IIIA	T1 N2 M0
	T2 N2 M0
	T3 N1 M0
	T3 N2 M0
Stage IIIB	Any T N3 M0
	T4 Any N M0
Stage IV	Any T any N M1

Conventional Staging

Conventional staging of lung cancer usually involves contrast-enhanced computed tomography (CECT) scan of the thorax and upper abdomen, a radionuclide bone scan, and a magnetic resonance imaging (MRI) scan of the brain.

Patterns of FDG Uptake in Lung Cancer

In general, lung cancer is very FDG avid.

Two lung tumors, pulmonary carcinoid and bronchioloalveolar cell cancer, are well known to be less FDG avid and are frequently listed as causes of false-negatives on FDG-PET.

Although it is important to remember that these tumors may be a cause of false-negative scans, many of them are positive on FDG-PET.

Bronchioloalveolar cell lung cancers with a prominent solid component on CT are more likely to be FDG avid.

Atypical carcinoid tumors are more likely to be FDG avid than typical carcinoid tumors. Thus, FDG uptake should not be used to exclude either of these tumors.

The FDG avidity of lung cancer is a valuable prognostic factor: the more FDG avid the cancer, the worse the prognosis.

A recent study showed that FDG uptake correlated with tumor proliferation as measured by Ki-67 immunostaining in lung cancers. Comparison of FDG uptake with F-18-fluorothymidine (FLT) showed that FDG had higher mean uptake as measured by SUV, while the FLT had a better correlation with tumor proliferation. Studies of FLT as a prognostic indicator have not yet been done. These two PET tracers may provide different insight into cancer metabolism.

Potential Role of PET–CT in Lung Cancer

- Assessment of the solitary pulmonary nodule (SPN)
- Staging of non-small-cell lung cancer (NSCLC)
- Assessment of mediastinal lymphadenopathy
- Identification of distant metastatic disease
- Detection of recurrent disease
- Radiotherapy planning
- Response to therapy assessment
- As a prognostic indicator
- Possible role in staging small-cell lung cancer
- Detecting lung carcinoma as the unknown primary with metastatic presentations (Figs. 12.7a, b and 12.8) with simultaneous disease staging

Case Record

The 49-year-old female diagnosed of multiple ring-enhancing lesions with irregular margins and varying sizes at the gray-white junction in the supratentorial parenchyma and the left cerebellum suggesting multiple brain metastases was referred for a whole-body PET–CT scan (Fig. 12.7a, b) to search for unknown primary tumor. PET–CT revealed an intense and homogenous FDG uptake in the large well-defined mass in the upper lobe of the left lung abutting the arch of aorta with no definite infiltration, measuring 2.9×2.7×4.5 cm (SUVmax 10.82 g/ml). FDG-avid multiple nodes noted in the ipsilateral mediastinum involving the prevascular (largest measuring 1.6 cm, SUVmax 4.77 g/ml), aortopulmonary window, and subcarinal and left hilar region. Multiple FDG-avid skeletal metastases were noted in the distal end of the right clavicle, manubrium sternum, proximal ends of the left humerus, and the right femur. The known brain metastases were not appreciable in the early scan (postinjection 1 h), which were well delineated in the dedicated delayed brain image (Fig. 12.7a).

Message

FDG PET/CT can serve as a one-stop shop for detecting the site of primary in patients with brain metastases; most frequently, they are encountered in the lungs. A whole-body FDG PET/CT would also provide disease staging in the same go in such a patient. The scan highlights the importance of delayed brain imaging in FDG-PET to identify and clearly delineate brain metastases.

Case Record

A 50-year-old male patient, biopsy-proven poorly differentiated adenocarcinoma from the abdominal nodes; baseline CECT revealed soft tissue density in the right lower lobe-medial basal segment of size 2.5×2.1×1.6 cm. USG abdomen revealed portal, peripancreatic, paraaortic nodes. Staging FDG PET/CT (Fig. 12.8) revealed FDG-avid soft tissue mass, measuring 2.0×1.5 cm in the lower lobe of the right lung (SUVmax 7.74), and FDG-avid left supraclavicular, pretracheal, subcarinal, and right hilar node, (SUVmax 8.88). FDG-avid conglomerulated nodal mass in the multiple peripancreatic, preaortic, and paraaortic nodes (SUVmax 13.0). FDG-avid single right lobe of liver (SUVmax 7.43).

Message

This is another example of the role of FDG PET/CT in detecting and delineating the primary lesion in a patient with presentation of abdominal nodes from unknown primary and also to do a whole-body disease assessment in a single go.

Assessment of the Solitary Pulmonary Nodule (SPN)

Solitary pulmonary nodule (SPN) is defined as a single spherical lesion of 3 cm or less in diameter completely surrounded by lung parenchyma without any associated atelectasis or lymphadenopathy.

Imaging characteristics favoring benign or malignant nature of the nodule has important implications for patient management.

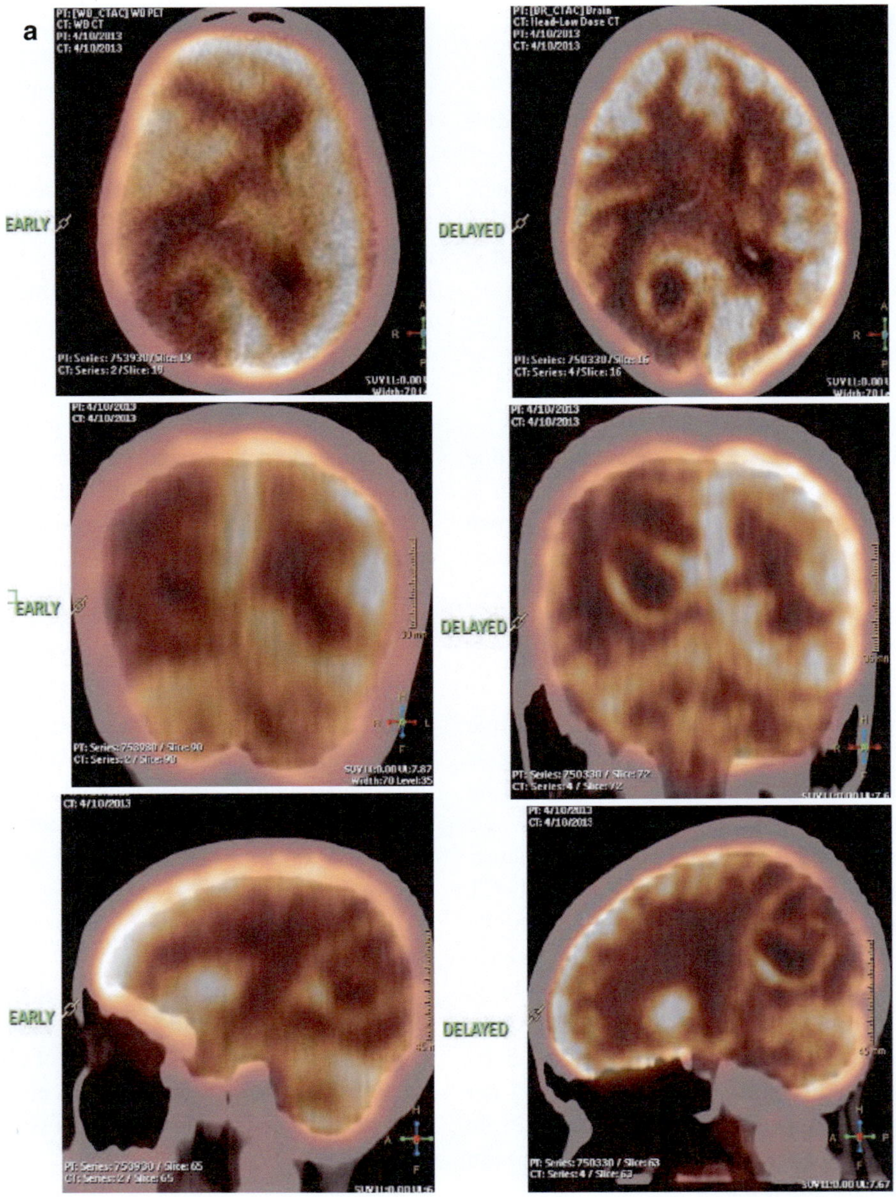

Fig. 12.7 (**a**, **b**) Case record: a 49-year-old female diagnosed of multiple ring-enhancing lesions with irregular margins and varying sizes at the gray-white junction in the supratentorial parenchyma and the left cerebellum suggesting multiple brain metastases was referred for a whole-body PET/CT scan (**a**, **b**) to search for unknown primary tumor. PET/CT revealed an intense and homogenous FDG uptake in the large well-defined mass in the upper lobe of the left lung abutting the arch of aorta with no definite infiltration, measuring 2.9 × 2.7 × 4.5 cm (SUVmax 10.82 g/ml). FDG-avid multiple nodes noted in the ipsilateral mediastinum involving the prevascular (largest measuring 1.6 cm, SUVmax 4.77 g/ml), aortopulmonary window, and subcarinal and left hilar region. Multiple FDG-avid skeletal metastases were noted in the distal end of the right clavicle, manubrium sternum, proximal ends of the left humerus, and the right femur. The known brain metastases were not appreciable in the early scan (postinjection one hour), which were well delineated in the dedicated delayed brain image (**a**)

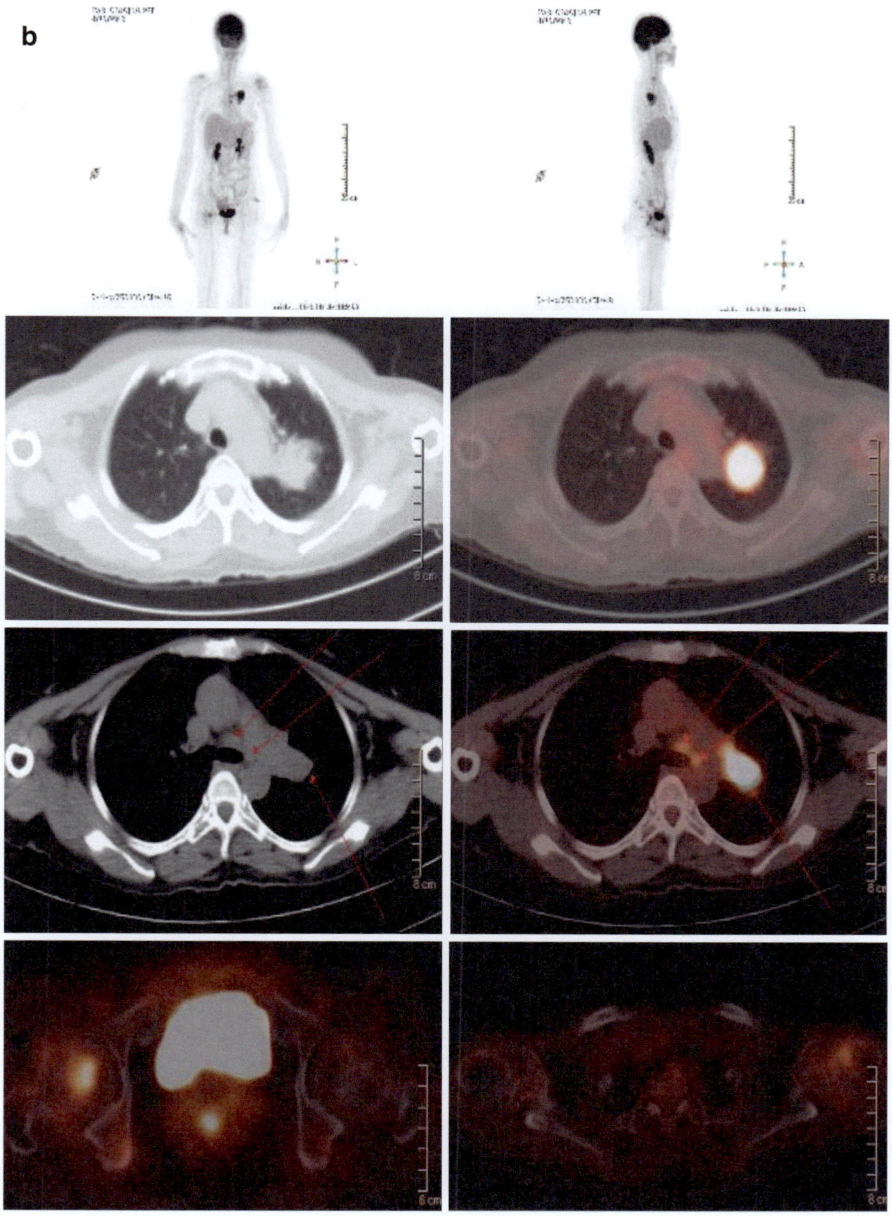

Fig. 12.7 (continued)

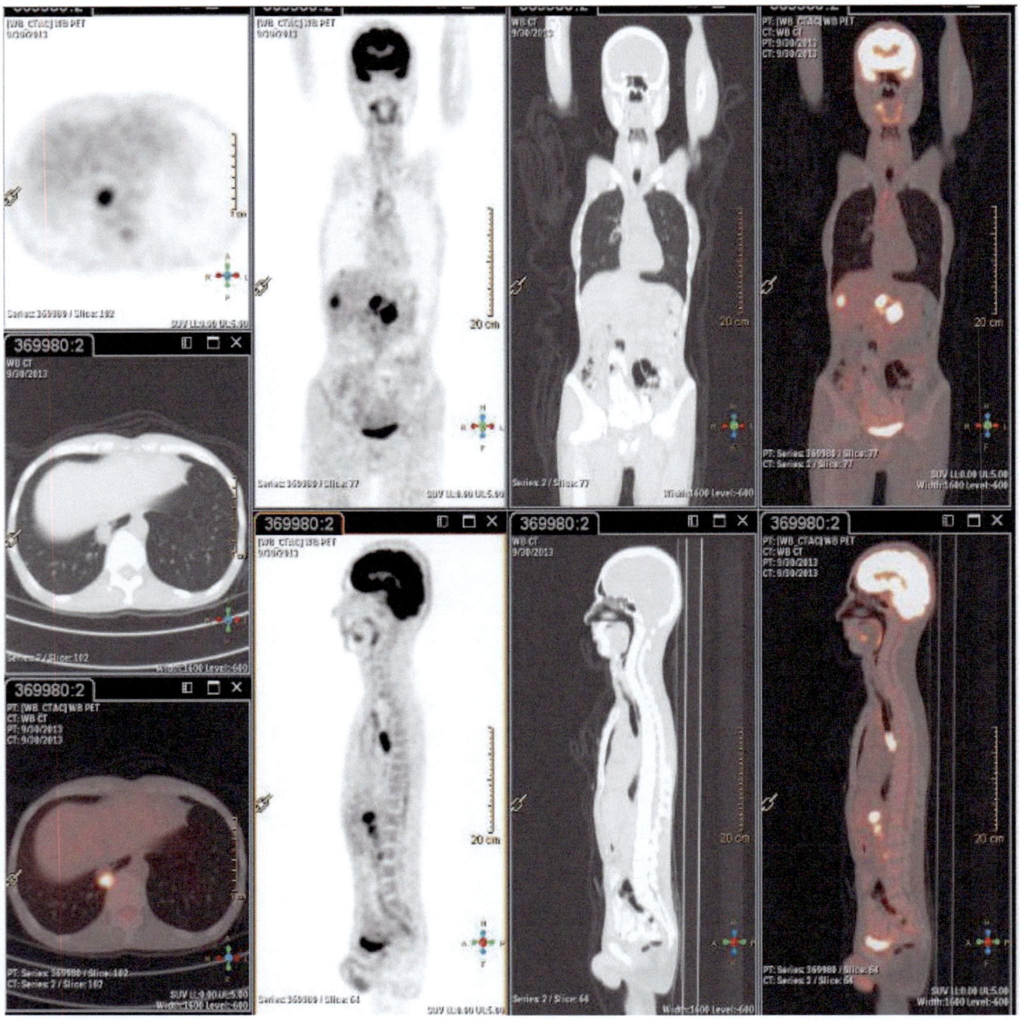

Fig. 12.8 Case record: a 50-year-old male patient, biopsy-proven poorly differentiated adenocarcinoma from the abdominal nodes; baseline CECT revealed soft tissue density in the right lower lobe-medial basal segment of size 2.5 × 2.1 × 1.6 cm. USG abdomen revealed portal, peripancreatic, and paraaortic nodes. Staging FDG PET/CT revealed FDG-avid soft tissue mass, measuring 2.0 × 1.5 cm in the lower lobe of the right lung (SUVmax 7.74) and FDG-avid left supraclavicular, pretracheal, subcarinal, and right hilar node(SUVmax 8.88). FDG-avid conglomerulated nodal mass in the multiple peripancreatic, preaortic, and paraaortic nodes (SUVmax 13.0). FDG-avid single right lobe of liver (SUVmax 7.43)

Findings Favoring a Benign Lesion

Conventional Imaging with chest X-Ray or CT Scan

- Size less than 2 cm.
- Stable appearance (since 2 years).
- Smooth margin.
- Diffuse calcification ("popcorn" calcification) is typical of hamartomas, and central calcification is typical of granulomas.
- If cavitation is present, smooth, thin walls (4 mm or less) favor benign.
- Satellite nodules seen at the periphery of a dominant smooth nodule suggest infectious granuloma.
- Nodule enhancement <15 HU.

Imaging with FDG-PET

- SUVmax <2.5 or visually less metabolically active than mediastinal blood pool (for nodules >1 cm), though considerable overlap exists

Findings Favoring a Malignant Lesion

Conventional Imaging with chest X-Ray or CT Scan

- Size greater than 3 cm.
- Interval change, that is, increase in size.
- Spiculated, irregular, or lobulated margin.
- Stippled or eccentric calcification.
- Cavitation with irregular thick walls (>15 mm) favors malignancy.
- Malignant lesions are relatively hypervascular.

Imaging with FDG-PET

- SUVmax >2.5 or visually more metabolically active than mediastinal blood pool (for nodules >1 cm)

False-Positive SPN

- Granulomatous conditions
- Sarcoidosis
- Inflammation
- Infection
- Adenomas
- Hamartomas
- Neurofibromas
- Pulmonary fibrosis

False-Negative SPN

- Bronchoalveolar cancer
- Scar adenocarinoma
- Carcinoid tumors
- Neuroendocrine tumors
- Lesions that are too small (<1 cm)
- Well-differentiated lesions

PET scanning has a sensitivity of 91–97 % and a specificity of 78–88 % for predicting the pathologic nature of SPN.

Integrated PET–CT permits anatomic definition to be added to help resolve metabolic ambiguity. The improved localization with PET–CT also allows radiologists and nuclear physicians to differentiate areas of normal physiological uptake from abnormal areas of increased uptake. Integrated FDG PET/CT has been found to be more useful in characterizing SPN with better sensitivity, specificity, and accuracy. Semiquantitative analysis of glucose metabolism (SUVmax) is also frequently performed, in addition to visual assessment, because of observer independence and reproducibility.

Competing Methods for Evaluating a Single Pulmonary Nodule

A histological sample may be obtained from the nodule by transthoracic needle aspiration, by bronchoscopy with direct biopsy, transbronchial needle aspiration (TBNA), or bronchoalveolar lavage (BAL) or by surgical excision.

Advantage of All of These Methods

They provide histological samples.

A peripheral lesion, which is adherent to the chest wall, is ideal for transthoracic biopsy.

An endobronchial lesion is ideal for bronchoscopy.

A highly suspicious peripheral lesion, which is to be resected regardless of the findings on FDG, considered primarily for resection.

Disadvantages of These Methods

There is moderate morbidity from each of the aforementioned procedures. Transthoracic needle aspiration is associated with pneumothorax, hemoptysis, and rarely air embolism. Methods using bronchoscopy are associated with hemoptysis, infection, and pneumothorax.

Biopsy methods also suffer from sampling error.

The complications associated with surgical excision depend on the method—video-assisted thoroscopic surgery (VATS) or thoracotomy. Depending upon the exact location of the lesion, a biopsy may be more or less dangerous. However, the advantage of all of these methods is that they provide a histological sample.

Comparison with Conventional Imaging

In a large fraction of cases, FDG PET/CT will be the method of choice for characterizing a single pulmonary nodule and follow-up CT to assure 2-year stability in size. If there are other smaller nodule(s), then it is important to follow-up these nodules.

Staging of Non-small-Cell Lung Cancer (NSCLC)

T Staging

CT is a satisfactory method for the assessment of the size and location of a primary pulmonary lesion, but it is less successful in the characterization of mediastinal nodes. For the assessment of T stage, the combined PET–CT has increased the accuracy of tumor detection, chest wall, and mediastinal infiltration as compared to PEt alone. One potential advantage of PET over conventional imaging is the evaluation of extension of the primary tumor to involve the pleura with a high positive and negative predictive value for the evaluation of malignant pleural effusions. Also, FDG-PET is more accurate than CT in determining the size of primary tumor (T1 and T2) when there is adjacent collapse or consolidation.

N Staging

Conventional imaging modalities (CT/MRI), using only dimensional criteria (>1 cm) to detect nodal involvement, have poor accuracy in differentiating benign from malignant nodal disease (sensitivity, 60–83 %; specificity, 77–82 %). The standard radiological practice is that the mediastinal nodes greater than 1 cm in diameter are considered abnormal and, therefore, more likely to represent metastatic disease. Small nodes may contain malignant cells and large nodes are often only reactive in nature; CT can therefore under- or overstage up to 40 % of cases (Fig. 12.1); thus, the accuracy of such a system is only approximately 60–79 %. Studies have shown that up to 75 % of metastatic nodes were within nodes considered to be of normal size. FDG PET/CT is reported to have a higher diagnostic accuracy than either CT or PEt alone for the N staging. A recent multicenter study has shown that FDG PET/CT has very high negative predictive value (91 %) and specificity (83 %), but limited positive predictive value (29 %).

The use of mediastinoscopy is reserved for those patients considered suitable for surgery by CT staging but who have enlarged mediastinal nodes that require evaluation. The procedure has its risks and may have a false-negative rate of up to 10 %.

Transbronchial needle biopsy can have a very high specificity; it has a poor sensitivity of not much more than 50 % and in turn can have a significant morbidity. Many studies concluded that nodal size is a poor determinant of metastatic involvement. A recent study examined lymph nodes of all sizes, in known lung cancer patients, and found that in those nodes with metastatic deposits, the nodal size was "normal" in 74 %. However, PET findings cannot replace histological confirmation of FDG-positive lesions by mediastinoscopy. FDG PET/CT cannot obviate the need for invasive procedures. PET–CT virtual mediastinoscopy has also been found to be a useful adjunct.

Teaching Point

- Without the aid of PET–CT, it can be difficult to distinguish active tumor from collapsed lung or necrotic tissue.
- The conventional staging of nodes by size criteria alone can often lead to the wrong conclusion. Metabolic imaging stages mediastinal

nodes with much greater accuracy than CT alone. The highest diagnostic accuracy has been with combined PET–CT with a generally accepted figure of approximately 90 % for mediastinal nodal involvement.

- PET has consistently been shown to be accurate in the distinction between N2 (operable) and N3 (inoperable) nodal disease.
- Not only is PET–CT superior than other non-invasive methods for mediastinal staging, it also has the advantage that whole-body images are obtainable, and therefore, an assessment can be made of potential distant metastatic disease.
- PET–CT has a higher accuracy for mediastinal nodal involvement compared to PET scan alone.

M Staging

Detection of a distant site of disease means the patient is stage IV and palliative chemotherapy is indicated. The common sites of metastases from NSCLC are to the adrenal glands, liver, bone, and brain.

Adrenal

PET and PET–CT are accurate in the assessment of metastatic disease to the adrenal glands. Evidence suggests that the removal of a solitary adrenal deposit at the time of resection of the lung primary results in an increased life expectancy.

Liver

FDG-PET appears more accurate than CT in detecting liver metastases because of its better specificity.

Bone

Lytic osseous metastases appear to have a greater uptake of FDG and as such are more readily detected (Fig. 12.2). PET–CT is both more sensitive and specific than bone scanning in lesion detection in lung cancer. Metastatic lesions start in marrow and FDG is positive much before bone scan becomes positive. In a recent meta-analysis, it was shown that the pooled sensitivity and specificity for the detection of bone metastasis in lung cancer using FDG PET/CT, FDG-PET, MRI, and bone scan were 92, 87, 77, and 86 %; and 98, 94, 92, and 88 %, respectively.

Brain

FDG-PET due to high physiological brain parenchymal uptake has its limitation in the detection of brain metastases. MRI brain is thus mandatory in the staging of lung cancer to rule out brain metastases.

Comparison with conventional imaging: Liver, adrenal, brain, and bony deposits are common with lung cancer, but many of the lesions are undetected in the course of conventional staging.

Anatomic imaging is often important in helping make important distinctions about issues such as invasion of structures, but FDG-PEt also has an important role.

Accurate staging is very important since it has such a major impact on both therapy and prognosis.

A meta-analysis of studies comparing FDG-PET and CT evaluation of mediastinal lymph nodes showed that FDG-PET had a sensitivity of 85 % and a specificity of 90 %, whereas CT had a sensitivity of 61 % and a specificity of 79 %. FDG-PET was more sensitive but less specific in enlarged lymph nodes (sensitivity of 100 %, specificity of 78 %) than in normal-sized nodes (sensitivity of 82 %, specificity of 93 %).

Integrated PET–CT improves the diagnostic accuracy of the staging of non-small-cell lung cancer.

Radiation Treatment Planning

FDG-PET may have an important role in radiation treatment planning in lung cancer.

In addition to determining if radiation therapy is appropriate and whether therapy will be given with curative or palliative intent, FDG-PET is useful for determining therapy ports. It can be used both to limit ports to spare normal tissue and to include additional involved regions. Treatment plans that include all the FDG-avid lesions or the FDG-avid portions of a complex mass will result in more effective local control with less unnecessary tissue being treated. PET–

CT should be used for RT planning in NSCLC because it more accurately images tumor extent than CT alone. PET–CT imaging can improve the accuracy of target volume delineation using anatomic biological contour (ABC) and determined directly on PET–CT images. PET–CT imaging has another positive effect on tumor volume delineation: Significantly reduced interobserver and intraobserver variability for tumor volume delineation.

Recurrent Laryngeal Nerve Palsy

The recurrent laryngeal nerves, especially the left recurrent laryngeal nerve, may be involved with lung cancer, causing recurrent laryngeal nerve palsy. In this case, the involved vocal cord does not function. The non-involved vocal cord must work harder than normal in order to make up for the loss in function of the involved vocal cord.

Consequently, the ipsilateral vocal cord will not show any FDG uptake, and the contralateral vocal cord will show increase uptake, which can be quite intense (Fig. 12.3).

Unilateral FDG uptake in the vocal cord region, especially with an appropriately placed mass, can be used to suggest the diagnosis of recurrent laryngeal nerve palsy. It should not be misinterpreted as spread of disease.

Monitoring Response to Therapy

FDG-PET is playing an increasing role in monitoring therapy for lung cancer.

A decrease in FDG uptake after therapy is very encouraging, although with lung cancer, one often has to settle for stable findings. One of the major theoretical advantages of FDG-PET compared with structural imaging techniques is that there is usually a more rapid change in cellular metabolism than in tumor size (Fig. 12.6).

A new region of uptake indicates that therapy is not effective or no longer effective. The level of FDG uptake as measured by the SUV is a prognostic factor both before and during therapy.

Remember that when comparing quantitative measures on sequential studies, the details of the analysis become even more important. Quantitative measurement should be made at the same time after injection, and when monitoring therapy with FDG-PET, it is important to recognize the confounding effects of radiation therapy. In the area of radiation change, there is typically low-level FDG uptake. Occasionally, there can be avid FDG uptake, especially when there is frank radiation pneumonitis. The FDG uptake often returns to normal after 6 months, but radiation change may persist for a longer period of time. There can be increased uptake in both the lung and the chest wall after radiation therapy. A local region with the radiation port with more marked FDG uptake, or a region with increasing uptake, especially if it is associated with a mass on anatomic imaging, should be considered suspicious for recurrence. Mild to moderate uptake in a region corresponding to typical radiation changes does not suggest recurrence. FDG PET/CT has been shown to be useful in monitoring response to the EGFR kinase inhibitor erlotinib in few studies.

Prognosis

The primary purpose of staging is to identify patients who may benefit from different therapies.

Staging is also a potent prognostic indicator. The 5-year survival for patients with non-small-cell lung cancer is 61 % if they have stage I disease and only 5 % if they have stage IIIB disease. Since FDG imaging allows for more accurate staging, it should make TNM staging an even more accurate predictor of prognosis.

In addition to staging, the FDG avidity of the tumor is an independent prognostic factor. It has been shown that FDG uptake correlates with tumor growth rate and FDG uptake correlates with Ki-67 immunostaining, a marker of tumor proliferation. It has been found that patients with an SUV greater than or equal to 7 had 6.3 times the 5-year mortality of those with an SUV less than 7. In a larger prospective study, a metabolic response to chemoradiation, as assessed by visual analysis of FDG-PET, was also much more powerfully correlated with survival than the response on CT determined from WHO criteria.

Most of the information about prognosis has dealt with non-small-cell lung cancer. But SUV may also provide prognostic information in small-cell lung cancer.

Detection of Recurrent Disease

PET–CT scan should not be carried out in less than 3 months after the therapy to allow any metabolically active macrophage activity within inflammatory lung to resolve preventing a false-positive report.

At least 3 months is suggested to allow post-operative change to improve.

FDG uptake correlates with the tumor growth and proliferation rates as well as the degree of tumor differentiation.

PET–CT in Therapy Response and Radiotherapy Planning

FDG uptake relates well to tumor growth and proliferation rates, thus, allows us to quantify the metabolic response of therapy within the tumor.

Results as early as 1 week after initiation of chemotherapy accurately reflect response, but imaging 6 weeks after chemotherapy has finished and at least 3 months after radiotherapy is generally regarded as more accurate. This helps avoid the problems with false-positive reports due to radiation-induced pneumonitis.

Tumors with higher pretherapy FDG uptake respond better to radiotherapy than those with low uptake, and additionally, persisting uptake after therapy is predictive of relapse.

Comparison with Conventional Imaging

PET–CT has been shown to be better than CT at assessing accurate tumor volume.

PET–CT is excellent at delineating tumor from distal atelectasis.

Small-Cell Lung Cancer and Mesothelioma

Diagnosis of small-cell lung cancer equates to a diagnosis of disseminated disease. This, however, is not always the case.

Principle role of staging in small-cell lung cancer is to determine if localized radiotherapy can be used in addition to chemotherapy.

The role of PET–CT in this instance is still under evaluation, but evidence suggests PET can be used to accurately upstage presumed limited disease and has been shown to be useful in the detection of paraneoplastic-associated small-cell lung cancers. The potential role of FDG-PET to assess early therapeutic response and disease prognostification has also been demonstrated.

In comparison with conventional imaging, PET–CT has been shown to be more sensitive in the detection of mesothelioma, in active pleural plaques. Circumferential pleural uptake of FDG is characteristic of this disease.

FDG PET/CT in Lung Cancer Restaging

As a rule, for lesions that are greater than 1 cm, an SUV greater than or equal to 2.5 and a visual intensity greater than that within the mediastinal blood pool are accepted criteria for malignancy.

Limitations

Lesion size is an important factor when a patient is undergoing FDG-PET evaluation for lung cancer. The threshold for lesion detection for most FDG-PET scanners currently in use is between 6 and 8 mm.

False-negative FDG-PET scans are seen in well-differentiated cancers such as bronchioloalveolar carcinoma (BAC), slow-growing neuroendocrine tumors such as bronchial carcinoid, and mucinous neoplasms.

Recurrent Lung Cancer/Second Primary

There may be considerable postoperative and postradiation therapy anatomic changes. These changes can make early detection of recurrent cancer a problem on anatomic imaging.

In patients who have been successfully treated for lung cancer, an important cause of mortality is the development of a second lung primary. In both recurrent and new primary lung cancer, FDG-PET imaging is valuable for detection of disease. Early detection is important, since new primaries are treated similarly to the initial primary and salvage therapies exist for localized recurrent lung cancer. A recent study revealed that the FDG-PET scan resulted in a major management change in 63 % of patients. Both the presence and the extent of disease were important prognostic factors.

Cost-Effectiveness

FDG PET/CT is a one-shop stop for staging with brain MRI, identification of distant metastatic disease, detection of recurrent disease, radiotherapy planning, response to therapy assessment, and as a prognostic indicator.

Newer Directions: Tracers Beyond FDG

FDG PET/CT is now an established modality in the management of lung cancer. A host of newer radiopharmaceuticals that target different aspects of tumor biology are being explored in lung cancers. These include the proliferation tracer 18F-fluorothymidine, which has been evaluated in few studies and found to be useful. Other tracers that provide information regarding hypoxia (18F-FMISO, 64Cu-ATSM), angiogenesis (RGD peptides), amino acid metabolism (11C-methionine), and choline metabolism (11C-choline, 18F-fluorocholine) have also been evaluated. An evolving area is the noninvasive assessment of epidermal growth factor receptor (EGFR) and EGFR tyrosine kinase overexpression in tumors by PET imaging that has the potential for in vivo a priori determination of EGFR-targeted drug efficacy. These agents might give better insight into tumor behavior, aggressiveness, and therapy-related toxicity, thereby helping in the formulation of individualized treatment strategies with targeted agents. However, substantial prospective assessment is needed before these agents come into routine use.

Concluding Remarks

Diagnosis, staging, and restaging of lung cancer are among the most extensively studied applications of FDG-PET. With the exception of bronchioloalveolar cell cancer and carcinoid, lung cancer is FDG avid. Lung cancer is one of the most commonly employed indications for FDG PET/CT in clinical practice, and the modality has a substantial impact on the management of lung cancer.

For Further Reading

1. Bradley J, Thorstad WL, Mutic S, Miller TR, Dehdashti F, Siegel BA, et al. Impact of FDG-PET on radiation therapy volume delineation in non-small-cell lung cancer. Int J Radiat Oncol Biol Phys. 2004;59:78–86.
2. Buck AK, Herrmann K, Schreyögg J. PET-CT for staging lung cancer: costly or cost-saving? Eur J Nucl Med Mol Imaging. 2011;38:799–801.
3. De Wever W, Ceyssens S, Mortelmans L, Stroobants S, Marchal G, Bogaert J, et al. Additional value of PET-CT in the staging of lung cancer: comparison with CT alone, PET alone and visual correlation of PET and CT. Eur Radiol. 2007;17:23–32.
4. Kim SK, Allen Auerbach M, Goldin J, Fueger BJ, Dahlbom M, Brown M, et al. Accuracy of PET-CT in characterization of solitary pulmonary lesions. J Nucl Med. 2007;48:214–20.
5. Lynch TB. Textbook of PET-CT in clinical practice. Springer-Verlag, Berlin/Heidelberg/New York; 2007. p. 16–47.
6. Mac Manus MP, Hicks RJ. Impact of PET on radiation therapy planning in lung cancer. Radiol Clin North Am. 2007;45:627–38.
7. Nahmias C, Wahl LM. Reproducibility of standardized uptake value measurements determined by 18F-FDG PET in malignant tumors. J Nucl Med. 2008;49: 1804–8.
8. Yamamoto Y, Kameyama R, Murota M, Bandoh S, Ishii T, Nishiyama Y. Early assessment of therapeutic response using FDG PET in small cell lung cancer. Mol Imaging Biol. 2009;11:467–72.

Application of PET and PET-CT in Cancer of the Gastrointestinal System

13

B.R. Mittal, Rahul Parghane, and J. Mohan Roop

Introduction

Esophageal and gastric cancers are highly aggressive malignancies with poor prognosis. Surgical resection is the primary curative treatment for early esophageal and gastric cancers. Multimodality treatment involving preoperative chemotherapy with or without radiation followed by surgical resection in suitable patients is increasingly becoming popular in patients who have locally advanced cancers.

Esophageal Cancer

Esophageal cancer is one of the most common malignancies, and its incidence is rising rapidly, particularly in developing countries. Accurate staging of the disease is essential, as surgery is reserved for early stages I, IIA, IIB, and occasionally stage III. Stage IV disease is considered inoperable and palliative therapy is given. The clinical staging of esophageal and gastric cancers is performed using endoscopy or endoscopic ultrasound (EUS) and CT. EUS is used to determine whether the tumor is localized and can be treated with surgery alone or the disease is locally advanced, requiring treatment with chemoradiotherapy with or without surgery. F-18 FDG-PET is now being increasingly used in the initial staging of patients who have esophageal cancer. The primary role for PET-CT in esophageal cancer is in the detection of distant metastases, assessing response to therapy, and demonstration of recurrent disease.

Diagnosis

F-18 FDG-PET imaging in primary tumor detection has been considered high (greater than 90 %) in patients who have esophageal carcinoma. However, the stage and size of the primary tumor at presentation affect the accuracy of detection. False-positive uptake of F-18 FDG may be caused by esophagitis or other inflammation. But esophagitis usually manifests as mild linear diffuse uptake, whereas that of esophageal cancer uptake is more focal and intense. False-negative results are more likely due to small or flat mucosal lesions as well as adenocarcinoma at or near the GE junction, possibly because of a diffuse growth pattern and/or mucinous histopathology.

B.R. Mittal, MD, FICNM, FAMS (✉)
Prof. & Head, Department of Nuclear Medicine and PET, Postgraduate Institute of Medical Sciences and Research, Chandigarh, India
e-mail: brmittal@yahoo.com

R. Parghane, MD
Senior Resident, Department of Nuclear Medicine and PET, Postgraduate Institute of Medical Sciences and Research, Chandigarh, India
e-mail: rahul_parghane@yahoo.co.in

J.M. Roop, MD
Junior Resident, Department of Nuclear Medicine and PET, Postgraduate Institute of Medical Sciences and Research, Chandigarh, India
e-mail: j.mohanroop@gmail.com

B.K. Das (ed.), *Positron Emission Tomography: A Guide for Clinicians*,
DOI 10.1007/978-81-322-2098-5_13, © Springer India 2015

The limitations of PET imaging in the detection of superficial esophageal cancers are greater because these tumors are typically small.

Initial Staging

As esophageal cancers are asymptomatic during initial stages, it is often detected at a later stage, leading to overall poor survival, although early disease has a better prognosis. Esophageal cancer with limited disease is usually treated with radical resection. For more advanced locoregional disease, chemotherapy and/or radiotherapy is advocated after surgery, depending on surgical and pathologic findings, and palliative techniques are used for advanced unresectable tumors or with distant metastases. Detection of involvement of locoregional lymph nodes has important prognostic implications but will not rule out surgery. The presence of distant metastases will preclude curative therapy and often is undetected by conventional imaging.

EUS combined with fine-needle aspiration (EUS-FNA) is considered the primary modality in the determination of depth of tumor invasion (T stage) and nodal status (N stage) preoperatively. PET has limited ability in determining depth of tumor invasion, and there is no clear correlation between T stage and intensity of uptake. FDG-PET may have a potential role in the determination of the length of the primary esophageal tumors. Accurate delineation of the upper and lower extent of viable esophageal tumor is important in radiotherapy planning, and the length of the tumor is also a strong independent predictor of prognosis in patients who have esophageal cancer. The sensitivity, specificity, and accuracy for detection of nodal metastases have been reported to be 52, 94, and 84 %, respectively, for FDG-PET and 15, 97, and 77 %, respectively, for CT. EUS has been shown to be superior to PET in the evaluation of nodal involvement (Fig. 13.1).

Prognosis and Response to Therapy

Several studies have evaluated the use of FDG-PET for predicting response shortly after initiation of therapy or to assess response following

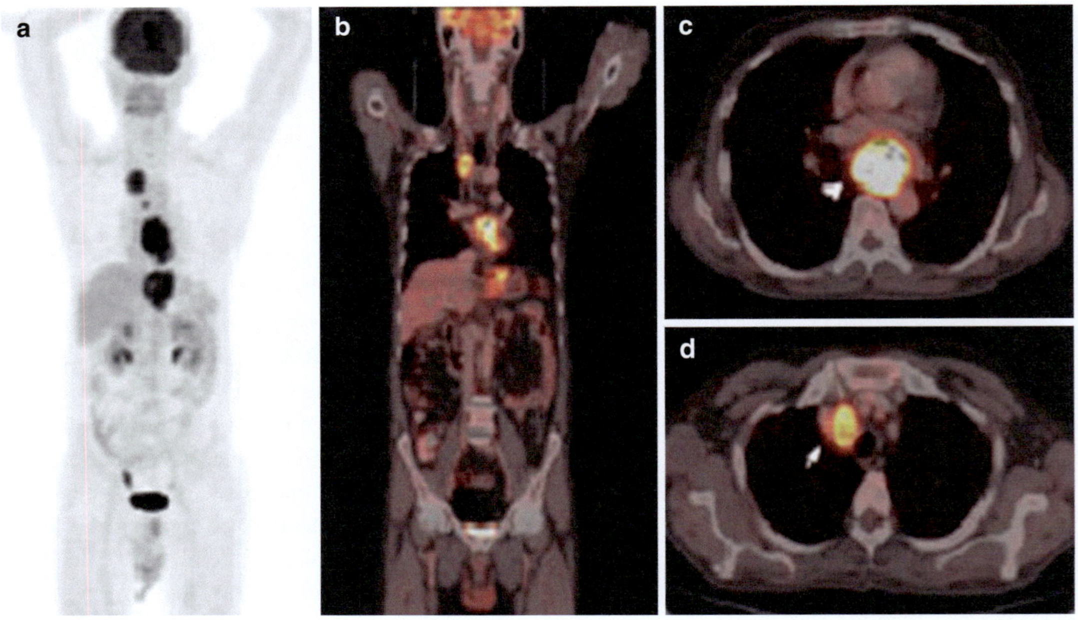

Fig. 13.1 Staging esophageal cancer in a 45-year-old man with adenocarcinoma of distal esophagus. Whole-body maximum intensity projection (**a**), coronal PET-CT fusion (**b**), and axial PET-CT fusion (**c**, **d**) images demonstrate intense FDG uptake within the thickened distal esophagus (*arrowhead*), consistent with primary esophageal cancer. Enlarged right paratracheal lymph node (*arrow*), which showed increased FDG uptake, suspicious for metastatic disease. Endoscopic biopsy of this lymph node was positive for metastatic disease

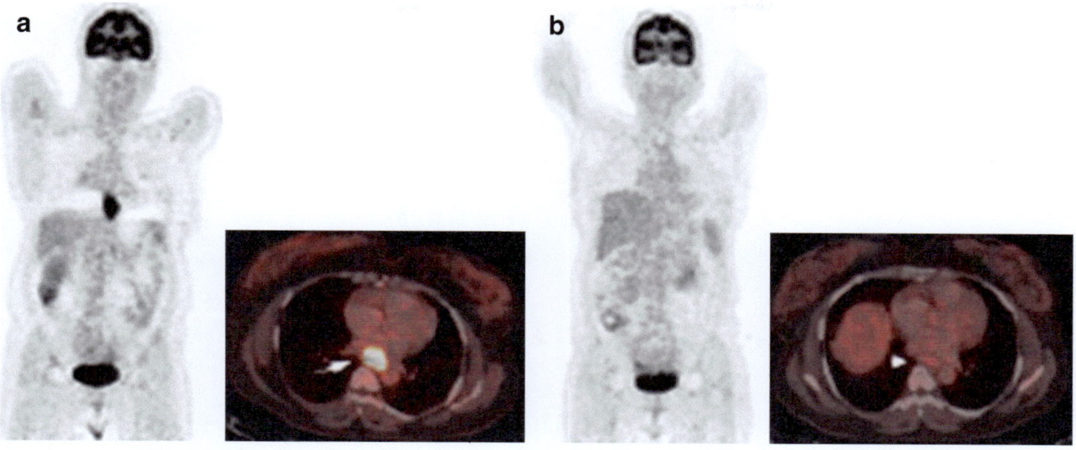

Fig. 13.2 Whole-body maximum intensity projection PET (**a**, **b**) and axial PET-CT fusion (*middle* and *right*) images of a patient with adenocarcinoma of the lower thoracic esophagus. Already 14 days after initiation of neoadjuvant chemotherapy, there is a marked decrease in tumor metabolic activity (>35 %) (see *arrow* and *arrowhead*). After completion of chemotherapy, no viable tumor cells were found in the resected specimen

completion of therapy in patients with esophageal cancer who underwent neoadjuvant therapy. FDG-PET has the potential to distinguish nonresponders from responders shortly after beginning induction therapy. A cutoff of 35 % value predicted clinical response with a sensitivity of 93 % and specificity of 95 %. It has been reported that a 52 % cutoff for decrease in SUV has 100 % negative predictive value and 72 % positive predictive value for response to therapy. Although response on PET cannot definitely predict a complete microscopic pathologic response, it does correlate with clinical response and survival. False-positive uptake due to radiation esophagitis may also be there. But by helping to separate responders from nonresponders, chemotherapy and surgical approach may be altered in these patients (Fig. 13.2).

Recurrence and Restaging

Recurrence of esophageal cancer is common after curative surgical resection. It typically occurs within 2 years after resection. Though locoregional recurrence of malignancy is not uncommon, most patients present with distant metastases. FDG-PET imaging shows high sensitivity in detecting recurrent malignancy after curative resection of cancer of the esophagus or gastroesophageal junction. FDG-PET has an overall sensitivity of 95 % for the detection of locoregional and distant metastases, though it did not improve the diagnostic accuracy for locoregional recurrences compared with conventional imaging. Whole-body PET-CT is the best imaging method for the detection of metastases as they can be clinically occult and in unusual and uncommon locations after induction therapy.

Gastric Cancers

Gastric cancer is usually diagnosed at an advanced stage, with most resected tumors having already spread to the regional lymph nodes. Gastric wall shows physiologic F-18 FDG uptake and is subsequently excreted into the lumen of the digestive tract, so it can be difficult to detect small gastric tumors. PET-CT would best be deployed for the detection of distant metastatic spread from gastric tumors rather than for the primary tumor. FDG-PET has a poor sensitivity ranging from 60 to 91 % for diagnosing gastric cancer. The intensity of uptake tends to be low in mucinous carcinomas and signet ring cell carcinomas than in other pathologic types. Endoscopic ultrasound is used to assess the local extension of the tumor, whereas abdominal ultrasound and CT are used for metastatic workup. Compared to these techniques, PET has no role for evaluating the T stage of the disease.

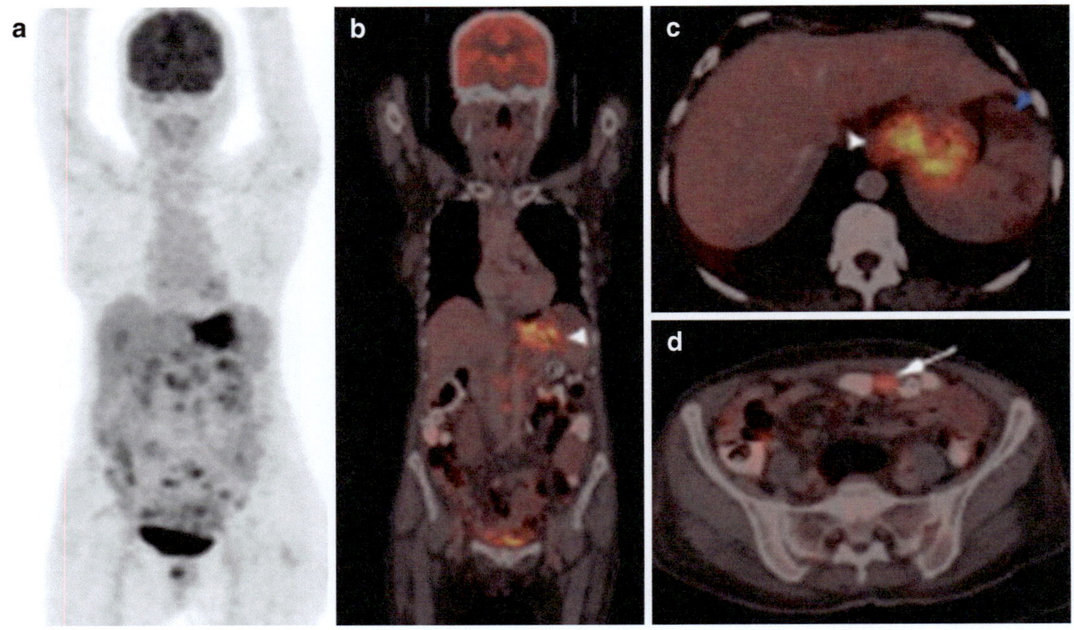

Fig. 13.3 Staging of gastric cancer in a 65-year-old woman with poorly differentiated adenocarcinoma showing metastases to spleen and omentum. Whole-body maximum intensity projection (**a**), coronal PET-CT fusion (**b**), and axial PET-CT fusion (**c**, **d**) images demonstrate intense FDG uptake in the wall of the cardia and fundus of the stomach (*arrowheads*). Axial PET-CT images showing hypodense lesions in spleen (*blue arrowhead*) and serosal/omental deposit, which showed increased FDG uptake (*arrow*)

FDG uptake is reported to be significantly more in well-differentiated adenocarcinomas than in poorly differentiated adenocarcinomas and signet ring cell carcinomas (SUV 13.2 ± 6.6 versus 7.7 ± 2.6, $P < 0.05$). It is also reported that FDG-PET is useful in detecting distant metastatic disease in the liver, lung, and lymph nodes, but not for detection of osseous metastases and peritoneal or pleural carcinomatosis.

Sensitivity of PET-CT for nodal staging appears to be extremely low. However, PET is more specific than CT, especially for assessing the proximal lymph node status. In patients selected for surgery, PET-CT may change the clinical management by detecting additional distal lesions.

A retrospective analysis of 33 patients with suspected recurrence of gastric carcinoma showed a sensitivity of only 70 % (14/20) and a specificity of 69 % (9/13) for FDG-PET. However, the mean survival for the PET-positive group (6.9 months) was significantly lower than the PET-negative group (18.5 months), suggesting that PET may serve as a prognostic rather than diagnostic tool in gastric carcinomas.

Overall, FDG-PET has poor sensitivity for detection of mucinous carcinomas, small-volume disease, and low-grade tumors. In addition to these, the normal, moderately intense physiologic FDG uptake in the stomach may obscure tumors showing low-level uptake. So, F-18 FDG-PET should be used as a problem-solving tool in selected patients with gastric cancer, and there is no sufficient data to recommend its routine use for staging, restaging, or treatment monitoring of this disease (Fig. 13.3).

Gastrointestinal Stromal Tumors (GISTs)

GISTs are uncommon tumors of the GI tract, well-demarcated spherical masses that appear to arise from the muscularis propria layer of the GI wall. Intramural in origin, they often project exophytically and/or intraluminally, and they may have overlying mucosal ulceration. GISTs are usually asymptomatic till they reach 5 cm in size. Almost 30 % of GISTs are malignant and metastasize to liver and peritoneum. Contrast-enhanced CT differentiates benign from malignant lesions. Impressive response to targeted molecular therapy with Gleevec has increased the interest in GISTs. Approximately 15 % of GISTs do not respond to Gleevec treatment, and, hence, it is important to assess response to Gleevec treatment.

Metastasizing lesions show increased FDG uptake compared to nonmetastasizing lesions. In another prospective study, it was reported that contrast-enhanced CT showed more lesions and PET had variable FDG uptake; the prognostic information was better seen in FDG-PET images than in CT findings. Another group showed similar findings in a retrospective analysis of 54 patients treated with Gleevec. FDG PET-CT is use for early prediction of disease response to therapy. It has also been noted that combined PET-CT imaging may provide additional information when measuring response in some cases.

Gastrointestinal Neuroendocrine Tumors

Neuroendocrine tumors are a wide range of neoplasms that have certain characteristics in common. Neuroendocrine tumors are defined not by site but by molecular characteristics. They can exist in any tissue of embryonic gut origin and therefore mostly appear in the gastrointestinal (GI) system. It is also possible to find neuroendocrine tumors in other tissues such as the lung, kidneys, and genitalia.

The most common forms arise in the mid- and hindgut. In most subjects, these can be benign and asymptomatic, with some reports suggesting that at postmortem 5 % of the population may have such a tumor. NET cells originate from neural crest and often secrete hormones that resulted in the term neuroendocrine tumors (NETs).

Carcinoid tumors occurring in the pancreas and cecum may produce serotonin, but the cellular characteristics and prognosis of the two tumors can be very different. Except for some gastric carcinoids, most foregut neuroendocrine tumors are found in the pancreas. In the mid- and hindgut, neuroendocrine tumors are normally carcinoid tumors, some of which may be nonsecretory. In foregut tumors, there is evidence that FDG-PET may be of use, especially if the tumors are somatostatin receptor negative. PET using F-18 DOPA and Ga-68 DOTA octreotate has also shown promise results for NET.

Hepatocellular Carcinoma

Hepatocellular carcinoma (HCC) is the most common primary tumor of the liver, with most cases associated with alcoholic cirrhosis and chronic viral hepatitis. Various imaging modalities usually used in the evaluation of HCC are ultrasonography, contrast-enhanced CT, and MRI. Various studies have been done to determine the potential usefulness of F-18 FDG PET-CT in HCC, similar to its use with other cancers. FDG PET-CT has been shown to be useful in staging, prognostication, evaluation of recurrence, and response assessment in HCCs despite high false-negative rates in detecting primary intrahepatic low-grade HCCs.

Tumors in the liver can be shown with good contrast even though there is normal physiological uptake of FDG by the liver. Image contrast improves with time because tumor lesions continue to accumulate FDG whereas in normal liver tissue it gets cleared off. Delayed PET imaging at about 90 min has been suggested to improve detection rate of malignant liver lesions.

FDG accumulation in HCC is variable, which may be related to the underlying varying degrees of glucose transporter, hexokinase II expression, and the effect of P-glycoprotein efflux pump on FDG retention. It has been shown that P-glycoprotein expression is significantly higher in well-differentiated HCCs than in poorly differentiated tumors and that there is an inverse relationship between tumor SUV and P-glycoprotein expression. Many studies have shown association between serum alpha-fetoprotein (AFP) level and F-18 FDG-PET positivity. In one study,

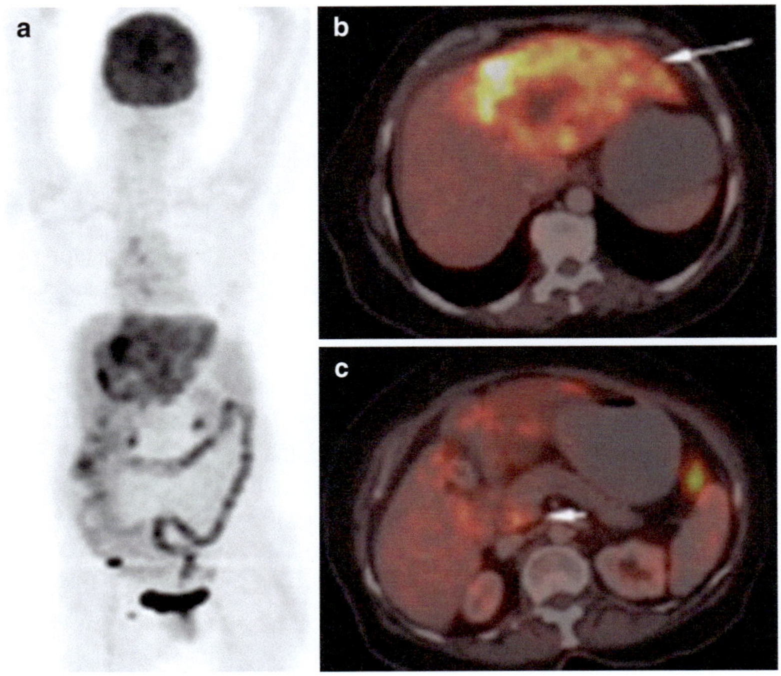

Fig. 13.4 A 60-year-old man with poorly differentiated hepatocellular carcinoma. Whole-body maximum intensity projection (**a**) and axial PET-CT fusion (**b**, **c**) images demonstrate intense FDG uptake in the hypodense lesion in the liver (*big arrow*) and enlarged portocaval lymph node (*small arrow*) with increased FDG uptake, suspicious for metastases

FDG-PET detected HCC in 55.6 % of patients with AFP <100 ng/mL and in 85.7 % of patients with AFP levels >100 ng/mL.

FDG PET-CT may also be useful in the assessment of response in patients with unresectable HCC who have undergone transarterial chemoembolization, external beam radiotherapy, and selective internal radiation therapy with Y-90 microspheres. Reduction in SUV has been associated with a favorable response to treatment. FDG PET-CT is also helpful in surveillance after therapy and detection of local recurrence in patients with unexplained elevated serum AFP after interventional treatment and negative conventional imaging. FDG-PET demonstrated sensitivity of 73.3 % and specificity of 100 % for detection of recurrent tumor in one study involving 26 patients with elevated AFP after either surgical resection or interventional therapy for HCC (Fig. 13.4).

Pancreatic Cancer

The most common (90 %) of pancreatic cancers are adenocarcinomas of ductal origin. Carcinomas of acinar cell origin, sarcomas, malignant lymphomas, and malignant islet cell tumors comprise the remainder of cancers originating in the pancreas. The pancreatic head is the most common location (70 %), with approximately 20 % of tumors arising in the body, 5 % in the tail, and 15 % involving more than one section. Most patients are asymptomatic until late in the course of disease, accounting for its dismal overall 5-year survival of less than 10 %.

Patients with adenocarcinoma of pancreas often have elevated tumor markers (CA 19-9) at diagnosis, which has been used for predicting prognosis and relapse in patients with known or suspected disease. However, CA 19-9 is not specific for pancreatic carcinoma. The uncommon islet cell tumors of the pancreas can present similar to adenocarcinoma of the pancreas due to mass effect if the islet cell tumor is a nonsecreting neoplasm. But, islet cell tumors that secrete peptide hormones (insulinoma, glucagonoma, gastrinoma, and VIPoma) will present with clinical consequences of the excess hormone production.

PET-CT in Diagnosis

FDG-PET has sensitivity ranging between 85 and 96 % in identifying pancreatic adenocarcinoma.

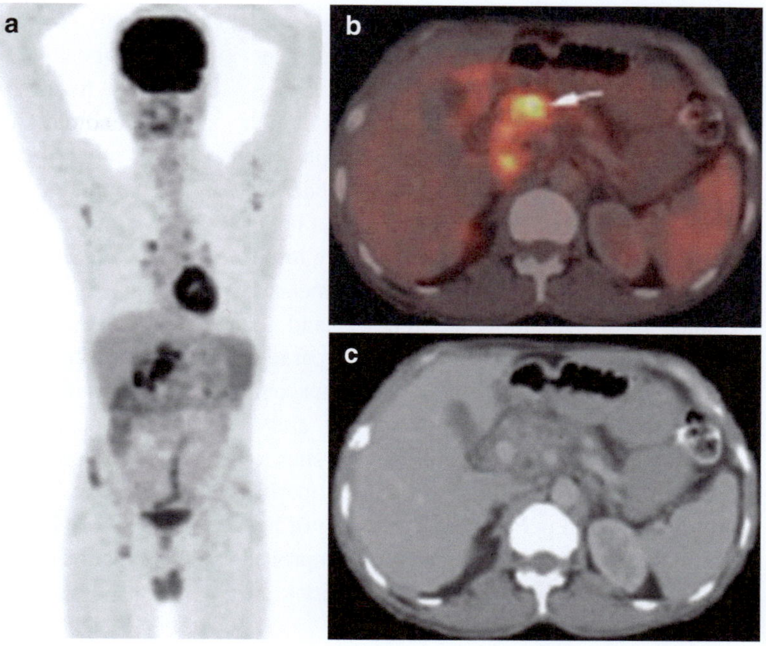

Fig. 13.5 F-18 FDG PET-CT images in a 55-year-old man with serous adenocarcinoma of the pancreas. Whole-body maximum intensity projection (**a**) and axial PET-CT fusion (**b**) images demonstrate intense FDG uptake in the hypodense lesion in the head of the pancreas (*arrow*) and corresponding axial CT image (**c**)

SUV uptake is useful for distinguishing between malignant and benign lesions. FDG-PET has been shown to be superior to contrast-enhanced CT alone in several studies for differentiating the benign mass of chronic pancreatitis from malignant pancreatic masses, with the overall sensitivity, specificity, and accuracy ranging from 85 to 100 %, 67–99 %, and 85–95 %, respectively, compared to sensitivity of 65–79 % and specificity of 44–62 % for CT.

Staging

FDG-PET has little role in determining T stage, due to the lack of anatomic detail and the inability to resolve the relationship of FDG activity with small pertinent structures. Extension beyond the pancreas and into adjacent organs is also based almost entirely on the CT images. PET-CT fused images have been shown to improve the identification of involved lymph nodes in patients with pancreatic cancer. However, it has limitations for detecting microscopic lymph nodal metastases, and it is common for a patient with a negative PET-CT examination for lymph node metastases to have multiple lymph nodes positive for metastatic disease on histopathologic examination. FDG-PET has been shown to be superior to other imaging modalities for determining the presence of metastatic disease.

Restaging

PET-CT is useful in diagnosing local recurrence, especially in postoperative cases. It has been reported to be useful in identifying recurrent disease in patients with rising CA 19-9 tumor markers, particularly when CT is negative or equivocal.

Therapy Monitoring

SUV uptake values of the primary tumor can be correlated to predicting effectiveness for chemotherapy in unrejectable cases. One study has suggested that a negative FDG-PET study after chemoradiation in a patient who had a positive pre-therapy scan may be a good prognostic indicator (Fig. 13.5).

Colorectal Cancer

It is the third most commonly diagnosed cancer and a major cause of mortality, particularly in Western countries. In most cases, colorectal cancer develops over a period of years in a preexisting adenomatous polyp. The most of colon cancers are adenocarcinomas. The tumor spreads sequentially from the primary site to adjacent lymph nodes and then up to the mesenteric lymph node chain. Low rectal and anal cancer spreads laterally to perineal nodes and may involve inguinal nodes rather than retroperitoneal nodes. Primary tumor is resected even in the presence of distant metastases since local symptoms are relieved and the quality of life is enhanced. In postoperative patients, an elevated serum carcinoembryonic antigen (CEA) level suggests recurrent and/or metastatic disease. The most common sites of metastases include the liver, lung, retroperitoneum, ovaries, and peritoneum. Resection of isolated metastases is associated with improved survival, while multifocal metastatic lesions are associated with less favorable prognosis.

PET-CT in the Initial Diagnosis

The diagnosis of colorectal carcinoma is based on colonoscopy and biopsy. FDG-PET is rarely specifically used for the diagnosis of colorectal cancer.

Staging

The pattern of metastatic spread of colorectal cancer depends on the lymphatic and venous drainage pathways. The sensitivity of FDG-PET for the detection of a primary colon carcinoma may be high, but its role in the preoperative staging is still debated except in high-risk patients for whom surgery can be avoided if metastases are identified. A recent study evaluated the usefulness of FDG-PET for staging patients with known or suspected primary colorectal carcinomas. In 48 patients, FDG-PET imaging identified all primary carcinomas, but it was poor for detecting local lymph nodes. FDG-PET was, however, superior to CT for detecting hepatic metastases, with sensitivity and specificity of 88 and 100 %, respectively, compared with 38 and 97 % for CT. False-positive findings on FDG-PET include abscesses, fistulas, diverticulitis, and, occasionally, adenomas. One study had shown that PET-CT altered the staging of colorectal carcinoma compared with conventional imaging in 31 % of cases.

In T staging, FDG-PEt alone is not suited for providing the precise information needed for local and regional staging, because of limited spatial resolution of PET. PET-CT is almost exclusively CT dependent for T staging of colorectal cancer. In serosal penetration and invasion of adjacent structures (T4 stage), PET scan has more sensitivity.

For M staging, role of PET-CT for the detection of recurrent or metastatic disease in patients with raised carcinoembryonic antigen is considerably better than multi-slice CT alone. Early detection of liver metastases gives the opportunity for neoadjuvant chemotherapy and resection, which may prolong survival for patients with colorectal carcinoma. FDG-PET is highly sensitive for the detection of liver metastases. The use of PET-CT for preoperative staging may improve the survival in patients with hepatic metastases deemed suitable for hepatic resection by ruling out the presence of other extrahepatic metastases. PET is accurate for the identification of hepatic metastases greater than 1 cm in diameter but is limited in its ability to demonstrate lesions smaller than 1 cm. It is also useful in patients with several hypodense or hypoenhancing liver lesions that are not clearly characterized by CT alone.

Other than the liver, the lung is another common organ for metastatic disease of colorectal tumors. Possible manifestations of lung metastatic disease include pulmonary nodules, lymphangitic carcinomatosis, and pleural involvement. A negative PET scan cannot exclude pulmonary metastases with certainty because of limited spatial resolution, while a positive PET will confirm suspicious abnormalities detected on the CT images. Bony lesions may be overlooked on CT scan, particularly if early metastatic disease is confined to the skeleton. FDG uptake in marrow of the vertebral bodies can appear focal on axial imaging and could be misinterpreted as metastatic disease. However, a repeating pattern, which is most evident on sagittal or coronal images, is characteristic of physiologic FDG uptake in vertebral marrow. Metastases from colon cancer can lead to peritoneal deposits, which are difficult to detect on CT alone without careful attention to CT technique and interpretation. The FDG-PET images can be very useful in locating peritoneal deposits 7–8 mm or greater in size and the CT images essential in confirming the presence of nodular deposits. Other sites for colorectal metastases include the adrenal glands. PET imaging of the adrenals is very sensitive for the detection of malignant adrenal lesions.

Restaging

Locoregional pelvic recurrence and liver metastases are the common sites of relapse after resection of colorectal cancer. Additional value of PET-CT in the detection of pelvic recurrence after surgical removal of rectal cancer is well reported. Isolated recurrence can occur at the site of prior resection of metastases, and PET-CT can be useful in precisely locating and characterizing such recurrences. It is helpful in determining the extent of local recurrence of rectal cancer, particularly in presacral region. Focal FDG uptake can guide to look for corresponding CT abnormalities such as a soft tissue mass or bone erosion on CT for detection of metastatic involvement.

Therapy Response Monitoring

PET-CT is now used as the standard test for detection of recurrence and monitoring treatment response. FDG-PET imaging has been used to monitor tumor response following chemotherapy. Complete resolution of abnormal FDG uptake in primary and metastatic lesions of colon cancer can be observed following chemotherapy. Studies have shown that FDG-PET can be used to predict response to chemotherapy and radiofrequency ablation in patients with hepatic metastases from colorectal cancer.

Radiotherapy Planning

Radiation therapy planning with FDG PET-CT scan is useful in patients with brain, lung, head, and neck, as well as colorectal cancer. PET-CT imaging is useful to plan laser-induced thermotherapy for liver metastases from colorectal cancer.

False-Positive FDG-PET Findings

Inflammatory conditions such as Crohn's disease, ulcerative colitis, typhlitis, and diverticulitis can all lead to increased colonic FDG activity, which leads to false-positive finding for colorectal cancer. A long segment of colon with diffusely increased FDG uptake should not be interpreted as inflammatory bowel disease unless a positive CT finding, as this finding is commonly seen in normal colon also. In the absence of specific corresponding CT findings, focal intense colonic activity on PET, however, certainly warrants further investigation such as colonoscopy.

False-Negative FDG-PET Findings

False-negative FDG-PET findings are noticed with mucinous adenocarcinoma of the colon (Fig. 13.6).

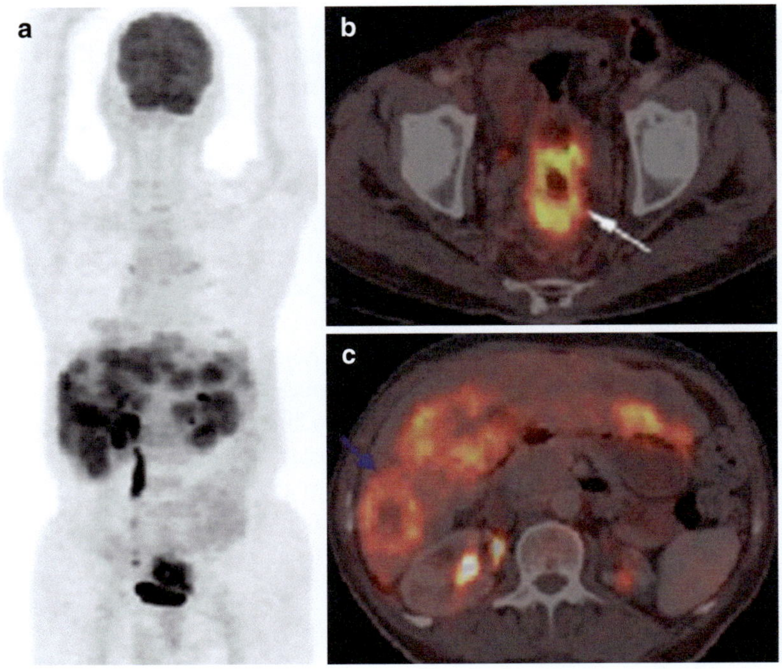

Fig. 13.6 F-18 FDG PET-CT images in a 61-year-old man with poorly differentiated adenocarcinoma of the sigmoid colon. Whole-body maximum intensity projection (**a**) and axial PET-CT fusion (**b**, **c**) images demonstrate intense FDG uptake in the thickened mural wall of the sigmoid colon (*white arrow*) and multiple lesions in the liver with FDG uptake (*blue arrow*) suggestive of liver metastases

For Further Reading

1. Bang S, Chung HW, Park SW, et al. The clinical usefulness of 18-fluorodeoxyglucose positron emission tomography in the differential diagnosis, staging, and response evaluation after concurrent chemoradiotherapy for pancreatic cancer. J Clin Gastroenterol. 2006; 40:923–9.
2. Chen J, Cheong JH, Yun MJ, Kim J, Lim JS, Hyung WJ, et al. Improvement in preoperative staging of gastric adenocarcinoma with positron emission tomography. Cancer. 2005;103:2383–90.
3. Dierckx R, Maes A, Peeters M, Van De Wiele C. FDG PET for monitoring response to local and locoregional therapy in HCC and liver metastases. Q J Nucl Med Mol Imaging. 2009;53:336–42.
4. Gayed I, Vu T, Iyer R, et al. The role of 18F-FDG PET in staging and early prediction of response to therapy of recurrent gastrointestinal stromal tumors. J Nucl Med. 2004;45:17–21.
5. Moehler M, Lyros O, Gockel I, Galle PR, Lang H. Multidisciplinary management of gastric and gastroesophageal cancers. World J Gastroenterol. 2008;14:3773–80.
6. Yoshioka T, Yamaguchi K, Kubota K, Saginoya T, Yamazaki T, Ido T, et al. Evaluation of 18F-FDG PET in patients with advanced, metastatic or recurrent gastric cancer. J Nucl Med. 2003;44:690–9.

Application of PET–CT in Genitourinary Malignancies

14

P.K. Pradhan and Raveena Mubalsha

18F-2-deoxy-D-glucose (FDG) has been the primary tracer for positron emission tomography (PET) in urologic malignancy. The primary clinical application of PET–CT in urology relates to the management of the urologic malignancies, including renal cancer, bladder cancer, prostate cancer, and testicular cancer.

Renal Malignancy

Renal cell carcinoma constitutes 3 % of adult malignancies. The worldwide incidence is increasing at an annual rate of 2 %. Renal malignancies include renal cell carcinoma (RCC), transitional cell carcinoma, squamous cell carcinoma (SCC), lymphoma, and metastatic neoplasm, usually lung cancer or melanoma.

Renal cell carcinoma, the most common malignant neoplasm of the kidney, is primarily a surgically managed disease. In advanced disease, medical management offers little in terms of improved survival or palliation. Surgical extirpation is the mainstay of treatment if there is no diagnostic imaging evidence of local spread beyond Gerota's fascia and distant metastasis. Metastases are present in about 50 % of patients at the time of presentation. Solitary metastasis may also be resected. RCC responds poorly to chemotherapy. Radiation therapy for RCC is used for palliation of metastatic sites, specifically the bone and brain. Immunotherapy with biologic response modifiers such as interleukin-2 and interferon alpha has the most impact on the treatment of metastatic disease. The 5-year survival is around 80–90 % for early stages of disease, while advanced disease carries a poor prognosis. Prognosis depends mainly on stage of the tumor at the time of presentation (Fig. 14.1)

Diagnostic imaging evaluation of a renal mass was performed as part of evaluation of hematuria, flank pain, or a palpable mass; however, incidental detection of malignant renal masses is also on surge due to widespread use of cross-sectional imaging. Characterization of renal masses from benign cysts, complex cysts, which can be malignant, and solid masses can be easily done on CECT.

Bosniak has classified the renal cysts based on morphological features on CECT. Bosniak category I are simple cysts, which do not require further evaluation. Category II complex cysts are usually followed up with anatomical imaging. Surgical removal and exploration are required in category III and IV cystic masses. Hence, PET–CT could potentially aid in the noninvasive evaluation and whole-body staging of Bosniak category III and IV renal masses.

P.K. Pradhan, MD, DNB (✉)
Professor, Department of Nuclear Medicine and PET, Sanjay Gandhi Postgraduate Institute of Medical Sciences, Lucknow, India
e-mail: pkp9918@gmail.com

R. Mubalsha
Senior Resident,
Department of Nuclear Medicine and PET, Sanjay Gandhi Postgraduate Institute of Medical Sciences, Lucknow, India
e-mail: mudalsharavina@gmail.com

B.K. Das (ed.), *Positron Emission Tomography: A Guide for Clinicians*,
DOI 10.1007/978-81-322-2098-5_14, © Springer India 2015

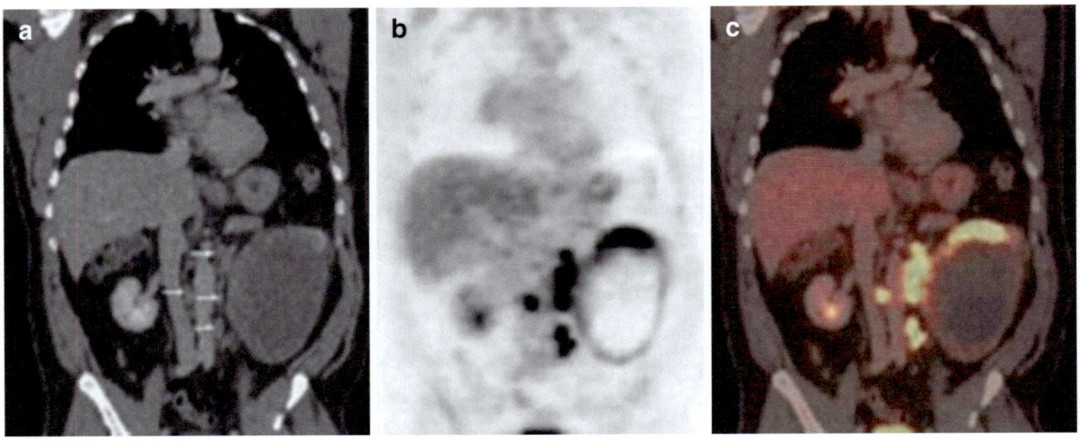

Fig. 14.1 FDG-avid large renal cell carcinoma of the left kidney. (**a**) Contrast-enhanced CT with enlarged lymph nodes (*arrows*). (**b**) FDG-PET with attenuation correction. (**c**) Fusion showing large renal cell carcinoma with central necrosis

PET–CT is mainly used for staging, response assessment, and surveillance in cases that have been operated. However, variants like clear cell carcinoma (CCC) are non-FDG avid. Other reasons for false negatives are small body size and interference from urinary activity as kidneys are the normal route of excretion. In oncocytomas, it is seen that FDG uptake is almost same as renal cell carcinoma. In one study with 11 Bosniak category III renal masses, 10 of 11 were benign, and the one containing RCC was false negative on FDG-PET. In another series of histopathologically proven 68 renal masses, it was found that the average standardized uptake value (SUV) for RCC to be 4.6, thus only slightly higher than adjacent nonmalignant renal parenchyma between 1 and 2 h after tracer injection.

In another series with 35 patients, it has been reported a sensitivity of 47 % and specificity of 80 % for FDG-PET characterization of renal masses. Hence, there may be inherent limitations in using the tracer FDG for evaluation of RCC in general, and renal masses in particular, because of the relatively modest FDG avidity of a significant fraction of malignant renal tumors. In the detection of locoregional and distant metastases, FDG-PET sensitivity and specificity have been estimated to be 63–77 % and 75–100 %, respectively. Positive predictive value appears to be high, in excess of 90 %, whereas generally, the negative predictive value is low to be clinically useful (e.g., a negative study does not exclude malignancy).

FDG PET–CT is superior to bone scan in evaluating osseous metastases as compared to bone scintigraphy. These observations are correlated with the heterogeneous expression of GLUT-1 in RCC, which may not correlate with the tumor grade or extent. FDG-PET can also alter clinical management in up to 40 % of patients with suspicious locally recurrent and metastatic renal cancer.

The normal route of excretion for FDG is urinary tract, so to avoid false negatives and false positives, newer tracers are being investigated, like amino acid analogs.

C-11-labeled acetate is one such tracer, which is retained by RCC but rapidly cleared from the renal parenchyma as carbon dioxide, and with no urinary excretion. Higher average SUV and tumor to renal cortex values are obtained within 10 min of tracer injection compared to FDG at 1 h postinjection, and the highest acetate tracer accumulation was found in granulocytic tumors. Although such tracers *of amino acid transport or lipid-related metabolism may have a role* in characterizing a small renal mass or response to therapy, detecting RCC in complex renal masses and metastatic disease requires high consistent tracer uptake. However, such consistency has not yet been demonstrated with this tracer.

The diagnostic accuracy of FDG-PET appears not to be improved by semiquantitative analysis, which is probably due to the fundamental variability of glucose metabolism in RCC. And delayed images 03 h postinjection with diuretic intervention helps us to delineate the tumor in a better way.

Thus, PET–CT has a role in staging (not in all histological types), response assessment, restaging, and surveillance in cases of renal cell carcinoma.

Bladder Cancer

Bladder cancer is the most common malignant tumor of the urinary tract and more common in patients aged 50–80. The incidence doubles in men >75 years of age versus younger men. It is more common in males as compared to females and presents usually as painless hematuria. The various risk factors being occupational exposure like in case of aniline dye workers, smoking, pelvic irradiation, and drugs like cyclophosphamide. Depth of tumor penetration into the bladder wall forms the basis for disease staging and is the most important prognostic factor.

Transitional cell carcinoma is the most common histopathology, up to 90 % of cases; 8 % are squamous cell and other varieties being adenocarcinomas, sarcomas, lymphomas, and carcinoid tumor types. Diagnostic procedures may include cystoscopy with biopsy, excretory urography, or retrograde pyelogram; pelvis ultrasound; and CT of the chest, abdomen, and pelvis. Superficial lesions may be treated with endoscopic resection, fulguration, or photodynamic therapy. For invasive tumors, cystectomy with urinary diversion is usually done. Radiotherapy may be employed as adjuvant therapy, in combination with other therapies, or as a palliative measure. There is no established systemic chemotherapeutic regimen for the treatment of metastatic bladder carcinoma. The 5-year survival is about 90 % for superficial disease and about 60 % for invasive disease. Systemic disease has a bad prognosis.

FDG-PET has been found to be modestly accurate in the diagnosis of bladder cancer, as urinary bladder is a route of excretion for FDG, and in the detection of pelvic lymph node and distant metastases (Fig. 14.2).

However, dual time point imaging where images are taken after 1 and 3 h postinjection, post-diuretic administration, produces optimal images. Primary bladder carcinoma and the lymph node metastases may exhibit an SUV in the range 1.7–6.2. For lymph node staging, a sensitivity of 67 % and a specificity of 86 % have been reported. Other PET radiotracers including C-11 methionine and C-11 choline may also be potentially useful in the imaging evaluation of bladder carcinoma.

Locoregional nodal staging is an important prognostic marker for proper management of bladder cancer patients. The size criteria of nodal involvement used with anatomic imaging is of limited accuracy. Hence, PET with FDG or other tumor-specific tracers could provide increased accuracy in N stage of bladder cancer, as has been demonstrated with several other malignancies. Distant metastatic disease, most commonly osseous, pulmonary, and hepatic metastases, is important in patients with invasive bladder cancer. PET offers improvement in detection of osseous or hepatic metastases analogous to that observed in other malignancies such as lung or esophageal cancer.

Limited pilot studies have demonstrated that metastatic bladder cancer is FDG avid and that involved local lymph nodes as small as 9 mm could be detected, whereas smaller involved nodes (less than 5 mm) were false negative.

A recent study has reported a sensitivity of 67 % and a specificity of 86 % for FDG-PET detection of pelvic lymph node metastases of bladder cancer. Osseous metastases of bladder cancer are readily detected on FDG-PET, but the relative accuracy of FDG-PET versus conventional bone scintigraphy has yet to be fully addressed. Bladder cancer appears to have relatively consistent avidity for FDG, and adding FDG-PET to conventional anatomic evaluation of locoregional and distant spread of bladder cancer such as with PET–CT may well prove clinically valuable.

As elsewhere in the urinary tract, alternative tracers that do not undergo urinary excretion, or can be imaged before the arrival of the excreted urinary tracer activity, have been investigated in

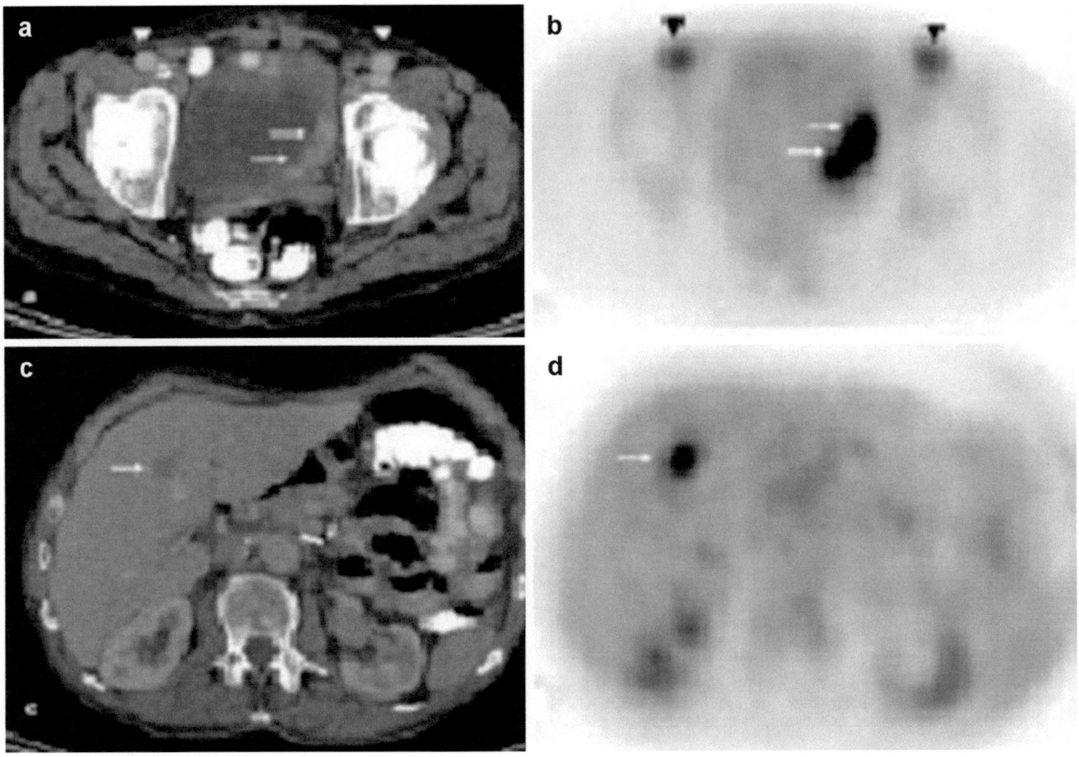

Fig. 14.2 FDG-avid left bladder wall thickening (**a**, **b**, *thin arrow*) corresponding to a primary urothelial neoplasm. The bilateral FDG uptake in the groins is physiologic in aortofemoral bypass graft (**a**, **b**, *arrowheads*). A solitary hepatic metastasis is also intensely FDG avid (**c**, **d**, *arrows*)

an attempt to obviate the confounding effects of urinary tract excretion. 11C-L methionine was used in a limited series to investigate the PET detection of primary bladder cancer. T4, most T3, and 2 of 4 of T2 primary bladder cancers were detected.

The T staging was not superior to anatomic imaging, and there were insufficient proven cases of nodal metastases to evaluate accuracy of local lymph node metastases. A preliminary study reported detection of 10 of 18 primary bladder cancers with 11C-choline PET. In two patients, pelvic lymph node metastases were visualized; however, again there were insufficient proven cases of nodal metastases in the series to evaluate the accuracy for local lymph node metastases. In addition to locoregional lymph node staging, differentiating post-radiation therapy scar from recurrent tumor in patients treated for locally advanced disease and assessment of neoadjuvant therapy response are areas warranting further investigation of PET with FDG and other tracers.

Prostate Cancer

As life expectancy increases, so will the incidence of this disease, creating what will become an epidemic male health problem. The commonest histology is adenocarcinoma. Digital rectal examination is considered the standard of reference for detection of prostate cancer. About 50 % of all palpable nodules are carcinomas. Neither prostatic acid phosphatase (PAP) nor prostate-specific antigen (PSA) is useful for screening prostate cancer, although elevated serum levels of these substances are usually suggestive of locally advanced or metastatic disease. The commonly used histopathological Gleason score, which ranges from a minimum of 2 to a maximum of 10, is based on both the tumor's glandular differentiation and its growth pattern and has been shown to be associated with the clinical stage of disease. The role of diagnostic imaging in the management of prostate cancer is both as rapidly evolving and as controversial as the clinical management of the disease. Although two decades ago staging before prostatectomy with bone scintigraphy was common, today the management of prostate cancer is varied, with far less reliance on surgery and the routine use of serum markers (prostate-specific antigen) to assess disease progression and response to therapy. Because the prostate itself is easily accessed via the rectal vault, very high-resolution anatomic imaging by ultrasound or MRI is possible. Biopsy of all sectors of the prostate gland, either randomly or assisted by ultrasound guidance, is routine, and hence tumor histological grade is readily obtained at initial diagnosis.

The potential roles for diagnostic imaging of prostate cancer include diagnosis of primary disease, determination of extracapsular spread, and detection of locoregional lymph node metastases and distant metastatic spread. Bone scintigraphy with SPECT/CT is very useful in detecting osseous metastases. However, bone scintigraphy cannot detect soft tissue or lymph nodal involvement, quite prevalent with metastatic spread of this disease.

Preliminary studies show that accumulation of FDG in prostate cancer is low, and also the uptake may overlap with the uptake in benign prostatic hyperplasia (BPH), in the normal gland, and in the operative site or local recurrence. However, animal and preliminary clinical studies have demonstrated that FDG-PET may be useful in the evaluation of advanced androgen-independent disease and in patients with high Gleason scores and serum PSA level, in the detection of active osseous and soft tissue metastases, and in the assessment of response after androgen ablation and treatment with novel chemotherapies. For metastatic disease, FDG localization in the lesions may display a standardized uptake value (SUV) of up to 5.7 at a positive predictive value of 98 %.

FDG-PET may be more useful than In-111 ProstaScint in the detection of metastatic disease in patients with high PSA or PSA velocity. FDG-PET appears to be more useful in the detection of soft tissue metastases than osseous metastases and also differentiates between active osseous diseases from the scintigraphically quiescent lesions. It also helps in restaging patients who have a rising PSA level despite treatment. Additionally, it has been suggested that FDG-PET has a better specificity but lower sensitivity for detecting osseous metastases in comparison to bone scintigraphy and is useful in differential diagnosis of flare reaction after endocrine therapy. FDG-PET has also shown a role in prognosticating patients undergoing radical prostatectomy. Lesions demonstrating higher FDG accumulation (i.e., high SUV) had poorer prognosis in comparison to those with low SUV.

In a series of 24 patients with organ-confined prostate cancer in which urinary tracer activity in the bladder was minimized, only 1 (4 % sensitivity) was detected. Tumor volume ranged from 1.2 to 10.4 ml with a mean of 6.9 ml. The failure of detection most likely reflects the low tumor to background achieved with FDG. Similar disappointing results were reported for the detection of local recurrence of prostate cancer in patients treated by prostatectomy, also attributed to the relatively low avidity of prostate cancer for FDG.

Novel PET radiotracers for imaging of prostate cancer are also in vogue. Acetate participates in cytoplasmic lipid synthesis, which is believed to be increased in tumors. Cellular retention of radiolabeled acetate in prostate cancer cell lines is primarily due to incorporation of the radiocarbon into phosphatidylcholine and neutral lipids of the cells. The lack of accumulation of acetate in urine is also advantageous to imaging prostate cancer in particular, because the prostate bed remains unobstructed by the adjacent high levels of radioactivity in the urinary bladder, commonly a problem with FDG. Although there can be a considerable overlap between the uptake level in primary cancer and in the normal prostate gland, generally, the uptake appears to be greater in the tumor than in the normal tissue. C-11 acetate may also be useful in detection of tumor recurrence in patients who had been treated previously with prostatectomy or radiation therapy.

Choline PET may also be useful in imaging prostate cancer. The biological basis for radiolabeled choline uptake in tumors is the malignancy-induced upregulation of choline kinase, which leads to the incorporation and trapping of choline in the form of phosphatidylcholine (lecithin) in the tumor cell membrane in proportion to the rate of tumor duplication. The tracer uptake has been noted to decrease in both the primary tumor and in the metastases after hormonal therapy and increase after relapse as measured by the increase in the serum PSA level. Both acetate and choline appear to be more or less equally useful in imaging prostate cancer in individual patients and are more advantageous than FDG. In some clinical circumstances, such as detection of the locally recurrent disease, the diagnostic potential of other tracers such as radiolabeled androgen analog F-18 fluoro-5a-dihydrotestosterone (FDHT) for imaging androgen receptors may be necessary. However, this is yet to be clinically established.

In addition, emerging roles include guidance of local therapy in patients with organ-confined disease and assessment of tumor response to systemic therapy in patients with advanced metastatic disease.

In prospective series of 67 patients, C-choline PET was 80 % sensitive and 96 % specific in the staging of pelvic lymph node metastases. 18 F-labeled choline derivatives have subsequently been synthesized and tested, including 18 F-fluoromethyl choline and 18 F-fluoroethyl choline. Fluoromethyl choline most closely matches the in vivo phosphorylation rate of choline and appears to be the preferred 18 F-labeled choline analog for PET imaging. Both soft tissue and bone metastases are readily identified with fluorocholine, with SUVs ranging from 2.5 to as high as 10, but on average roughly 4.5 in untreated prostate cancer. In addition to high liver, pancreas, and bowel activity, fluorocholine undergoes urinary excretion. The rapid tumor uptake and blood pool clearance, however, does permit early imaging of the prostate and adjacent tissues before arrival of urinary tracer in the bladder. In comparison to FDG-PET, 18 F-fluorocholine PET was generally better in detection of primary lesions and osseous and soft tissue metastases on initial clinical evaluation.

As with acetate and choline PET tracers, amino acid tracers have been investigated both as a probe of an alternate metabolic pathway and as a strategy to avoid the confounding effects of urinary tracer in the bladder. In patients with progressive metastatic prostate cancer, L-methionine uptake in metastatic lesions was consistently higher than FDG uptake and demonstrated progressing metastatic lesions more consistently than did FDG on PET imaging. Some success was also reported in using 11C-L-methionine PET to direct biopsy in patients with rising serum prostate-specific antigen and negative routine biopsies.

Thus PET–CT could be valuable in the management of advanced prostate cancer.

Role of PET–CT in Testicular and Gynecological Malignancies

Testicular Cancer

Role of imaging is vital in testicular and gynecologic malignancies as they are important cancers with significant morbidity. The incidence of testicular malignancy is increasing and is a cancer of relatively young men. Ovarian, uterine, and cervical cancer are among the most common tumors in women, with high morbidity and mortality rates. Despite early screening tests for cervical cancer, mortality from this tumor remains significant. Outcome in terms of long-term survival is based on staging. Imaging techniques have been fundamental in this process, and positron emission tomography (PET) is developing to be an important tool in pretreatment assessment, assessing early response to therapy, restaging, and in cases of tumor recurrence (Fig. 14.3).

Testicular cancer (seminoma and non-seminoma, NSGCT) is a relatively rare tumor affecting only 1 % of men, but it is the commonest tumor in young males (aged 15–35), and its incidence is increasing. Biological behavior and potential for metastasis for the two tumors are different; however, the overall prognosis is good (Fig. 14.4).

Because of advances in chemotherapy, cure is now possible for the majority of patients with

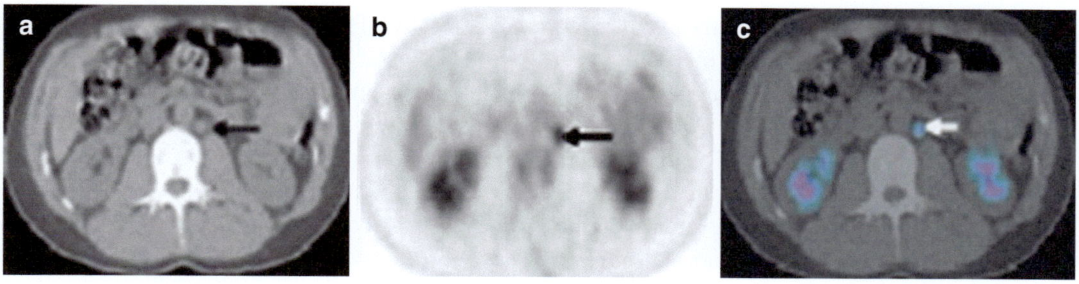

Fig. 14.3 A 33-year-old man with a history of testicular cancer treated 18 months back, now presented with rising tumor markers. There was no evidence of disease on conventional imaging. PET–CT imaging revealed FDG-avid lymph node lying close to the L2 vertebra, as shown on the transaxial image of CT (**a**), PET image (**b**), and PET–CT fusion image (**c**) (see *arrows*)

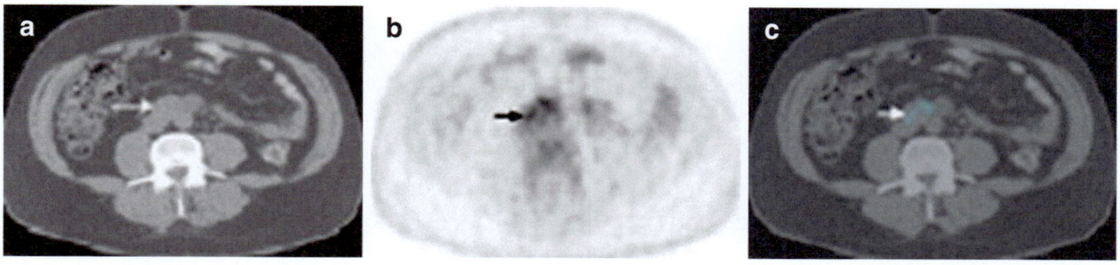

Fig. 14.4 A patient with known metastatic testicular cancer had an FDG-PET scan performed before chemotherapy (**b**). The transaxial images of CT (**a**), PET showed an FDG-avid nodal lesion (*arrow*). One week later following chemotherapy, there has been a rapid decrease in uptake in the tumor (**c**) (see *arrow*), indicating an early response to chemotherapy. The patient responded well to the current course of chemotherapy

minimal metastatic disease. If tumor spread could be reliably assessed, some patients with NSGCT stage I (no evidence of metastases) could be clinically observed rather than undergo prophylactic chemotherapy.

Testicular cancers usually present as an asymptomatic lump and urgent orchidectomy is warranted. Usual histological diagnosis being seminoma and non-seminomatous germ cell tumor (NSGCT), 10 % can have mixed tumors as well. The tumors spread to the para-aortic region initially, although hematogenous spread is more common in NSGCT and metastases are seen in the lung, brain, liver, and bone also. At the time of diagnosis, all patients are staged by clinical examination and computed tomography (CT) scans of the chest and abdomen and pelvis. Tumor markers are measured for prognosticating and allowing monitoring of treatment response and assessment of recurrence.

In seminoma, conventional practice has been to perform retroperitoneal radiotherapy even in stage I disease, and about 15 % of patients at presentation have disease confined to the abdomen. Retroperitoneal and pelvic radiotherapy is a common practice and has a good rate of achieving local control. So, if there is nodal involvement, PET–CT can act as a guide for radiotherapy planning.

Anatomic staging techniques including CT, ultrasound, and lymphangiography have all been used to stage testicular cancer. The most widely used now is CT, which is routinely performed as part of the initial staging protocol. All staging procedures have limitations, and even for CT, false-negative rates of 59 % have been reported. The false-negative rates for lymphangiography and ultrasound are 64 and 70 %, respectively. The diagnosis of nodal metastases by CT is based on detection of nodal enlargement, with a 1-cm upper limit for normal lymph node size. Before nodal enlargement, the entire volume of lymph node may be replaced by malignant cells, whereas a large lymph node may contain only benign reactive cells. As a result, the false-positive rate of CT is also high at 40 %. This inaccuracy has led to search for more accurate imaging methods, including metabolic imaging with PET.

As 18 F-2-deoxy-D-glucose (FDG)-PET relies on metabolic uptake criteria rather than size criteria. It has the potential to identify small-volume disease in a lymph node that is normal in size; this may have a direct effect on patient management. Unnecessary treatment and morbidity is avoided in stage I tumors by more accurate classification of patients at high or low risk. Treatment plan is also changed in stage III and IV tumors if there is distant metastasis.

FDG-PET can clearly identify more sites of disease in patients with established metastatic disease than seen on CT.

There are only a few studies that have addressed the issue of improving the initial staging using PET. The sensitivity ranged between 70 and 87 % and the specificity between 94 and 100 %. The three major initial studies confirmed overall better sensitivity, positive predictive value (PPV), and negative predictive value (NPV) for PET than for CT. Both CT and PET missed small (approximately 1 cm) retroperitoneal lymph node metastases. However, one limitation of these studies was that not all patients had histological confirmation of findings on PET and assessment of true negativity or positivity.

FDG-PET potentially has the ability to detect small-volume tumors in solitary residual masses, to identify a specific mass as the site of relapse in patients with multiple masses, to detect other unsuspected sites of tumor, and to determine the site and extent of disease in patients with raised tumor markers.

In a recent study including 51 patients with seminomas and post chemotherapy residual masses, PET detected residual tumor in all masses greater than 3 cm and in 95 % of masses less than 3 cm, with PPVs and NPVs of 100 and 96 %, respectively, for PET versus 37 and 92 %, respectively, for CT. These findings showed that PET was a reliable predictor of residual tumor. Two problems emerged in the studies of residual masses. First, FDG-PET can miss some small-volume active disease. Overall, the numbers of false-negative PET studies were small, and the NPV was high.

Another useful role of PET–CT is in a scenario where there are rising tumor markers, and this may be the first indicator of disease recurrence. However, they are neither sensitive nor specific for tumor detection, and marker-negative relapse may occur even where the initial tumor was marker positive. Also, some patients with residual masses posttreatment may show modest elevation of markers even though the masses contain no active tumor, and a return of markers to normal, posttreatment, does not guarantee disease remission. Thus, PET–CT can help in lesion localization in such cases. In a recent study of 70 patients with known markers who underwent PET imaging, it was found that patients with raised tumor markers, including those with a residual mass, PET identified the site of disease in all patients but 1. In the group with raised tumor markers and no residual masses, PET demonstrated the tumor in all patients. In the group with raised tumor markers and residual masses, there was one false positive. Negative PET scans in the presence of raised tumor markers presented more of a problem as there were five patients with false-negative findings in this group. In three of these cases, all imaging was normal, and subsequent PET scans were the first studies to identify the site of recurrence. This finding suggests that, in the presence of raised tumor markers and negative imaging findings, the most appropriate follow-up procedure is repeating the PET study. These findings have important implications for the management of patients. It has been found that the ability of PET to find unsuspected disease resulted in management changes in 57 % of patients. Management changes involved changes from local therapy like radiotherapy/surgery to chemotherapy or surveillance. Many of their patients had multiple recurrences and had chemotherapy-resistant tumor, and in such cases local control of active sites may be the only chance of cure. In the first relapse, determination of whether there are one or multiple sites will help to determine the type of consolidation treatment.

PET–CT also plays a role in assessing response to treatment. A recent study evaluated the value of FDG-PET imaging compared to tumor markers and CT/MR in 23 patients with relapsed testicular cancer after two or three cycles of induction chemotherapy before high-dose chemotherapy. The outcome of high-dose chemotherapy was correctly predicted by PET–CT scan/serum tumor marker in 91, 59, and 48 % of patients, respectively. In those patients who showed response to induction chemotherapy according to CT scans or serum tumor marker evaluation, a positive PET study correctly predicted treatment failure. In addition, PET identified patients most likely to achieve favorable response to subsequent high-dose chemotherapy. It was suggested that FDG-PET is a valuable addition to the prognostic model of low-, intermediate-, and high-risk patients, particularly in the low and intermediate groups, for further selection of patients who would benefit from high-dose chemotherapy.

Cervical Cancer

Cervical carcinoma is a leading cause of death in females in India. The advent of routine Papanicolaou (Pap) smears has led to a substantial reduction in the incidence and greater than 70 % decrease in mortality of cervical cancer over the last 50 years. About 80 % are squamous cell carcinomas and about 20 % adenocarcinomas, in addition to other rare types of tumors. Uncommon subtypes are adenosquamous carcinoma, lymphoma, adenoma malignum, and small cell carcinoma, the latter tending to be locally invasive as well to have distant metastases. The International Federation of Gynecology and Obstetrics (FIGO) has defined a staging system for carcinoma of the cervix that uses a combination of clinical and radiologic findings. Although prognosis is related to stage, other factors are equally important in determining prognosis, including the extent of lymph node involvement. Only limited data is available on the use of imaging modalities, including FDG PET–CT in the primary detection of these malignancies.

FIGO staging of cervical cancer is done in accordance with the FIGO clinical criteria, which allows information obtained from physical examination, lesional biopsy, endocervical curettage, hysteroscopy, cystoscopy, proctoscopy, barium enema, intravenous pyelogram (IVP), and radiographs of the chest and skeleton, to be used for staging purposes. This criterion allows uniform staging evaluation of patients with cervical cancer. Cervical cancer spreads in a predictable pattern from the primary tumor to pelvic, para-aortic, supraclavicular lymph nodes, and then to non-nodal distant sites. Extrapelvic disease without pelvic lymph nodal involvement is rare. Lymphangiography and surgical staging have become obsolete, although surgical staging is still advocated as the gold standard to assess pelvic and para-aortic lymph node metastasis. The best methods for defining the status of lymph nodes are CT and MRI scanning. The less-invasive nature of CT/MR as well as the extra anatomic information provided on tumor extent has made the imaging procedures preferable; however, as far as lymph nodal assessment is concerned, CT and MRI have lower sensitivity and specificity as it depends on size-based criteria in comparison to functional imaging.

These problems with conventional imaging have led to evaluation of FDG-PET as an alternative for the staging of cervical carcinoma and for the evaluation of lymph nodes in particular. MR imaging had accuracies for detecting parametrial, pelvic sidewall, bladder, and rectal involvement of 88, 95, 96, and 100 %, respectively. The overall staging accuracy was 88 %.

FDG-PET has a superior ability to detect regional and distant metastatic disease in patients with cervical cancer as compared with CT and MR imaging, which has made FDG PET–CT a useful investigation. A recent study evaluated the usefulness of tumor volume measurement with PET in patients with advanced cervical cancer treated by radiation therapy. Conclusion made from this study was that tumor volume can be accurately measured by PET; the tumor volume separates patients with a good prognosis from those with a poorer prognosis; a subset of patients with relatively small tumors and no lymph node involvement does remarkably well; and tumor volume does not correlate with the presence of lymph node disease.

The same group of investigators has evaluated a treatment planning method for dose escalation to the para-aortic lymph nodes based on PET with intensity modulated radiotherapy (IMRT) for cervical cancer patients with para-aortic lymph node involvement. They subsequently determined the guidelines regarding the selection of appropriate treatment parameters. A group of researchers developed a simple, rapid, and highly reproducible system for visual grading of characteristics of the primary tumor in patients with cervical cancer at the time of diagnosis. Another group evaluated the outcome of patients with FIGO clinical stage IIIb cervical carcinoma as a function of site of initial regional lymph node metastasis as detected by FDG-PET. They concluded that the cause-specific survival in this group was highly dependent on the extent of lymph node metastasis as identified on FDG-PET. FDG-PET imaging helps in predicting response in patients with cervical carcinoma pre- and post radiotherapy. A recent study evaluated 152 patients with cervical cancer who underwent radiotherapy and/or chemotherapy with pre- and posttreatment FDG-PET imaging. The 5-year survival of patients with positive FDG-PET scans (at previous or new sites) was of 32 % and with negative FDG-PET post-therapy scan was 80 %. They concluded that persistent or new FDG uptake on the post-therapy scan was the most significant prognostic factor for developing metastatic disease and for predicting death from cervical cancer. The main cause of false positives in FDG PET–CT is interference of activity from urinary bladder, which might cause difficulty in image interpretation or artifacts. Imaging patients pre- and post-void without furosemide intervention was found, however, to have 100 % sensitivity for tumor detection on post-void images.

As with other tumors, timing of the acquisition of the PET images post-administration of FDG may be important, and dual time point intervention is usually done and well documented in studies.

Several non-FDG tracers have been tried like imaging using both 11C-choline and 11C-methionine and evaluated in carcinoma of the cervix. Absence of significant renal excretion with subsequent little or no radioactivity in the urinary bladder makes this tracer more suitable than FDG for evaluation of gynecologic malignancies.

Despite significant advances, cervical cancer continues to be a major worldwide public health concern. Although no randomized trials have directly compared FDG imaging to surgical staging, there is substantial evidence supporting its clinical utility in the management of patients with cervical cancer. The robust data has led the Center for Medicare and Medicaid Services to make it a covered oncologic indication.

Uterine Cancer

Endometrial carcinoma is the most common gynecologic malignancy, with approximately 40,000 new cases diagnosed in the United States each year. The two main subtypes of endometrial cancer are endometrioid (type I) and non-endometrioid (type II) tumors. 80–90 % of patients are associated with endometrioid histology. Patients are usually perimenopausal and have risk factors associated with increased estrogen exposure such as nulliparity, chronic anovulation, and obesity. The tumors are confined, as a rule, to the uterus and have a good prognosis. On the other hand, non-endometrioid subtypes are seen in older multiparous women, usually without increased estrogen exposure. The most common forms are uterine papillary serous carcinoma and clear cell carcinoma. In view of high propensity for myometrial and vascular invasion as well as peritoneal carcinomatosis, these types of cancers carry a poorer prognosis than endometrioid carcinoma. Painless bleeding is the most frequent presenting symptom of endometrial cancer. Effective steps for the evaluation of patients' postmenopausal bleeding (PMB) are transvaginal sonography (TVS), endometrial biopsy (EMB), and hysteroscopy. After the detection of malignancy, tumor bulk as well as local and distant spread can be assessed with this imaging modality before surgical staging (Fig. 14.5).

Routine surveillance imaging has not proved to be effective, and likely will not be until a cura-

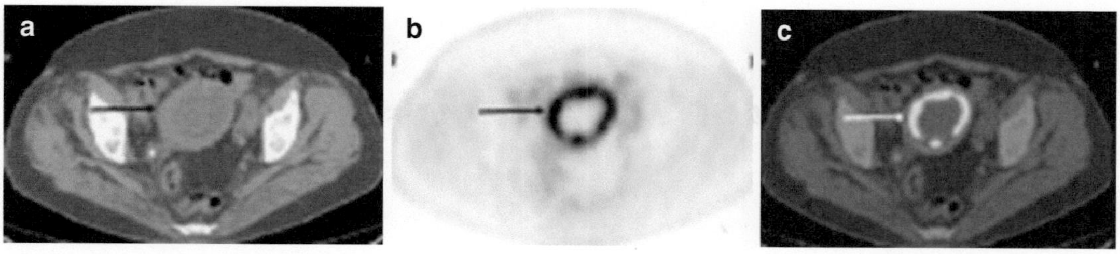

Fig. 14.5 A 61-year-old woman with stage IV endometrial cancer had completed chemotherapy and radiotherapy and being considered for surgery. PET–CT was performed to exclude metastasis. There was evidence of disease outside the uterus. The transaxial CT images (**a**), PET image (**b**), as well as fusion image (**c**) showed persistent FDG uptake in the primary endometrial site indicating residual tumor (see *arrows*)

tive treatment for this recurrent metastatic disease is found. Seven percent patients show vaginal recurrence, which presents as vaginal bleeding. So, clinical examination and MRI/CT are the important tools in this scenario.

FDG imaging has been shown to play an important role in the decision-making process for women with known recurrent endometrial cancer. For women with an isolated site of recurrence, surgery and/or radiotherapy may be either curative or provide effective palliation, but with multifocal recurrent disease, only palliative chemotherapy is indicated. A well-documented study on PET versus CT performance in 90 women with recurrent endometrial or cervical cancer found that PET improved the sensitivity and specificity for assessing the extent of disease when compared with CT. These investigators also noted that in 42 % of patients, PET results led to a change of management.

Studies using 18 F-17b-estradiol (FES) and FDG-PET showed that FES-PET are more useful in monitoring hormone therapy, especially in endometrial hyperplasia than FDG-PET. This differential monitoring of PET signals certainly could provide valuable insights into the management of recurrent disease or fertility-sparing interventions where hormone receptor status could inform the decision to treat a woman with progestin or antiestrogen therapy rather than chemotherapy. These treatment decision strategies, however, are only theoretical and have not yet been investigated. Endometrial cancer is a common malignancy that usually has a good prognosis. Given the favorable outcomes generally seen, there is no apparent benefit to extensive surgical or radiologic staging of these women. The utility of FDG PET–CT is confined to clarifying the extent and location of recurrent disease, thus assisting in the individualization of salvage therapy decisions.

Ovarian Cancer

About two-thirds of patients present with advanced FIGO Stage III or IV disease; ovarian cancer accounts for a greater number of deaths than all other gynecologic malignancies. Epithelial tumors account for approximately 90 % of ovarian cancers and can have serous, mucinous, endometrioid, clear cell, and undifferentiated histologies. Serous carcinoma represents approximately 80 % of all ovarian cancers and is histologically graded as low or high grade. Low-grade serous carcinomas arise from borderline tumors, whereas high-grade tumors do not have a definite precursor lesion, are more frequent, and have a poorer prognosis. Borderline tumors lack stromal invasion and occur at a younger age group than invasive cancer. Primary ovarian mucinous carcinoma is uncommon and is diagnosed after excluding metastatic disease to the ovary (Fig. 14.6).

Imaging is used to characterize an adnexal mass and assess for metastatic disease following the diagnosis of malignancy. USG is the first-line approach for lesion characterization, with MR imaging as a problem-solving tool. CT or MR imaging can be used to stage patients for metastatic disease. Adnexal lesions are common findings on imaging procedures, and the key is to distinguish benign from potentially malignant lesions.

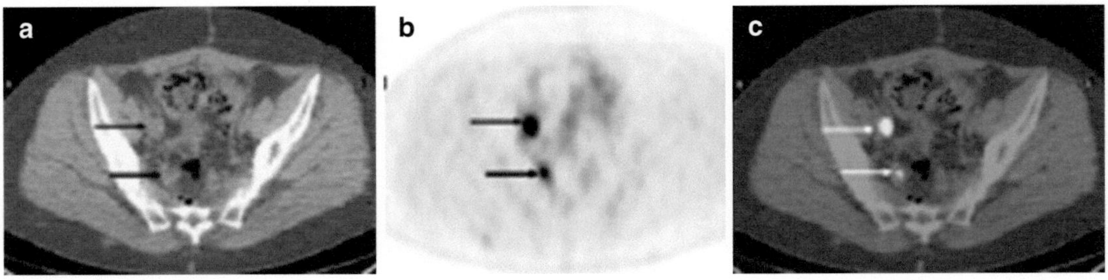

Fig. 14.6 A 40-year-old woman was diagnosed with ovarian cancer and was treated with radical surgery. On follow-up, she was found to have rising tumor marker (CA-125). MRI was stable without evidence of recurrence. The transaxial images of CT, (**a**) PET (**b**), and PET–CT fusion image (**c**) showed FDG-avid lesion in the pelvis that were proven to be recurrent disease (see *arrows*)

There are several possible diagnostic tasks for FDG-PET imaging in ovarian cancer, including noninvasive characterization of an ovarian mass as malignant or benign early diagnosis and defining the extent of localized ovarian cancer; staging and initial treatment planning (which often includes debulking surgery) after the diagnosis; predicting whether response will occur and determining whether the disease is responding to treatment; and determining whether there is residual tumor (restaging) or recurrence after the treatment. There are critical diagnostic points because the treatments for ovarian cancer can be aggressive and difficult for patients to tolerate. FDG-PET has been evaluated, to some extent, in each of these settings, but the role of PET in detecting recurrent disease and in monitoring response to tumors is the major area of focus. Multimodality screening using serum CA-125 and pelvic USG or TVS is used. Ovarian cancer primarily recurs in the peritoneal cavity and retroperitoneal lymph nodes. Patients are monitored for recurrence with periodic physical examinations, serum CA-125 level measurements, and USG examinations. Additional imaging (CT, MR, and FDG-PET or PET–CT) is commonly performed when there are signs or symptoms suggestive of recurrence.

However, both morphologic (CT and MR imaging) and metabolic FDG-PET imaging are limited in their ability to detect small-volume (<5–10 mm) disease. Rising CA-125 levels may precede the clinical detection of recurrence in 56–94 % of cases, with a median lead time of 3–5 months. Recent data suggest FDG-PET to be more sensitive than CA-125 levels, in some instances.

There is increased use of PET–CT for evaluating response to neoadjuvant, adjuvant, or standard chemotherapy or radiotherapy, but the literature is just evolving and studies are small. In a prospective study of 33 patients with advanced-stage ovarian cancer (FIGO stage IIIC and IV), receiving neoadjuvant chemotherapy before cytoreductive surgery, FDG-PET of the abdomen and pelvis was obtained before treatment and after the first and third cycles of chemotherapy. A significant correlation was observed between FDG-PET metabolic response after the first (threshold of 20 % decline in standardized uptake value [SUV]) and third cycle of chemotherapy (threshold of 55 % decline in SUV) and overall survival. The investigators concluded that FDG-PET seems to be a promising tool for early prediction of response to chemotherapy.

PET–CT is the established hybrid imaging method for ovarian carcinoma, but combinations of PET and MR imaging are being explored and this technology is evolving rapidly. The combination of PET and MR imaging using software fusion techniques has been examined in 31 patients with ovarian cancer. The investigators concluded that anatomic localization was superior for PET–MR imaging fusion than for PET–CT. However, diagnostic accuracy was not formally evaluated. Nonetheless, although PET–MR imaging could play a growing role in ovarian cancer imaging in the future, it is still in its infancy.

Uterine Sarcomas

Uterine sarcomas are mainly classified into three subtypes: leiomyosarcoma, carcinosarcoma, and endometrial stromal sarcoma. They constitute less than 8 % of uterine carcinomas. In a small study of five patients, it has been reported that FDG-PET detected leiomyosarcoma with an accuracy of 100 % compared with an accuracy of 80 % for MR imaging. However, FDG-PET was unable to detect metastases in subcentimeter lymph nodes. Because of their aggressive tumor biology, leiomyosarcoma typically recurs within 2 years of initial treatment. FDG PET–CT has a better detection rate as compared to CT for extrapelvic recurrences.

Vaginal Cancer

It constitutes less than 3 % of gynecological malignancy. It is usually diagnosed by Pap smear and staged according to FIGO classification system. Squamous cell carcinomas constitute 80 % of the histological variants. The usual presentation is bleeding. Tumor usually spreads by local invasion, lymphatic dissemination to inguinal and pelvic lymph nodes, and hematogenous spread to the lungs. Depending upon the anatomical involvement, like in cases where upper third of the vagina is involved, drainage is into the pelvic lymph nodes, internal and external iliac chain, and obturator nodes, whereas tumors of the lower third drain into deep pelvic lymph nodes, femoral, and inguinal nodes. A comparison of CT and FDG-PET in 23 patients with stages II and IV vaginal carcinoma found that FDG-PET identified all metabolically active primary tumors with a sensitivity of 100 % and detected metastatic lymph nodes in 35 % of patients in comparison to 17 % by CT. Lymph node involvement usually indicates poor prognosis.

Vulvar Carcinoma

It accounts for less than 4 % of the malignancies. Usual age of presentation is 70 years, and the most common histology is Squamous cell carcinoma. The lymphatic drainage is usually to the inguinal lymph nodes. The presence of lymph node metastases decreased 5-year survival from 97 % (stage I) to 50 % (stage III). As inguinal femoral lymph nodal dissection carries chances of increased morbidity, preoperative evaluation helps in proper management of these patients. Pre-op evaluation of patients with FDG-PET with respect to nodal metastases to the groin reported an overall sensitivity of 80 %, specificity 90 %, PPV 80 %, and NPV of 90 %. Given the low sensitivity and high NPV, a negative scan result does not preclude surgical resection. Given the high specificity, FDG imaging identifies metastatic pelvic lymph nodes and can be used to plan preoperative chemoradiation or spare extensive groin dissection. A sensitivity of 75 % and specificity of 62 % have been reported for inguinofemoral nodal metastasis by using the new agent (11C) tyrosine. A number of new agents are being investigated for staging of vulvar cancer.

For Further Reading

1. Aide N, Cappele O, Bottet P, Bensadoun H, Regeasse A, Comoz F, et al. Efficiency of [(18)F] FDG PET in characterizing renal cancer and detecting distant metastases: a comparison with CT. Eur J Nucl Med Mol Imaging. 2003;30:1236–45.
2. De Hullu JA, Pruim J, Que TH, et al. Noninvasive detection of inguinofemoral lymph node metastases in squamous cell cancer of the vulva by L-[1e11C] tyrosine positron emission tomography. Int J Gynecol Cancer. 1999;9(2):141e6.
3. De Jong IJ, Pruim J, Elsinga PH, Jongen MM, Mensink HJ, Vaalburg W. Visualisation of bladder cancer using C-11-choline PET: first clinical experience. Eur J Nucl Med Mol Imaging. 2002;29:1283–8.
4. Grigsby PW, Siegel BA, Dehdashti F, Rader J, Zoberi I. Posttherapy [18F] fluorodeoxyglucose positron emission tomography in carcinoma of the cervix: response and outcome. J Clin Oncol. 2004;22:2167–71.
5. Hain SF, O'Doherty MJ, Timothy AR, et al. Fluoro deoxyglucose positron emission tomography in the evaluation of germ cell tumours at relapse. Br J Cancer. 2000;83:863–9.
6. Ho KC, Lai CH, Wu TI, et al. 18F-fluorodeoxyglucose positron emission tomography in uterine carcinosarcoma. Eur J Nucl Med Mol Imaging. 2008;35(3):484–92.

7. Kitajima K, Murakami K, Yamasaki E, et al. Performance of integrated FDG-PET/contrast enhanced CT in the diagnosis of recurrent uterine cancer: comparison with PET and enhanced CT. Eur J Nucl Med Mol Imaging. 2009;36(3):362–72.
8. Lamoreaux WT, Grigsby PW, Dehdashti F, et al. FDG-PET evaluation of vaginal carcinoma. Int J Radiat Oncol Biol Phys. 2005;62(3):733–7.
9. Lassen U, Daugaard G, Eigtved A, et al. Whole body FDG-PET in patients with stage I non-seminomatous germ cell tumours. Eur J Nucl Med Mol Imaging. 2003;30:396–402.
10. Price DT, Coleman RE, Liano RP, Robertson CN, Polascik TJ, DeGrado TR. Comparison of [18F] fluorocholine and (18F) fluoro deoxyglucose for positron emission tomography of androgen dependent and androgen independent prostate cancer. J Urol. 2002;168:273–80.
11. Risum S, Hogdall C, Loft A, et al. The diagnostic value of PET-CT for primary ovarian cancer- a prospective study. Gynecol Oncol. 2007;105(1):145–9.
12. Safaei A, Figlin R, Hoh CK, Silverman DH, Seltzer M, Phelps ME, Czernin J. The usefulness of F-18 deoxyglucose whole-body positron emission tomography (PET) for re-staging of renal cell cancer. Clin Nephrol. 2002;57:56–62.
13. Singh AK, Grigsby PW, Dehdashti F, Herzog TJ, Siegel BA. FDGPET lymph node staging and survival of patients with FIGO stage IIIb cervical carcinoma. Int J Radiat Oncol Biol Phys. 2003;56(2):489–93.
14. Toth G, Lengyel Z, Balkay L, Salah MA, Tron L, Toth C. Detection of prostate cancer with 11C-methionine positron emission tomography. J Urol. 2005;173:66–9.

15 Application of PET in Cancer of the Endocrine Organs

Birendra Kishore Das

The role of functional imaging for endocrine abnormalities has increased over the last few decades. Endocrine tumors can be overlooked on conventional anatomic imaging because of small tumor size as well as limited or equivocal clinical data. It can also be difficult for conventional imaging to differentiate disease recurrence from postsurgical changes. Furthermore, anatomic imaging does not provide information on tumor activity. Endocrine malignancies are uncommon, comprising 1–2 % of all tumors affecting adults, and 4–5 % of tumors affecting children. Some endocrine tumors are functional and secrete active substances that trigger physiological symptoms that prompt patient presentation. A smaller fraction of patients with endocrine neoplasms present with symptoms associated with mass effects that are more typically seen in patients with large nonfunctional tumors, which are commonly detected incidentally (so-called incidentaloma). The increasing use of whole-body multidetector CT@ (MDCT) scans has led to an appreciable rise in the identification of such tumors.

B.K. Das, MD, ANM, FAR, FAMS
Director and Chief of Nuclear Medicine,
Utkal Institute of Medical Sciences,
Jaydev Vihar, Bhubaneswar, India
e-mail: bkishoredas@yahoo.com

The Role of PET-CT in Thyroid Disorders

Thyroid nodules are extremely common among the adult population, ranging from 4 to 7 % within the whole population and a female-to-male ratio of 4:1. The occurrence of malignancy in these nodules is fairly rare. However, correctly categorizing identified nodules is critical for management decisions. Current strategies for characterizing these nodules include thyroid scintigraphy as well as thyroid ultrasound. Iodine-123 scintigraphy or even Tc- 99m thyroid scan allows analysis of nodules and classification of "hot" versus "cold" nodules. These terms describe uptake of radiotracer by the nodule and the surrounding thyroid gland. Hot nodules, which display high uptake of I-123 or Tc-99m, are rarely malignant, while cold nodules, which display low uptake or absence of tracer, represent malignancy in 10–15 % of cases. Cold nodules in younger patients can harbor malignant tissue up to 35–40 %. This test does provide useful data but may require biopsy to delineate a more accurate diagnosis.

Thyroid carcinomas are classified as either differentiated (i.e., papillary or follicular) or undifferentiated (i.e., anaplastic), depending on the histologic and cytopathologic findings. Typically, patients with papillary or follicular carcinomas receive iodine-131 (I-131) total body scans in their evaluation. Figure 15.1 shows a

B.K. Das (ed.), *Positron Emission Tomography: A Guide for Clinicians*,
DOI 10.1007/978-81-322-2098-5_15, © Springer India 2015

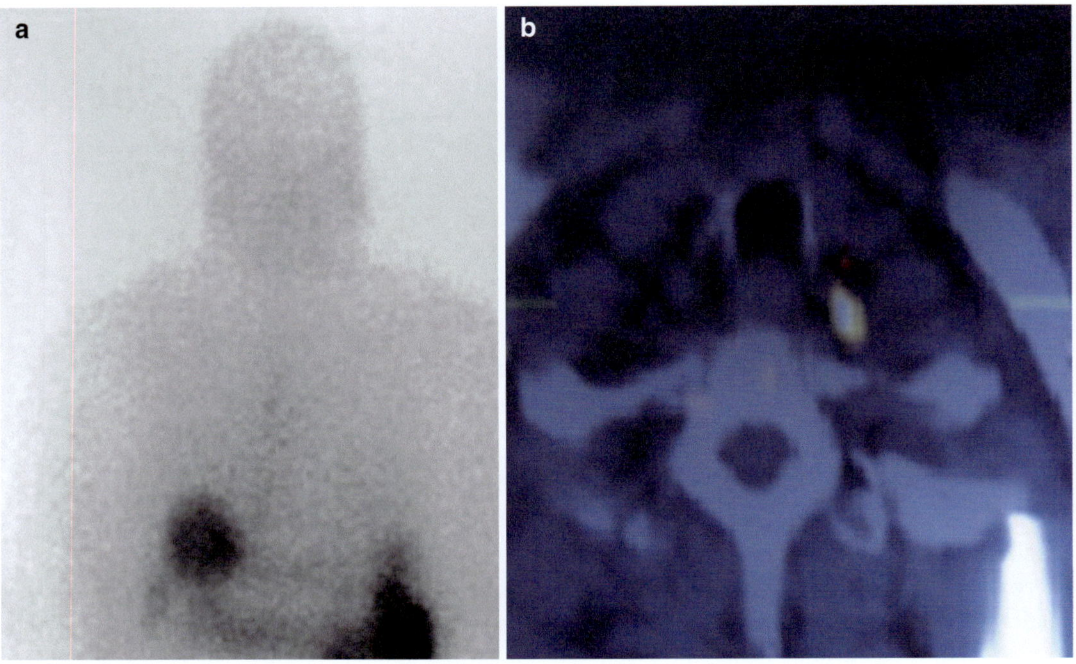

Fig. 15.1 Radio iodine scan (**a**) showing no remnant tissue in the thyroid bed. However, FDG-CT image shows an 8-mm lesion in the left thyroidectomy bed (**b**)

patient who underwent total thyroidectomy for papillary thyroid cancer and was being evaluated for increasing thyroglobulin levels after surgery. This patient had a negative I-131 scan along with an 18F-FDG PET-CT scan that showed increased metabolic activity within the thyroid bed approximately 6 months after surgery.

There have been numerous reports of patients with negative I-131 scans and positive 18F-FDG PET scans and vice versa. This phenomenon was attributable to dedifferentiation of the cancer. Figure 15.2 shows a patient who underwent an 18F-FDG PET-CT evaluation after a total thyroidectomy for biopsy-proven papillary thyroid cancer. This particular patient was treated with I-131 on two separate occasions and on follow-up 2 years later was noticed to have increasing thyroglobulin levels. He had an I-131 scan that was negative and was referred for 18F-FDG PET-CT evaluation for metastatic disease. The FDG studies showed presence of malignant tissue at different sites.

These and other studies noted that more differentiated cancers are better imaged by I-131, while dedifferentiated thyroid cancers are better imaged by 18F-FDG PET. Therefore, the role of 18F-FDG PET-CT is presently applied to those patients who have a negative I-131 scan, a negative chest CT (looking for metastasis), and rising serum thyroglobulin levels following initial thyroidectomy. These patients are thought to have recurrent or metastatic disease that is undetectable by conventional imaging but biochemically apparent because of increasing thyroglobulin measurements. A recent study reported 18F-FDG PET-CT sensitivity of 68.4 %, which is slightly lower than previous reported data, for detecting recurrent or metastatic thyroid cancer. However, several studies revealed higher sensitivities for 18FFDG PET-CT with higher thyroglobulin levels, with a sensitivity approaching 72 % when thyroglobulin levels exceed 10 ng/mL. Other studies have reported 18F-FDG PET (using visual fusion with CT) and dedicated 18F-FDG PET-CT sensitivities for recurrent or metastatic thyroid cancer to be 95–100 %.

The advantage of 18F-FDG PET and PET-CT in thyroid cancer imaging can be further verified by reviewing data on patient management affected by 18F-FDG PET-CT results. In many

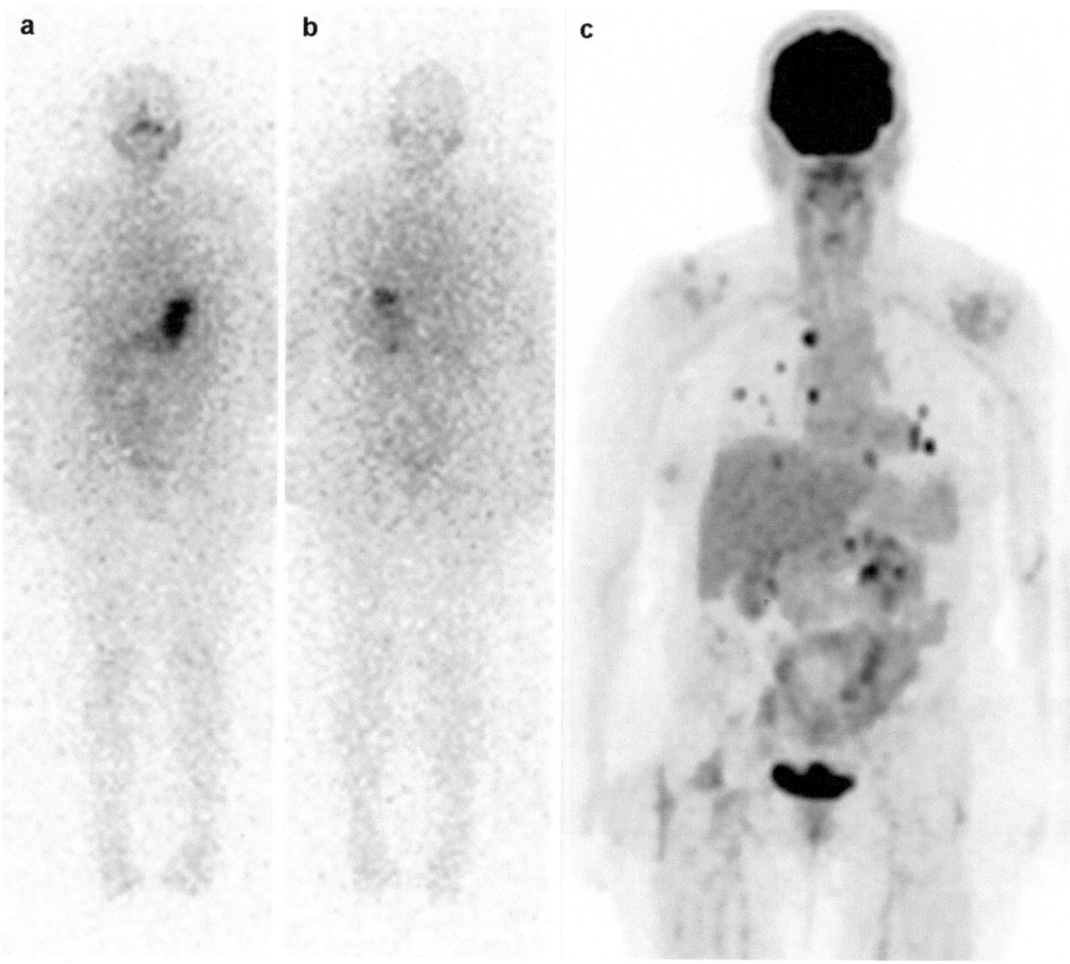

Fig. 15.2 Anterior (**a**) and posterior (**b**) radio iodine whole body scan of a thyroid cancer patient after NTT and two times high-dose RI therapy. There is no abnormal uptake suggestive of any metastatic lesions. However, TG was high. FDG PET (**c**) image shows multiple metastases

cases, finding distant metastases using 18F-FDG PET (with CT comparison) led to alteration of the initial plan and management in some surgical cases. These distant metastases were identified on a 18F-FDG PET staging scan. Using FDG PET-CT modification of the treatment has been necessary in more than 40 % of cases.

The Role of 18F-FDG PET-CT in Adrenal Tumors

Up to 79 % of adrenocortical cancers produce some hormone or active agent that may lead to clinical symptoms. Adrenal pheochromocytomas account for approximately 80 % of catecholamine-secreting neoplasms and generally measure 4–5 cm at presentation. Since functioning adrenocortical tumors can be elucidated using hormonal assays, it is the nonfunctioning tumors that require additional testing and imaging to determine management for optimal patient care.

Imaging of patients with suspected pheochromocytoma usually begins with CT or magnetic resonance imaging (MRI) to assess the tumor. However, the variable appearance of pheochromocytomas on these modalities sometimes makes it quite difficult to establish an accurate diagnosis. These tumors can range from solid to cystic, fatty to necrotic, and homogenous to heterogeneous. Functional imaging provides localization of the

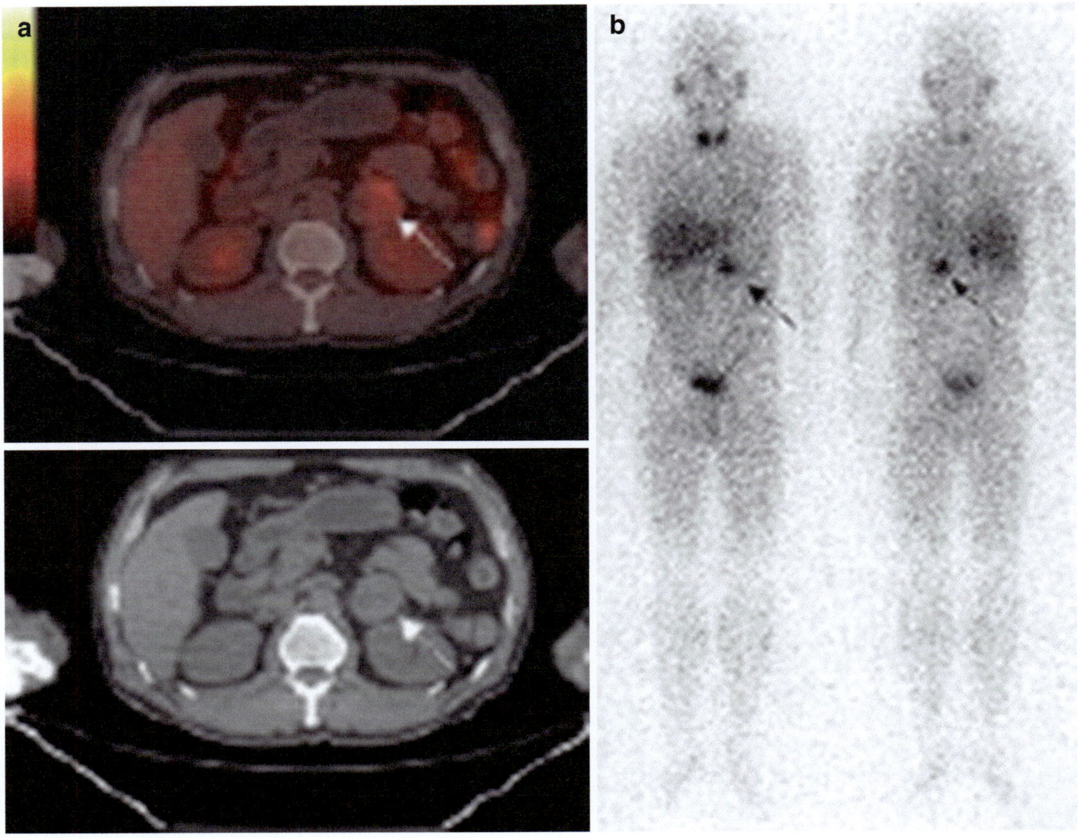

Fig. 15.3 Pheochromocytoma of the left adrenal seen in FDG-PET-CT (**a**) and I-131 MIBG whole body scan showing the tumor (see *arrow*) in (**b**) of the same patient.

tumor to any part of the body, especially extra-adrenal sites or previously postsurgical areas that can have distorted anatomy. Functional imaging with either I-123 metaiodobenzylguanidine (MIBG) scintigraphy or 18F-FDG PET-CT can prove useful for localization. However, due to the anatomic correlation that CT fusion offers, PET scanning is superior to that of MIBG alone (Fig. 15.3).

Several studies have shown variable uptake of MIBG and 18F-FDG in pheochromocytomas. Some pheochromocytomas fail to accumulate MIBG but are able to accumulate increased 18F-FDG on PET-CT. Some pheochromocytomas also accumulate MIBG without increased metabolic activity seen on 18F-FDG PET-CT. Studies have shown that most malignant pheochromocytomas are better visualized with 18F-FDG, while more benign pheochromocytomas are better detected with MIBG.

A recent study showed also superiority of 18F-FDG PET-CT in detecting metastatic foci in patients with paragangliomas pheochromocytomas (Fig. 15.4).

With the increased utilization of whole-body PET-CT scans for diagnosis and clinical staging in oncology, the incidence of unsuspected adrenal gland abnormalities is increasing. It has been reported that up to 1 % of patients being evaluated with abdominal CT and as many as 2–5 % of patients being scanned with contrast-enhanced CT have been found to have adrenal incidentalomas. The majority of these incidentalomas are benign and do not produce any biologically active metabolites.

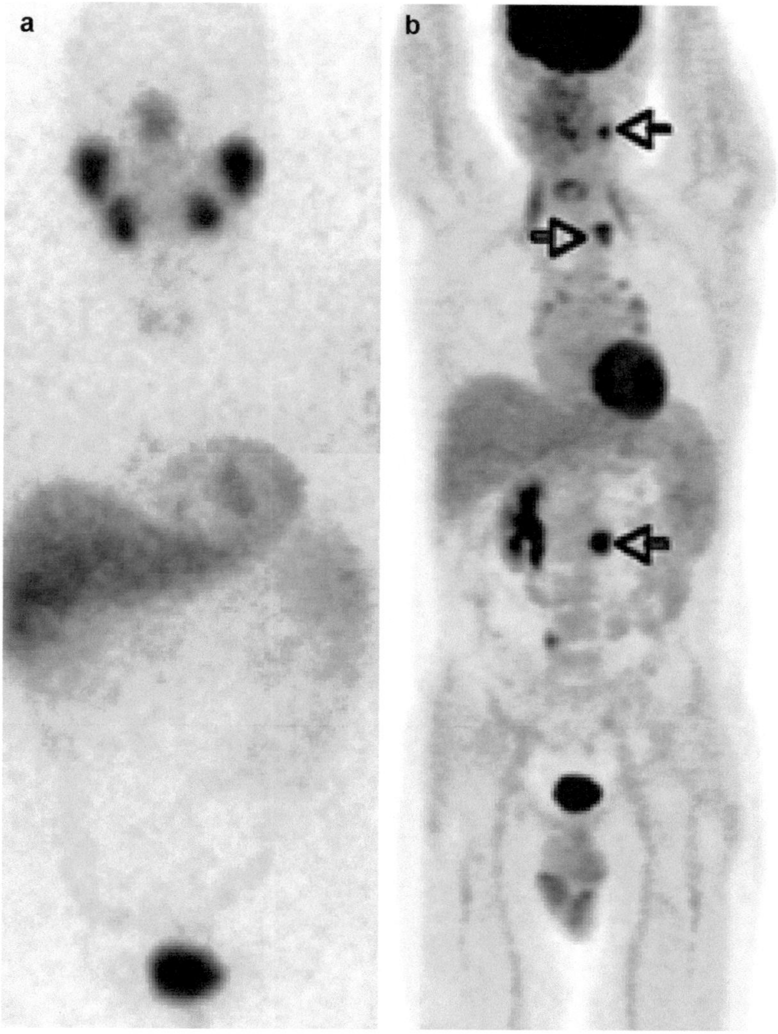

Fig. 15.4 Demonstrates a case of metastatic pheochromocytomas clearly seen in 18F-FDG PET image (**a**) with negative 123-I-MIBG scan (**b**). In this case, PET scan is shown to be more accurate

For Further Reading

1. Poeppel TD, Binse I, Petersenn S, et al. 68Ga-DOTATOC versus 68Ga-DOTATATE PET-CT in functional imaging of neuroendocrine tumors. J Nucl Med. 2011;52:1864–70.
2. Rufini V, Calcagni ML, Baum RP. Imaging of neuroendocrine tumors. Semin Nucl Med. 2006; 36:228–47.
3. Virgolini I, Ambrosini V, Bomanji JB, Baum RP, Fanti S, Gabriel M, et al. Procedure guidelines for PET-CT tumor imaging with 68Ga-DOTA-conjugated peptides: 68Ga-DOTA-TOC, 68Ga- DOTA-NOC, 68Ga-DOTA-TATE. Eur J Nucl Med Mol Imaging. 2010;37:2004–10.

16 Role of PET-CT in Malignant Involvement of Bone and Bone Marrow

Alok Pawaskar and Sandip Basu

Introduction

Traditionally, benign and malignant diseases of the skeletal system have been evaluated with radiological investigations like X-ray, CT scan, or MRI. These have established themselves as modalities of choice because of wide availability and excellent anatomical details they provide. X-ray and CT scan have been able to provide excellent images of cortical bone while MRI depicts details of bone marrow and associated soft tissue like no other imaging modality can. In the last decade or so, there has been tremendous interest and advances in applications of positron emission tomography (PET) in the field of oncology. Particularly, 2-fleuro-2-deoxy-glucose (FDG) PET-CT has become a backbone of oncological imaging for various cancers, including the bone and bone marrow tumors.

The bones comprise of cortical, trabecular, and marrow components (Fig. 16.1).

Involvement of bones by malignancy may be in the form of primary bone tumors, bone marrow disorders, or metastatic infiltration from other solid organ tumors. The purpose of imaging modalities in skeletal oncology has been to identify disease early, to understand complete extent of the disease, to assess any complications associated with malignant bone involvement (e.g., pathological fractures, cord compression, etc.), to monitor response to therapy, and to guide biopsy if warranted.

Conventional X-ray imaging shows the abnormalities of cortical and trabecular bone. The change in bone density creates the characteristic

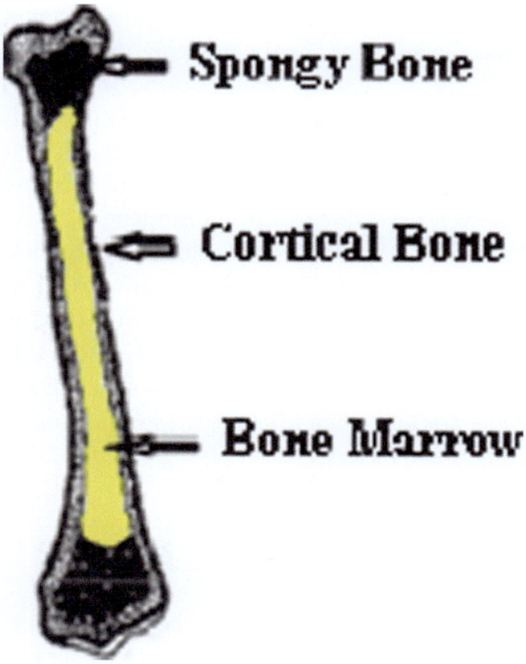

Fig. 16.1 Normal components of a long bone

A. Pawaskar, DRM, DNB (✉)
Consultant, Department of Nuclear Medicine, Radiation Medicine Centre, Bhabha Atomic Research Centre, Tata Memorial Hospital Annexe, Mumbai, India
e-mail: alokpawaskar@rediffmail.com

S. Basu, MD
Professor, Radiation Medicine Centre, Bhabha Atomic Research Centre, Tata Memorial Hospital Annexe, Mumbai, India
e-mail: drsanb@yahoo.com

B.K. Das (ed.), *Positron Emission Tomography: A Guide for Clinicians*, DOI 10.1007/978-81-322-2098-5_16, © Springer India 2015 patterns which enable one to diagnose various bone lesions like primary bone tumors or lytic, sclerotic, or mixed type of skeletal metastasis. The main limitation of X-ray in imaging the malignant bone involvement is that there needs to be considerable bone destruction before the abnormality is detected by X-ray. It has been estimated that 30–75 % reduction in bone density is required to visualize abnormality on X-ray. This may lead to significant delay in diagnosis or monitoring the changes in response to therapy.

The computed tomography (CT) scan has good anatomical resolution and better soft tissue contrast. It gives details of cortical and trabecular bone very well. However, again there is need for significant cortical bone destruction by malignancy to be visible on CT scan. This obviously leads to delayed detection of skeletal lesions reducing the sensitivity for detection of early bone lesions. In addition, malignant lesion detection may be difficult in presence of severe osteoporotic and degenerative changes. It is also not very sensitive for detection of bone marrow lesions, although occasionally these can be demonstrated by contrast enhancement or altered attenuation.

On the other hand, MRI is an optimal imaging modality for bone marrow imaging. It can even differentiate between hematopoietic (red) marrow from non-hematopoietic (yellow) marrow. It has very good spatial and contrast resolution. It can detect early malignancy in the bone marrow before it has caused cortical destruction. At the same time, it is better for visualization of soft tissue, brain, and spinal cord lesions. However, it is unable to visualize cortical destruction as good as CT scan does. Further, it may be difficult to differentiate between active disease and scar, necrosis or fracture when MRI is used for monitoring response to therapy.

Bone Scan

As far as nuclear medicine is concerned, bone scan for many decades has made significant difference in management of many cancers. It has worked as a tool for skeletal survey for detection of skeletal metastasis. It has enabled us to visualize an entire skeletal system within reasonable time and cost. Technetium (Tc) -99m-Methylene diphosphonate (^{99m}Tc-MDP) is the tracer used for bone scan using gamma camera in general nuclear medicine. Its uptake in the bone is dependent upon perfusion and osteoblastic activity of the bone. Bone scan is very sensitive to change in osteoblastic activity of the bones with as little as 5–10 % change in lesion to normal bone being sufficient to detect abnormality on bone scan. However, MDP is not a tumor-specific tracer and increased accumulation may also be seen in the benign lesions. Correlation with X-ray/CT scan is often necessary to confirm the finding on bone scan. As MDP uptake is primarily in the cortical bone, bone marrow lesions are missed on bone scan. Lytic lesions having typically less reactive bone formation may also be missed out on bone scan. During follow-up after therapy, bone scan may show reduction in uptake or resolution of lesions noted on baseline scan. However, if healing of the lesion occurs by increased osteoblastic activity, bone scan is unable to differentiate between active disease and healing process. Sometimes, osteolytic lesion may be missed on baseline bone scan and after treatment may heal by osteoblastic reaction. This new area of increased tracer uptake can be misinterpreted easily as new lesion indicating disease progression. Correlation with clinical status, tumor markers, or other imaging modalities may be useful in these cases.

Single photon emission computed tomography (SPECT) study may add to information provided by planar bone scan especially in the region of spine. SPECT gives three-dimensional image of the tracer uptake and hence has better resolution as well as localization of the lesion. This leads to better sensitivity and specificity of SPECT compared to planar bone scan. It was reported to detect 20–50 % more lesions in the spine compared with planar bone scan. With advances in imaging technologies, gamma cameras with capability of performing "whole body SPECT" in reasonable time are available. Fusion imaging gamma cameras with SPECT and CT mounted on same gantry are available as well.

These machines are not only able to perform SPECT but also perform CT scan with patient on the same bed without moving between the SPECT and CT studies. Best possible fused SPECT-CT images are available for review. This definitely makes life easy for nuclear medicine physician as well as patient. Although sensitivity of bone scan is very high for detection of osteoblastic lesions, specificity is poor due to nonspecific uptake of Tc-99m-MDP in nonneoplastic lesions. Hence, additional radiological imaging is often needed for definitive diagnosis. This adds to the time, cost, and anxiety for the patient. With availability of SPECT-CT, the specificity of bone scan has improved significantly making bone scan even more valuable investigation.

Positron Emission Tomography (PET)

Positron emission tomography (PET) imaging has better spatial and contrast resolution. It is capable of whole body tomographic imaging and quantification of tracer uptake. Using PET tracers it is possible to detect functional changes in the bone and bone marrow much earlier than structural changes detected by anatomical imaging like CT scan/MRI. Currently ^{18}F-fluoride and $^{18F\text{-}FDG}$ are the PET tracers used in clinical practice for assessment of cancers of bone and bone marrow.

^{18}F-Fluoride PET-CT Scan (F-18 Bone Scan)

^{18}F-Fluoride was first introduced as a bone-imaging agent by Blau at al. in 1962. Its uptake mechanism is similar to that of $^{99mTc\text{-}MDP}$. After diffusion through capillaries into bone extracellular fluid, fluoride ions exchange with hydroxyl groups in hydroxyapatite crystal bone to form fluoroapatite, which is deposited mainly at the surface, where bone remodeling and turnover are greatest. Similar to ^{99m}Tc-MDP, accumulation of ^{18}F-fluoride in malignant bone lesions reflects the increased regional blood flow and bone turnover. Bone uptake of ^{18}F-fluoride is twofold higher than that of ^{99m}Tc-MDP. The higher capillary permeability of ^{18}F-fluoride and its faster blood clearance result in a better target-to background ratio. Better spatial resolution of PET and pharmacokinetics of ^{18}F-fluoride lead to a better quality bone scan done in shorter time frame. The little osteoblastic reaction associated with osteolytic lesions may not be seen on ^{99m}Tc-MDP bone scan. Hence sensitivity of ^{18}F-fluoride PET bone scan is better in these cases compared to ^{99m}Tc-MDP bone scan.

Like ^{99m}Tc-MDP bone scan, abnormal F-18 uptake may also be seen in benign bone diseases and it is not very specific tracer for imaging of malignant disease. Here again just like in ^{99m} Tc-MDP bone scan, fusion imaging with PET-CT machines which have become a norm these days helps to a great extent. Not only CT attenuation data is used for attenuation correction, but CT images help in localization as well as characterization of the lesion detected by F-18. This leads to better specificity in addition to increased sensitivity provided by this bone scan (Fig. 16.2).

FDG PET-CT Scan

Whole body PET-CT with FDG has gained wide popularity in oncology for diagnosis, staging, restaging, and therapy monitoring in various malignancies. FDG being glucose analogue is taken up by GLUT receptors located on the cell membrane just like glucose molecules. However, once inside the cell, FDG does not undergo further metabolism after being phosphorylated. It gets trapped in the cells having low dephosphorylation rate which is the case with malignant cells. Hence malignant cells having higher glucose consumption are easily detected on FDG PET-CT scan. Unlike F-18 bone scan, FDG PET-CT detects both skeletal as well as soft tissue disease.

Normal bone marrow shows low-grade diffuse FDG uptake (Fig. 16.3).

However, in early marrow disease, there is focal increased FDG uptake indicating presence of disease much earlier than it can be picked up

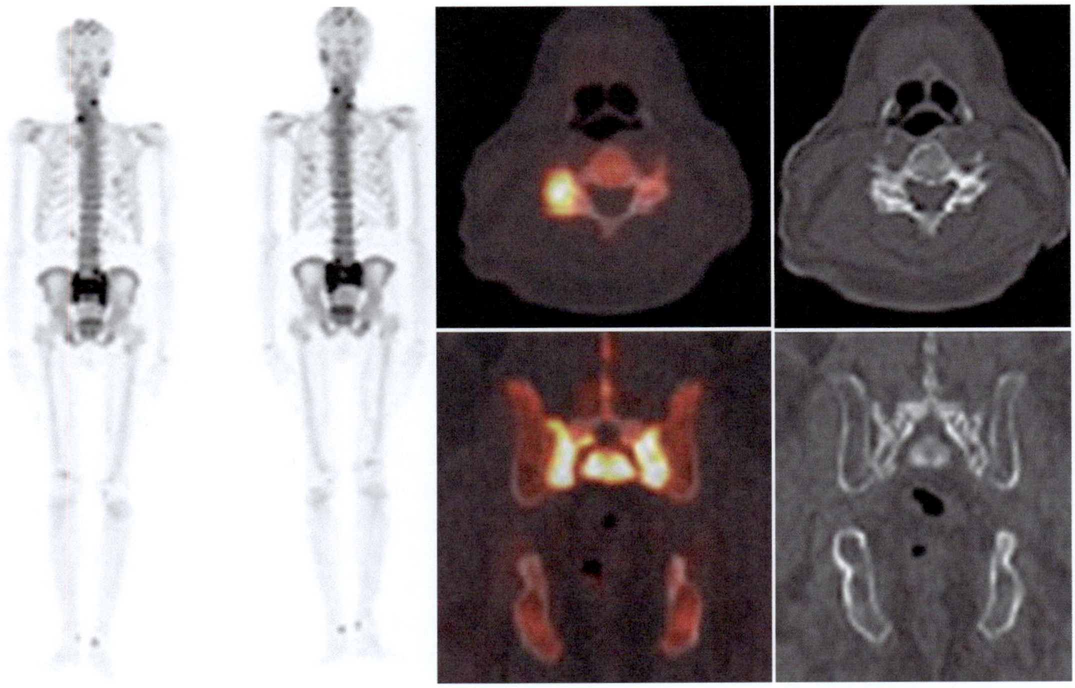

Fig. 16.2 Whole body bone scan (*left images*) and ^{18}F-fluoride PET bone scan of the sacro-iliac joints along with CT scans and fusion images (*right images*)

on bone scan or CT scan (Fig. 16.4). Also FDG PET appears to be more sensitive to osteolytic lesions than osteoblastic bone lesions. Some of the diseases like prostatic carcinoma with skeletal metastases, where skeletal metastases are predominantly osteoblastic; ^{99m}Tc-MDP/F-18 bone scan would be positive while FDG PET-CT may not show significant tracer uptake. This is due to the difference in the mechanism of localization of these tracers. While ^{99m}Tc-MDP/F-18 bone scan shows "osteoblastic reaction" of the body to the malignant disease, FDG PET-CT actually shows 'malignant disease' having increased glucose metabolism. Hence osteolytic bone disease is seen prominently on FDG PET and osteoblastic bone disease is seen predominantly on bone scan. However, with fusion of PET with CT, both types of lesions may be detected with high sensitivity and specificity as CT scan can easily detect osteoblastic lesions negative on FDG PET. Absence of FDG uptake in these lesions may further indicate their slow-growing nature and good prognosis.

FDG PET is less hampered by nonspecific uptake in incidentally found benign bone lesions compared with bone scan or ^{18}F-fluoride PET. However, false-positive increased FDG uptake may occasionally be detected in benign bone lesions, especially histiocytic or giant cell-containing lesions, including osteoblastoma, brown tumor, aneurysmal bone cyst, and sarcoidosis. Tissue histiocytic and giant cells are the cells in monocytes–macrophage lineage and play a central role in the host response to injury and infection. Their energy is predominately supplied by means of intracellular glucose metabolism. Although the standardized uptake value (SUV) of FDG in malignant bone lesions is generally higher compared with benign bone lesions, there is a overlap.

Primary Bone Tumors

Differentiating Benign from Malignant Bone Tumors

X-ray, CT scan, and MRI have traditionally diagnosed and differentiated benign bone tumors from malignant ones. These modalities are able to precisely localize the tumors, define tumor extent, and show associated complications like necrosis and fractures. However, these modalities typically lack the information about the activity and metabolism of the tumor. FDG PET fills in this lacuna. Initially, there was lot of enthusiasm about being able to differentiate benign from malignant lesion based on FDG uptake (SUV values). However, over the time, it is found that there is significant overlap of uptake of FDG in benign and malignant bone tumors. For example, several authors have revealed benign tumors with a high SUV, causing a high false-positive rate in trials to differentiate malignancy from benignancy with FDG-PET. In two other studies with a SUV cut-off of 1.9, the sensitivity of FDG-PET to correctly diagnose malignancy was 72.7 and 84.6 %, with a specificity of 66.0 and 80.0 % and an accuracy of 68.0 and 81.8 %, respectively.

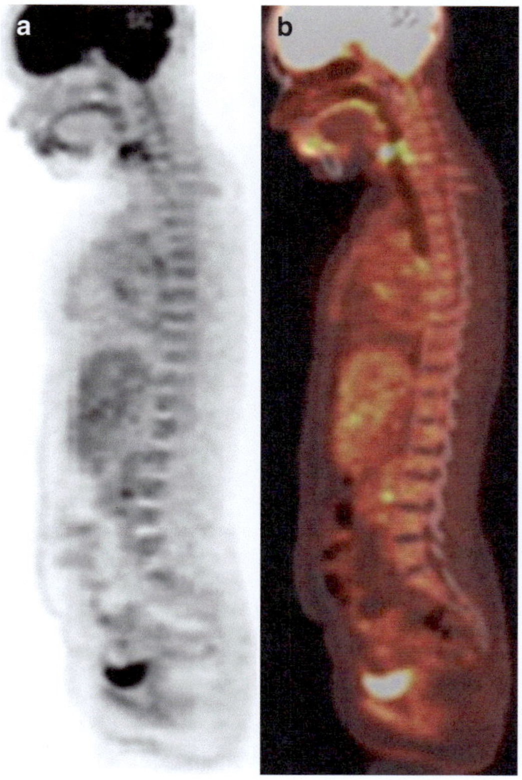

Fig. 16.3 Lateral view of FDG-PET whole body scan (**a**) and PET-CT fusion image (**b**) showing normal bone marrow FDG uptake

Benign bone tumors with high FDG uptake by FDG-PET are giant cell tumor, chondroblastoma, Langerhans cell histiocytosis, fibrous dysplasia, and osteoid osteoma, to name a few. It is still uncertain as to why some benign bone tumors show high glucose uptake? However, the point to be noted is that most of the above benign tumors are from giant cells or osteoclasts, which are speculated to originate from macrophages. Kubota et al. reported that FDG was highly accumulated not only in tumors but also in macrophages and granulation tissues surrounding the tumors, which might explain why some benign tumors show high FDG uptake.

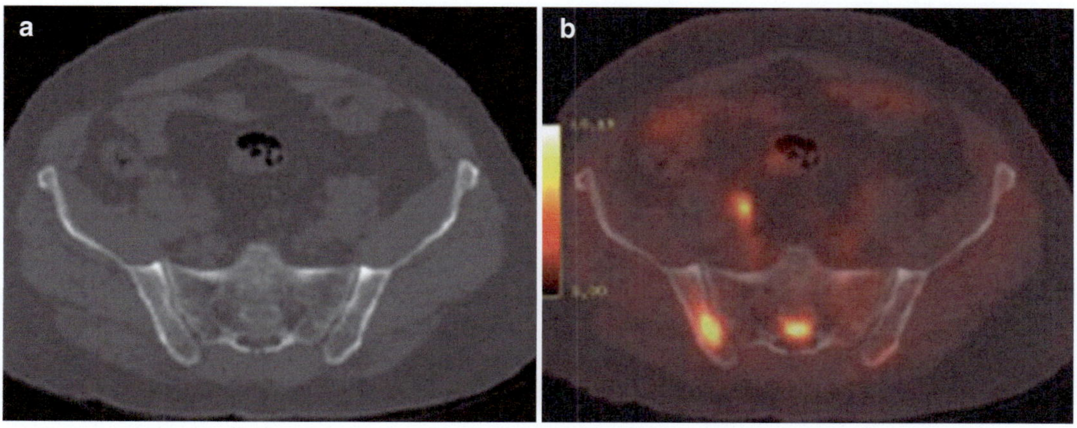

Fig. 16.4 Early bone marrow disease showing high FDG uptake (**b**) before changes appear in CT scan (**a**)

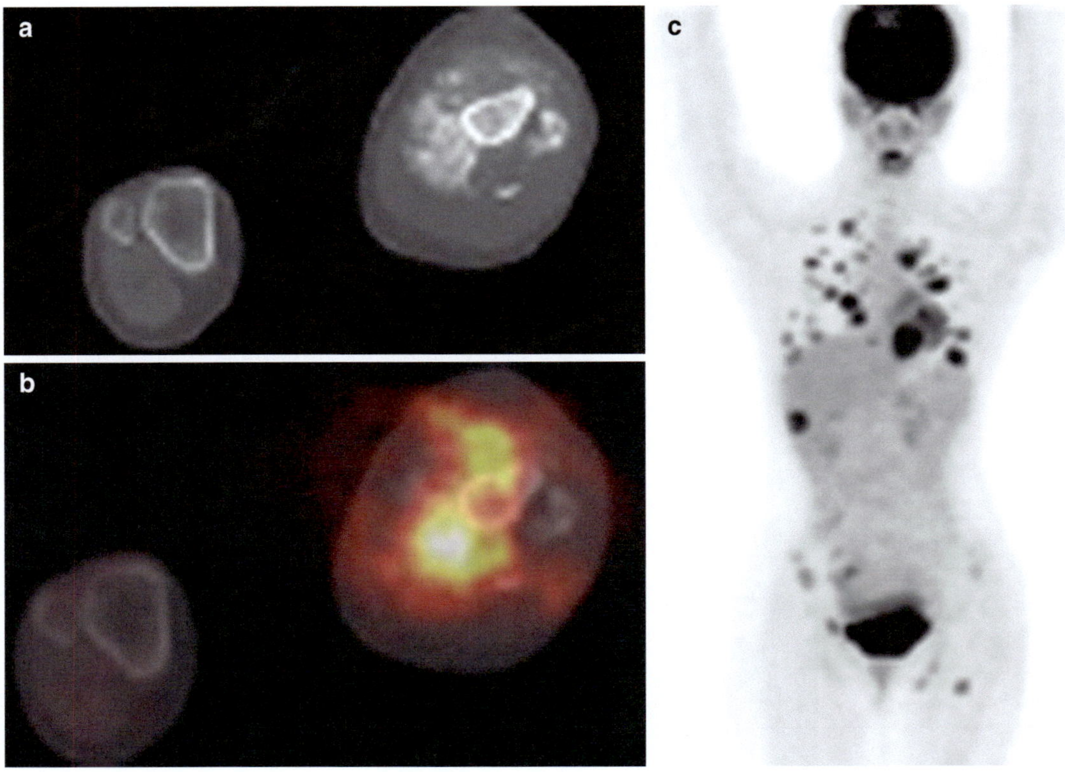

Fig. 16.5 Soft tissue sarcoma of a pediatric patient in CT (**a**) and PET-CT fusion image (**b**). FDG whole body image showing extensive metastatic disease (**c**)

Grading and Staging

Like many other tumors, FDG uptake in the primary bone tumors is generally proportional to grade of these tumors; with low-grade tumors having low FDG uptake and high-grade tumors having more FDG uptake. However, there is significant overlap in uptake values among the tumors of different grades. Folpe et al. examined the relationship between FDG-PET SUVs and histopathological findings. A significant difference was revealed between grade I bone and soft tissue sarcomas and grade II and III using SUV by FDG-PET, although there was little difference between benign tumors and grade I sarcomas. They also reported that high SUV was related with hypercellularity, high mitotic activity, the MIB labeling index, and p53 overexpression.

As for tumor staging, a prospective multicenter trial revealed that FDG-PET was superior to conventional imaging modalities, including ultrasound CT, MRI, and bone scintigraphy, to detect lymph node involvement and bone manifestation in pediatric sarcoma patients, although CT could depict lung metastases more reliably (Fig. 16.5).

Evaluation of Response to Therapy

Evaluating response to chemotherapy is very important in the treatment of osteogenic sarcoma because the degree of necrosis by chemotherapy is one of the most important prognostic factors, and a poor response to chemotherapy increases the local failure rate after limb salvage operations. Schulte et al. showed that a decreased ratio of post- and pre-therapeutic tumor-to-background in FDG-PET correlated with the amount of tumor necrosis by chemotherapy and FDG-PET could discriminate therapy responders from nonresponders in all seven but two patients with a tumor-to-background

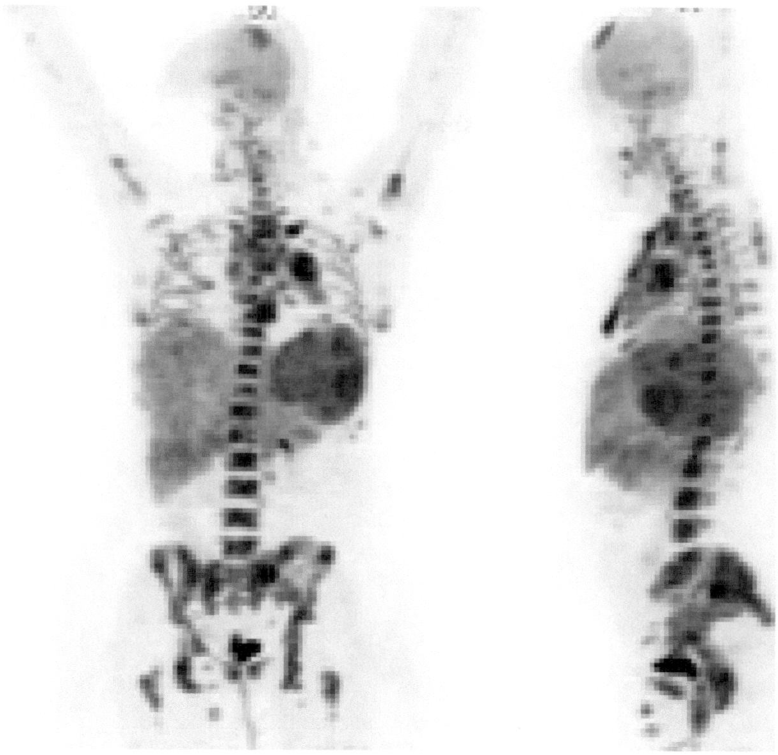

Fig. 16.6 FDG whole body scan showing disseminated disease in a case of lymphoma

cutoff level of 0.6. Hawkins et al. reported that an SUV less than 2.5 after chemotherapy was predictive of progression-free survival in Ewing's sarcoma family of tumors. Jones et al. also described that FDG accumulation in soft tissue and musculoskeletal sarcomas decreased after neoadjuvant therapies, although complete absence of FDG uptake could not be achieved. They speculated that the remaining FDG uptake seems to correspond with a pseudocapsule or infiltrating granulation tissues and fibrosis.

Bone Marrow Tumors

Lymphoma

FDG PET is now an established modality of choice in imaging lymphoma. It is useful for staging, restaging, and therapy monitoring of lymphoma patients. Compared to CT scan which depends on size criteria, FDG PET has been proven to be more sensitive for detection of disease in lymph nodes, spleen, as well as bone marrow. The metabolic changes typically precede the structural changes and hence are detected earlier by FDG PET scans. Further with availability of PET-CT, the localization and attenuation correction has improved significantly. PET-CT is very useful in differentiating localized from extensive disease which has major impact on management of lymphoma (Fig. 16.6).

However, PET has low sensitivity for detection of central nervous system involvement because of physiological high FDG uptake in brain. Uptake of FDG in the tumor quantified as standardized uptake value (SUV) has prognostic implications. Lesser the SUV, less aggressive is the disease and better the prognosis. Higher SUV means aggressive disease with poor prognosis. FDG PET has better sensitivity than bone scan for detection of early marrow infiltration. It is also more sensitive than CT for detecting bone involvement (Fig. 16.7). The bone marrow involvement by lymphoma is often patchy. Routine bone marrow biopsy is done without any imaging guidance; hence may miss bone marrow lesion leading to false-negative bone marrow

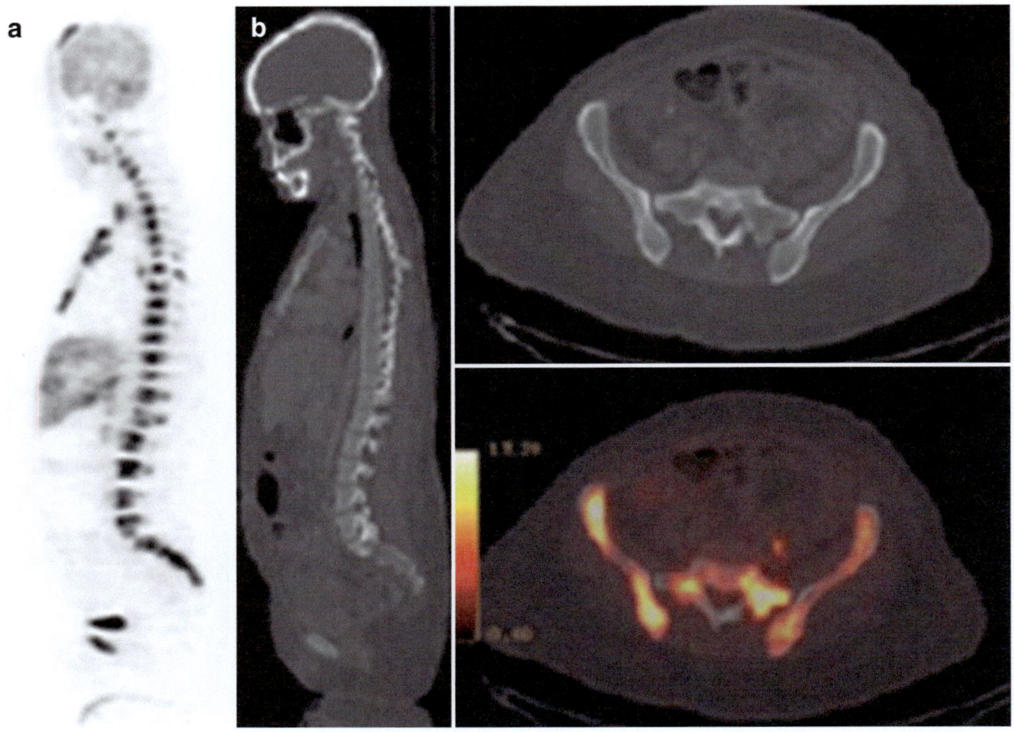

Fig. 16.7 FDG whole body scan (**a**) showing extensive involvement of the bone marrow. CT of the same patient does not reflect the extent of the disease whereas PET-CT fusion image (**b** *right below*) shows involvement of the bone

biopsy in case normal marrow is sampled. FDG PET may be useful to guide site of bone marrow biopsy to reduce false negative biopsies. Hence, FDG PET is proven to be complementary to bone marrow biopsy; however, it cannot replace it.

Multiple Myeloma

The diagnosis of multiple myeloma is primarily from finding of "M" band proteins in blood and/or urine or by biopsy from the skeletal lesion. However, skeletal survey with X-ray or whole body MRI or PET-CT provides vital information about number of skeletal lesions which determines the prognosis and treatment for a particular patient. Patients with solitary lesion rarely need systemic therapy. The number of focal lesions at baseline is a key finding, inversely and significantly related to both overall and event-free survival. During follow-up when there is evidence of tumor progression or relapse on PET-CT, these should be corroborated with standard laboratory tests. Restaging of the patients after therapy should be done 2–4 weeks after completion of therapy. However, if progression or relapse while on treatment is suspected, FDG PET-CT can often provide verification and direction for biopsy or change in treatment.

Presence of extramedullary lesion is important because it identifies the patient as high risk. The extramedullary disease is suggestive of multiple myeloma which is poorly differentiated, non-secretory, rapidly progressive, and resistant to standard treatment. However, it is important to differentiate between extramedullary disease and breakout lesions. Breakout lesions are the lesions in which the tumor has broken the cortex and is extending into the soft tissue. These lesions behave similar to bone marrow lesions and are amenable to therapy. These lesions can be imaged with MRI or PET-CT. However, PET-CT has advantage of being whole body imaging over conventional region MRI. However, whole body MRI, if available, may be equally informative. In hypo or nonsecretory patients who have low

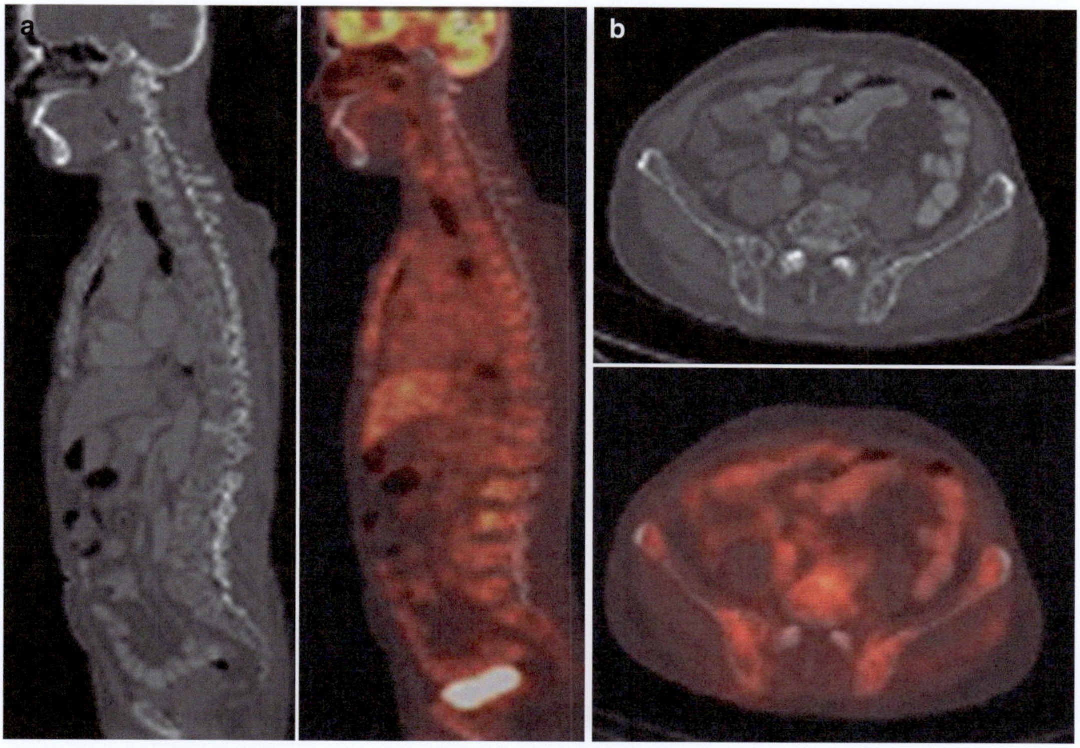

Fig. 16.8 CT and PET-CT fusion whole body image (**a**) as well as spot view of pelvis (**b**) showing no localized abnormality

levels of "M" band protein at diagnosis, FDG PET-CT is particularly useful for monitoring disease status and response to therapy.

FDG PET-CT is especially useful in patients with hematologic malignancies for detection and localization of occult infection which is common during the course of treatment. Detection of site of infection is often difficult in these patients and FDG PET-CT is very useful in these settings. After completion of treatment, negative PET study augurs well for patient undergoing stem cell transplantation (Fig. 16.8). However, if PET-CT detects the disease during treatment /follow-up usually some intervention is warranted.

Metastatic Bone Disease

Bone metastases are the most common malignant bone tumor. Skeletal involvement occurs in 30–70 % of all cancer patients, with breast cancer being the leading cause for bone metastases in women and prostate cancer in men followed by lung cancer. Bone involvement by cancer occurs most commonly by hematogenous spread. Over 90 % of bone metastases are found in the distribution of the red active marrow, which is located in the axial skeleton in adults. The bone undergoes constant remodeling and based upon balance between osteoblastic and osteolytic processes, there may be lytic, sclerotic, or mixed type of metastasis. Bone metastases of bladder, kidney, and thyroid cancer and lesions of multiple myeloma are invariably lytic. Blastic lesions are frequently seen in prostate and breast cancer, occasionally in lung, stomach, pancreas, and cervix carcinomas, and infrequently in colorectal cancer.

Breast Cancer

The skeleton is the most common site of distant metastases in breast cancer. Bone is the first site of metastasis in 26–50 % of patients with

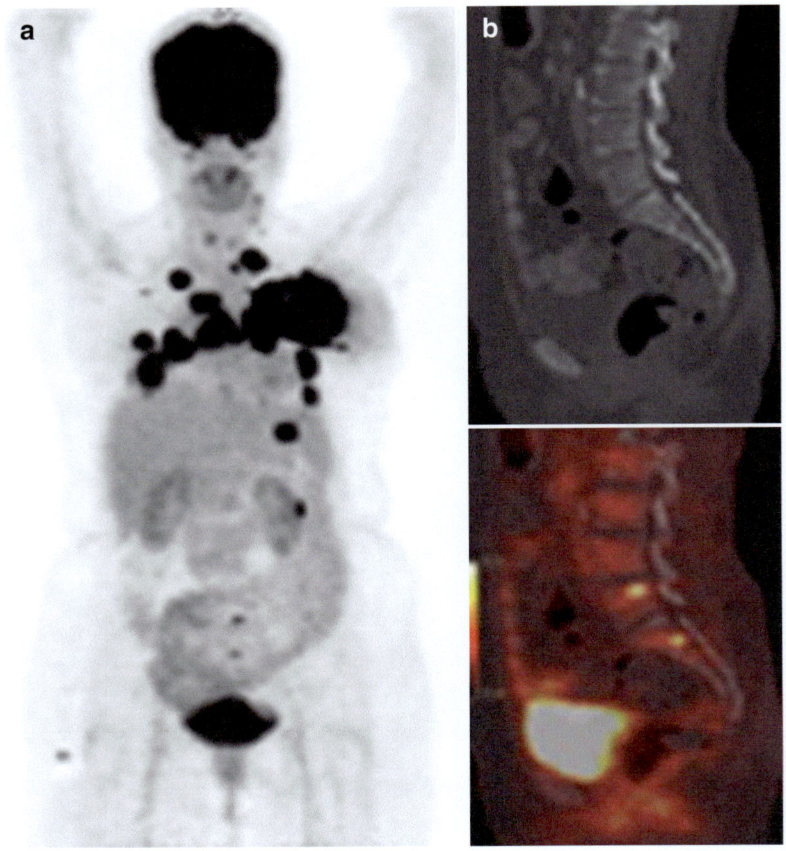

Fig. 16.9 Multiple metastases in FDG whole body scan (**a**) along with small metastases in lower end of the spine (**b**) which makes a whole body bone scan redundant

metastatic breast cancer, and it may develop during the course of the disease in 30–85 % of these patients. The risk for bone involvement, as assessed by bone scan, depends on the stage of the disease and varies between 0.8 and 2.6 % in early stages (I and II) and between 16.8 and 40.5 % in advanced stages (III and IV). Therefore, routine screening of asymptomatic patients with early-stage breast cancer is not recommended anymore. Bone scan, which is the most commonly used modality for detection of bone metastases, is indicated in patients with advanced disease or when bone involvement is clinically suspected. F-18 PET bone scan performs better than planar bone scan and should be performed specially in high-risk group patients.

FDG PET is useful for detection of not only skeletal metastases but also soft tissue metastases. As discussed earlier, FDG PET is more sensitive than bone scan for detection of lytic metastases and as PET-CT imaging has become a norm these days, high sensitivity of CT for detection of osteoblastic metastases adds to overall sensitivity of FDG PET-CT which may obviate need for separate bone scan (Fig. 16.9). However, some types of breast cancer, primarily well-differentiated histologic subtypes including some of the tubular and lobular ones, are less FDG avid and so are their metastases.

Prostate Cancer

Staging of newly diagnosed prostate cancer has significant bearing on treatment. Patients with low-risk prostate cancer are unlikely to have metastatic disease on bone scan. In a meta-analysis of 23 articles on the detection rate of bone metastases by bone scan in patients with newly diagnosed prostate cancer, the rates were 2.3 % of patients with prostate-specific antigen (PSA) levels of 10 ng/mL, 5.3 % of patients with levels

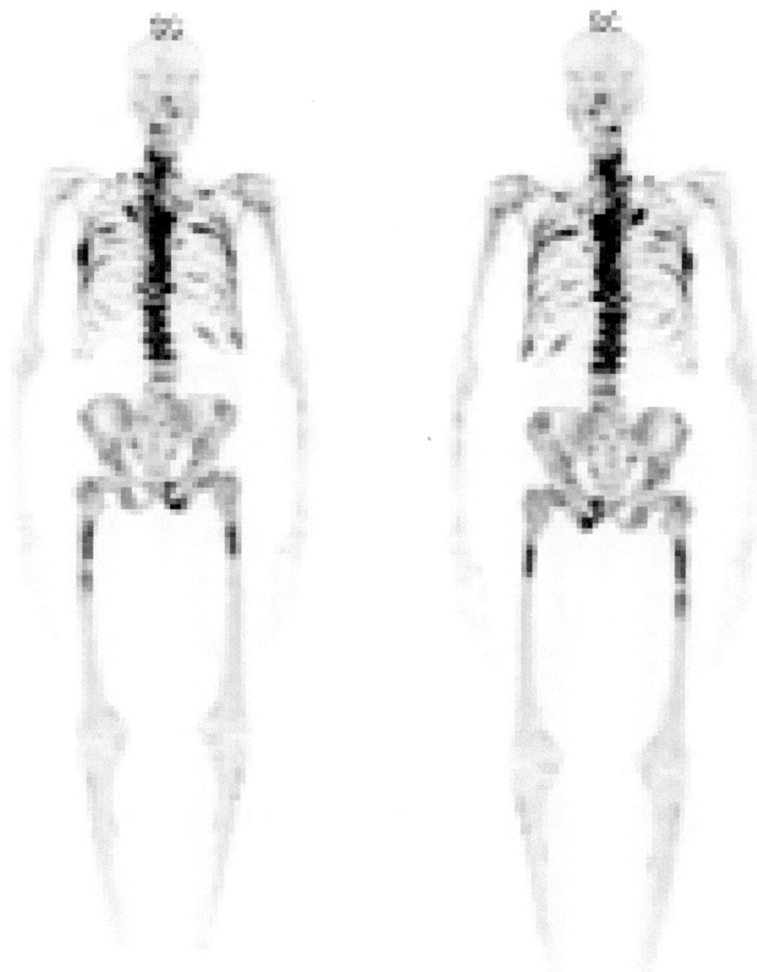

Fig. 16.10 Whole body F-18 PET scan showing extensive metastatic involvement of the entire skeletal system in a case of prostate cancer

between 10.1 and 19.9 ng/mL, and 16.2 % of patients with PSA levels between 20 and 49.9 ng/mL. Bone metastases were detected in 6.4 % of patients with organ-confined cancer and in 49.5 % of patients with locally advanced disease. Detection rates were 5.6 % in patients with a Gleason score of 7 and 29.9 % in patients with a Gleason score of 8. The likelihood of positive bone scan in asymptomatic patients with serum PSA levels of 20 ng % was found to be approximately 0.8 %. Based on these data, patients are referred for bone scan mainly if they are considered to be at high risk for bone metastases, with high PSA levels, a locally advanced disease, or a high Gleason score (Fig. 16.10). F-18 bone scan being more sensitive may replace the planar bone scan over the years. However, as of today its use is limited in cases where planar bone scan remains negative in a patient with rising serum PSA levels.

The role of FDG PET seemed to be limited in prostatic malignancy as both the soft-tissue sites of disease and bone metastases are usually FDG negative or show only a low-intensity uptake. However, FDG PET-CT may have complementary role to bone scan as it detects soft tissue metastases in selected patients. Recently, there are reports suggesting role of FDG PET in advanced or aggressive or nonsecretory prostatic primary in staging and therapy monitoring. Other PET tracers suggested for assessment of prostate cancer include ^{11}C- or ^{18}F-labeled choline and acetate, ^{11}C-methionine, and ^{18}F-fluorodihydrotestosterone.

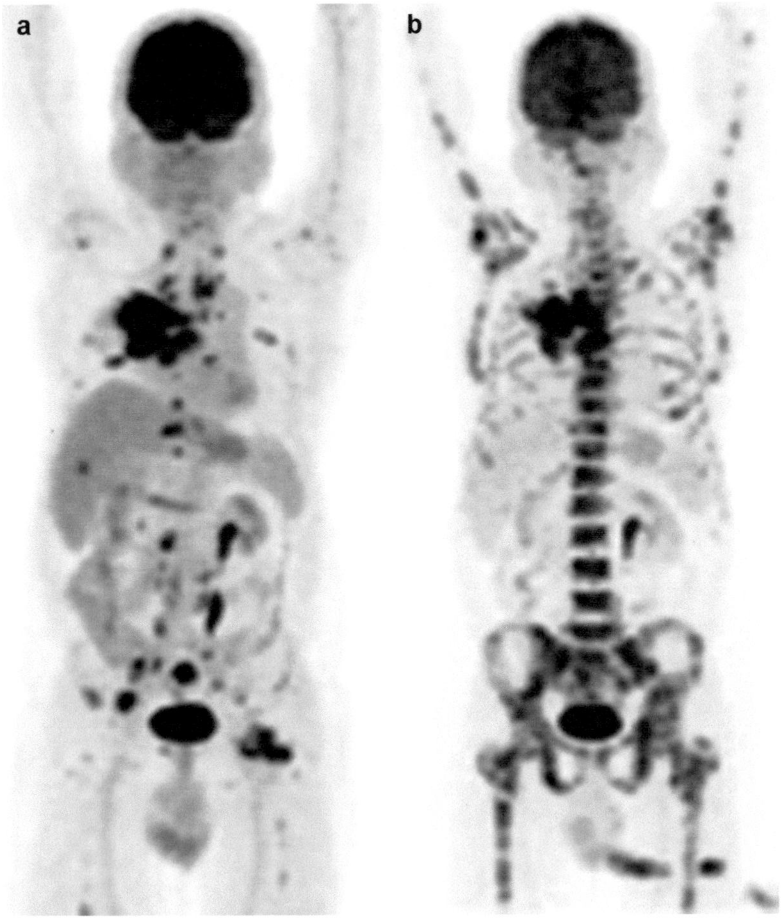

Fig. 16.11 FDG whole body scan showing both soft tissue (**a**) as well as bone involvement (**b**) making the need of a bone scan redundant

Lung Cancer

Bone metastases are diagnosed at initial presentation in 3.4–60 % of patients with NSCLC. Detection of skeletal metastasis changes patient management from curative surgery to palliative chemo-radiotherapy. Bone pain is usually considered an indicator of skeletal metastases, but up to 40 % of lung cancer patients with proven bone metastases are asymptomatic. FDG PET-CT recently reported to be of value in assessing the presence of soft tissue and bone spread in patients with NSCLC (Fig. 16.11), obviating the need to perform separate bone scan.

Concluding Remarks

PET-CT being the best possible "marriage" between functional and structural imaging modalities; there is progressive accumulation of data supporting its use in oncology and other fields of medicine. As far as skeletal system is concerned, PET is able to detect early marrow disease and CT is able to show detailed structure of bone with associated complications. Hence, combined PET-CT imaging hardly leaves any lacuna as far as imaging of skeletal malignancy is concerned. As PET-CT becomes widely available, the cost may further come down leading to optimal use of this technique for better patient management. Apart from F-18 and FDG, there are various new tracers which are being explored in oncology. These will further widen the scope of application of PET-CT in future. At present, management of most of the cancers remains inadequate without the use of FDG PET-CT for sure.

For Further Reading

1. Bartel TB, Haessler J, Brown TL, et al. F18-fluorodeoxyglucose positron emission tomography in the context of other imaging techniques and prognostic factors in multiple myeloma. Blood. 2009;114:2068–76.
2. Cheng G, Kwee TC, Basu S, Alavi A. Critical considerations on the combined use of ^{18}F-FDG and ^{18}F-fluoride for PET assessment of metastatic bone disease. Eur J Nucl Med Mol Imaging. 2013;40:1141–5.
3. Eubank WB, Mankoff DA. Current and future uses of positron emission tomography in breast cancer imaging. Semin Nucl Med. 2004;34:224–40.
4. Even-Sapir E. Imaging of malignant bone involvement by morphologic, scintigraphic, and hybrid modalities. J Nucl Med. 2005;46:1356–67.
5. Schoder H, Larson SM. Positron emission tomography for prostate, bladder, and renal cancer. Semin Nucl Med. 2004;34:274–92.
6. Watanabe H, Inoue T, Shinozaki T, et al. PET imaging of musculoskeletal tumours with fluorine-18 alpha-methyltyrosine: comparison with fluorine-18 fluorodeoxyglucose PET. Eur J Nucl Med. 2000;27:1509–17.

Role of PET in Diagnosis of Infection and Inflammation

17

Birendra Kishore Das

Inflammation and infection play a significant role in many disease processes. Development in molecular imaging in recent years provides new insight into the diagnosis and treatment evaluation of various inflammatory diseases and disease processes associated with acute or chronic infection.

Inflammation acts as the initial host defense against invasive pathogens and other inciting stimulus. It plays an important role in tissue repair and elimination of harmful pathogens. The inflammatory response is essential for host defense and inappropriate inflammatory reaction or delay in the resolution of inflammation will damage adjacent normal cells in the tissue. Microbial infection, most commonly caused by bacteria and viruses, is not the only cause of inflammation which can also be triggered by sterile stimulus involving physical, chemical, or metabolic noxiae such as burns, trauma, and dead cells. Similar to infection, the sterile inflammatory process also deals with the recruitment of neutrophils, macrophages, and the production of pro-inflammatory cytokines and chemokines. There are reports which suggest that various human diseases, including stroke, Alzheimer's disease, atherosclerosis, and many autoimmune diseases, are related to sterile inflammation.

B.K. Das, MD, ANM, FAR, FAMS
Director and Chief of Nuclear Medicine,
Utkal Institute of Medical Sciences,
Jaydev Vihar, Bhubaneswar, India
e-mail: bkishoredas@yahoo.com

Molecular imaging can visualize, characterize, and measure the biological processes at the molecular and cellular levels in humans and other organisms. Many imaging techniques are incorporated in the molecular imaging process, including magnetic resonance imaging (MRI), positron emission tomography (PET), single photon emission computed tomography (SPECT), optical imaging, ultrasound, etc. Each technique has its own unique applications, advantages, and limitations. Compared with other imaging modalities, PET has features of highest sensitivity and specificity. Therefore, PET has become one of the most frequently used molecular imaging techniques. Moreover, combination of PET with CT and MR provides additional anatomical details to the lesions, allowing very high sensitivity in molecular and anatomical imaging.

^{18}F-FDG (2-deoxy-2-^{18}F-fluoro-D-glucose) is the most extensively used PET imaging tracer and has been applied successfully in tumor detection, staging, and therapy evaluation, as well as in cardiovascular and neurological diseases (see Chaps. 1, 6, 7 and 10). In inflammatory diseases, ^{18}F-FDG PEt also has its value, particularly in atherosclerosis and some arthritis diseases. High glucose metabolism and consequent high FDG accumulation are not unique phenomena for only malignant cells. Benign processes including inflammatory disorders also show increased FDG uptake, which bring sometime false-positive results in tumor detection. As the key indicators and core participants in

B.K. Das (ed.), *Positron Emission Tomography: A Guide for Clinicians*,
DOI 10.1007/978-81-322-2098-5_17, © Springer India 2015

inflammatory foci, infiltrating inflammatory cells utilize glucose at a much higher level than peripheral non-inflammatory cells. The increased glucose metabolism of inflamed foci due to oxidative burst in the inflammatory cells become an important and most frequently used target in PET imaging of inflammation. CT or MRI is often combined with ^{18}F-FDG PET to increase the diagnostic accuracy.

Clinical Applications

Fever of Unknown Origin (FUO)

Labeled leukocyte imaging using ^{99m}Tc-HMPAO or ^{111}In-oxine has been the radionuclide procedure of choice for diagnosing most infections, due to the ability of radiolabeled granulocytes to migrate to the foci of infection. High sensitivity and specificity values have been reported for this technique in patients with FUO or a focal infection, while a negative study virtually excludes an infection. In cases of high suspicion of an infection, based on an increased leukocyte count, an elevated erythrocyte sedimentation rate and C-reactive protein level, this procedure remains the diagnostic technique of choice.

FDG is a radiopharmaceutical that accumulates in infections, malignancies, and inflammatory diseases. This nonspecificity is extremely valuable in FUO, which is caused, primarily, by these three conditions. The disadvantage of FDG's low specificity (lower than that of labeled WBCs) is that it cannot always discriminate between infection and neoplastic disease. Several recent studies, retrospective and prospective, have drawn attention to the added value of FDG-PET over conventional techniques. Most studies also stress the technique's high negative predictive value in the assessment of FUO.

In children with FUO or unexplained signs of inflammation without fever, the FDG-PET-CT technique was found to be a valuable diagnostic tool in these subjects in whom a nontraumatic method of depicting inflammation in the whole body is obviously particularly useful. In critically ill, mechanically ventilated patients suspected of having an infection or inflammatory process, FDG-PET-CT yielded overall accuracy of 91 %. A normal FDG-PET-CT scan ruled out an infection requiring prolonged antibiotic therapy or drainage. In patients with human immunodeficiency virus (HIV) and FUO, FDG-PET-CT emerged as a valuable tool and was helpful for diagnosis, especially when CT anatomical landmarks were added to PET findings (Fig. 17.1).

For the time being, the probability of infection remains the most useful criterion for choosing between FDG-PET-CT and WBC imaging. In the presence of low and medium probability, FDG PET appears to be more useful, whereas WBC imaging should be performed when the probability of infection is high. There are, however, increasing evidence to suggest that this technique is probably destined to become the preferred diagnostic procedure in the future, especially when a definite diagnosis cannot easily be achieved.

Infection of Bone (Osteomyelitis and Spondylodiscitis)

For acute osteomyelitis, WBC imaging and if necessary combined with bone marrow imaging or MoAb imaging, is very reliable, with overall accuracy of approximately 90 %. FDG-PET-CT, used in combination with conventional methods, may have limited value in the diagnosis of uncomplicated cases of acute osteomyelitis; conversely, it may play an important role in patients with chronic osteomyelitis, particularly those with previously documented osteomyelitis and suspected recurrence, or presenting with symptoms of osteomyelitis for more than 6 weeks. FDG-PET-CT can also be used to monitor response to antimicrobial treatment and to develop criteria for deciding when treatment can safely be stopped.

The performance of WBC imaging is poor in spondylitis and spondylodiscitis: Up to 50 % of all patients with spondylodiscitis show photopenic lesions due to encapsulation of the infection, and therefore relatively hampered migration of leukocytes; as a result, the specificity of the method

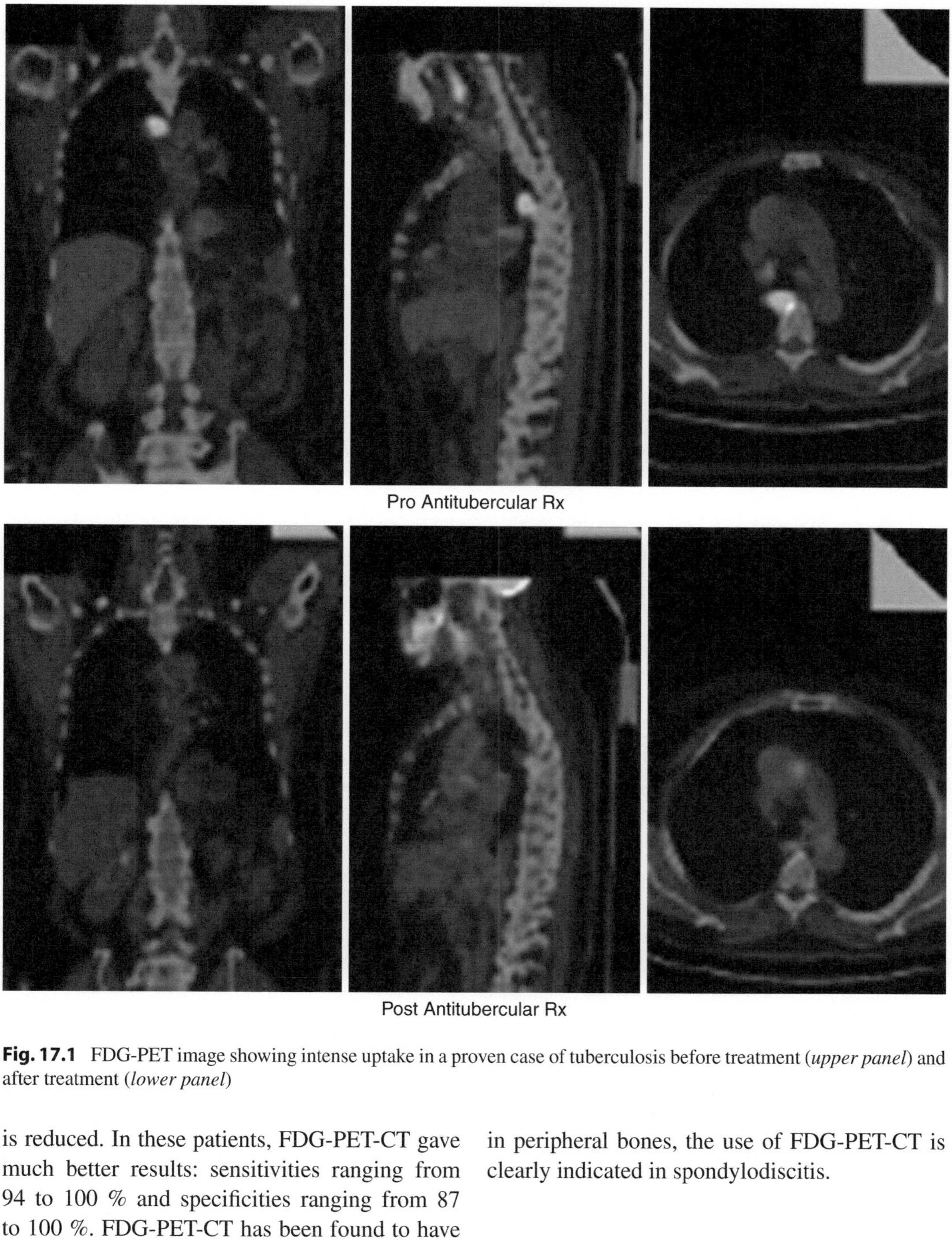

Fig. 17.1 FDG-PET image showing intense uptake in a proven case of tuberculosis before treatment (*upper panel*) and after treatment (*lower panel*)

is reduced. In these patients, FDG-PET-CT gave much better results: sensitivities ranging from 94 to 100 % and specificities ranging from 87 to 100 %. FDG-PET-CT has been found to have strong impact on the clinical management (initiation or prolongation of antibiotic therapy or recourse to surgical intervention) of patients with infectious spondylitis. Whereas WBC and MoAb imaging remain the gold standard imaging techniques in patients with suspected osteomyelitis in peripheral bones, the use of FDG-PET-CT is clearly indicated in spondylodiscitis.

Diabetic Foot Infection

The role of WBC imaging in the diagnosis of diabetic pedal osteomyelitis has been extensively investigated, with sensitivities ranging from 72 to 100 % and specificities ranging from 67 to 100 %.

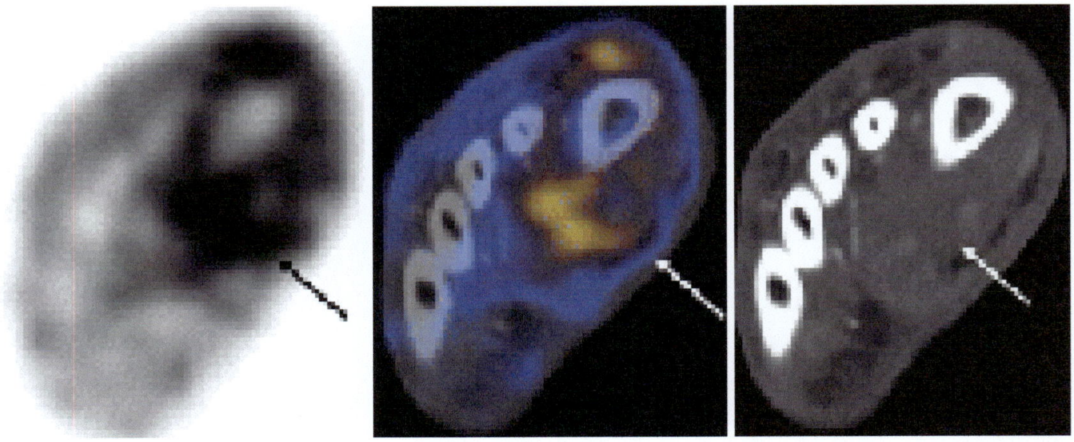

Fig. 17.2 FDG PET (*left*), PET-CT (*center*), and CT (*right*) slices (transaxial) show ^{18}F-FDG uptake at medial aspect of right forefoot, involving only soft tissues with sparing of metatarsal bones (*arrows*). Extensive soft-tissue infection involving muscles and planter fascia and no osteomyelitis were found at surgery

However, with poor spatial resolution and lack of bony landmarks makes the differentiation of soft tissue from bone infection difficult. Much better results are achieved using SPECT/CT. FDG-PET-CT was found to be highly sensitive in excluding osteomyelitis in the diabetic foot, and to complement MRI, particularly in cases with positive MRI findings (Fig.17.2). Conventional imaging like MRI or bone scanning also lacks specificity as a means of distinguishing osteomyelitis in the diabetic foot from Charcot's neuroarthropathy. Preliminary data suggest that FDG-PET-CT could play a role in assessing complicated and uncomplicated diabetic osteoarthropathy, being able to provide accurate assessment of patients with metal implants who may not be suitable candidates for MRI, and to correctly distinguish osteomyelitis from neuroarthropathy. However, FDG-PET-CT cannot yet replace WBC imaging, particularly when WBC is acquired using SPECT/CT modality and bone marrow imaging is added for Charcot's foot.

Hip and Knee Prosthesis Infection

WBC imaging combined with bone marrow imaging is currently the radionuclide imaging procedure of choice for diagnosing prosthetic joint infections. The accuracy of this technique has been found to be almost 90 %. MoAb imaging also shows very high accuracy, thus limiting the need for alternative modalities. Theoretically, FDG-PET-CT has the potential to detect infection in hip and, to a lesser extent, knee prostheses. It is not affected by artifacts due to metal implants and it provides images with higher resolution than those produced using conventional nuclear medicine techniques (Fig.17.3). However, non-infectious reactions around the neck of the prosthesis are common months and even years after surgery, and these may influence the diagnosis. Increased FDG uptake around the neck and/or head should not be interpreted as a finding suggestive of infection. Studies comparing WBC imaging with FDG-PET scanning in prosthetic joint infections show better results with WBC imaging, with more sensitivity and more specificity, thus making WBC/MoAb imaging the gold standard technique for evaluating prosthetic infections. The question of whether FDG-PET-CT can provide satisfactory diagnostic accuracy in these patients is still debated. Well-designed prospective studies may in future provide established criteria for infection and aseptic loosening, leading to a possible role for FDG-PET-CT in the evaluation of prosthetic joint infections.

Vascular Graft Infections

Vascular graft infections are uncommon but severe complications can occur long after surgery.

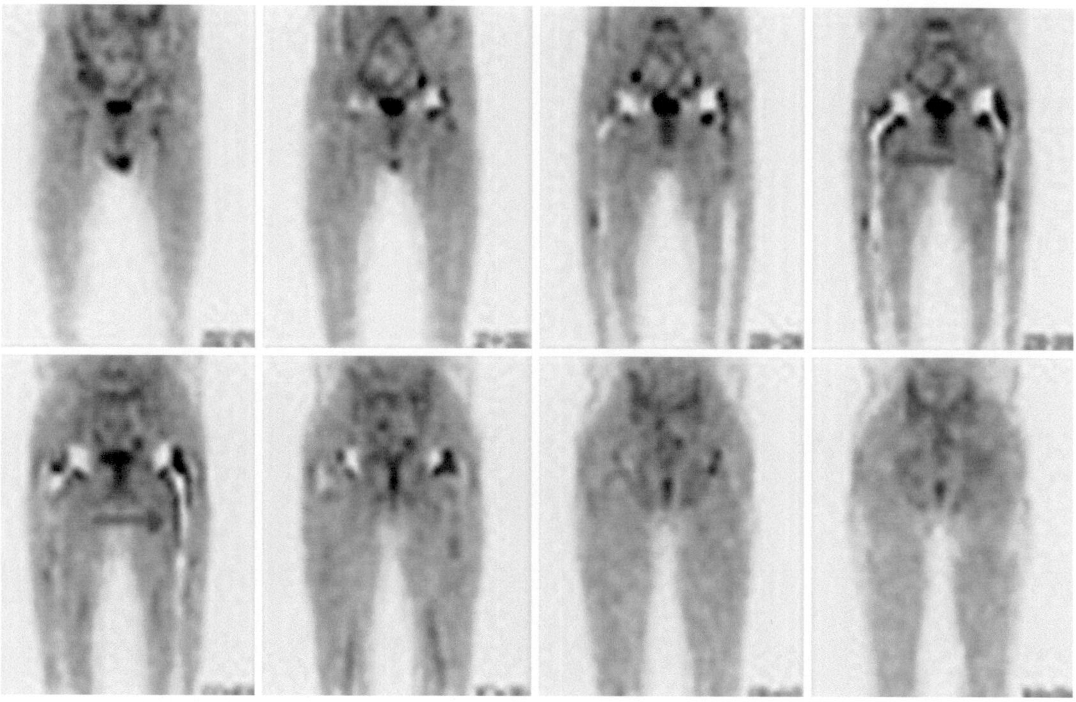

Fig. 17.3 FDG-PET showing infection of the hip prosthesis (*arrows*)

Sensitivities of WBC imaging for diagnosing graft infections range from 53 to 100 %, and specificities from 50 to 100 %. False-positive results are often reported and have been associated with perigraft haematomas, thrombosed grafts, bleeding, and recent surgery. The role of FDG-PET-CT in the assessment of vascular graft infections suggests that this modality is a reliable tool in this clinical setting. Indeed it is not uncommon for FDG-PET-CT to detect infection of vascular grafts even when the angio-CT results are negative. Comparison of the two modalities shows FDG-PET to be superior to CT scanning alone. Combining FDG-PET with CT increases the test specificity and therefore the diagnostic accuracy because the precise anatomical localization of increased FDG uptake, thanks to PET-CT, allows accurate differentiation between graft and adjacent soft tissue infection, leading to more accurate diagnosis and thus an optimal therapeutic strategy. This added value of PET-CT over PEt alone has been described by several investigators. In general, it is agreed that FDG-PET-CT is the modality of choice in vascular graft prosthesis.

Vasculitis

FDG-PET-CT in the evaluation of large-vessel vasculitis has been established to have a sensitivity value ranging from 77 to 92 % and specificity ranging from 89 to 100 %. FDG-PET-CT has proven utility in the initial diagnosis of patients suspected of having vasculitis, particularly those who present with non-specific symptoms, in the identification of areas of increased FDG uptake requiring biopsy, and in the evaluation of the extent of disease (Fig.17.4).

Cardiovascular Inflammation

As an inflammatory disease, the onset, progression, and destabilization of atherosclerosis involves multi-participants within the immune response, including activation of endothelial cells, infiltration of various cells, release of inflammatory cytokines, and macrophage apoptosis. Due to the high morbidity and mortality rates of atherosclerosis, early detection and full

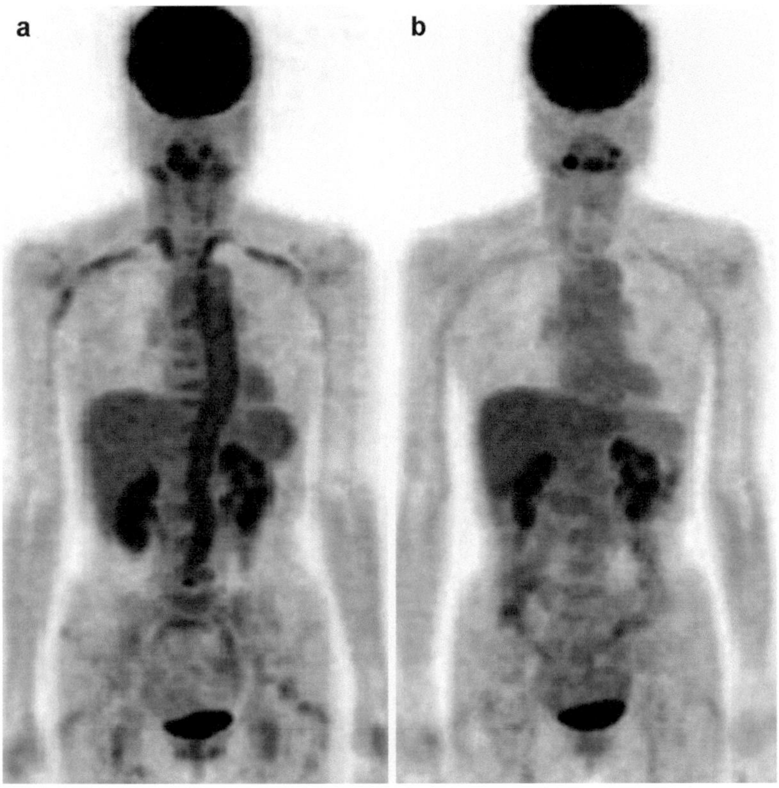

Fig. 17.4 FDG- PET image showing intense uptake in the aorta and the subclavian arteries caused by vasculitis (**a**). Follow-up PET image (**b**) after treatment showing complete disappearance abnormalities seen in the initial image

characterization of atherosclerosis is of great importance. So far, ^{18}F-FDG is the most extensively used probe for atherosclerosis imaging. Many preclinical and clinical studies have established the correlation not only between local FDG uptake and plaque macrophage density but also between high metabolic activities of macrophages within plaques and cardiovascular risk factors. However, the partial volume effect, high physiology uptake of ^{18}F-FDG in the myocardium or brain, and motion artifacts from cardiac movement all make visualization of small atherosclerotic plaques in these areas rather difficult. In fact, besides high glucose metabolism, many inflammation biomarkers have been evaluated for atherosclerosis PET imaging, including (described later) choline metabolism, TSPO, SSTR, VAP-1, MMPs, integrin receptors, and VCAM-1. Some of the PET probes, such as radio-labeled PK11195 (binding to TPSO), choline (targeting to the phosphatidylcholine catabolism of macrophages and monocytes), TATE/TOC (binding to SSTR), are superior to conventional ^{18}F-FDG due to their low myocardium biodistribution. Consequently, the images achieved high target-to-background ratio, facilitating the analysis of small coronary plaques. In addition, PET imaging of MMPs could assess the plaque-promoting activity of macrophages rather than their density in vulnerable plaques, and integrin receptor-targeted imaging could detect CD68-positive macrophages in the vulnerable plaques. PET imaging of these biomarkers, together with other conventional angiography, opens up the opportunity for better diagnosis and prognosis of atherosclerosis.

Neuroinflammation

Recently, accumulating evidences have revealed that many chronic neuroinflammatory diseases are caused by activated microglia in the CNS. As the resident immune cells in the CNS, microglial cells are activated in the acute neuroinflammation phase and protect brain tissue from further injury through migration, proliferation, and production of neurotoxic factors. However, in chronic neuroinflammation, microglia activation causes long-term cerebral damage by inducing autoimmune reaction. The activation of microglia is observed in various CNS diseases such as stroke, multiple sclerosis, Alzheimer's disease, and Parkinson's disease. There are several known neuroinflammation-related targets including TSPO, CB_2R, and COX-2. Among them, TSPO is the most popular target for PET imaging which has already undergone clinical application, while CB_2R and COX-2 are still in the preliminary stage as imaging targets.

Although some limitations do exist, TPSO-targeted PET imaging of neuro-inflammatory disease has provided some helpful information in disease diagnosis and prognosis. For example, in AD patients, TPSO PET enabled the discovery of the relationship between Aβ accumulation and microglia activation during disease process and can detect an age-related increase in microglia activation in normal human brains and in AD progression. However, based on current research, no definite conclusion can be drawn between the results of amyloid plaque imaging using ^{11}C-PIB and inflammation imaging targeting TPSO. Microglia activation was found to be a potential driving force in the development of Parkinson's disease (PD) with dementia and could be detected via TPSO PET at the early phase in PD patients. In stroke, TSPO PET imaging was able to find the temporal dynamics of microglia activation in patients, which was correlated with clinical outcome. In traumatic brain injury (TBI), the imaging of microglia via TSPO was found to be present up to 17 years after TBI, indicating the possible benefit of long-term interventions for post-TBI patients. However, some discrepancies exist among different studies, and this might be due to the lack of standardized analysis of imaging results and certain limitations of radiotracer for PET neuroinflammation imaging, such as low binding affinity and low target-to-background ratio. Therefore, the development of new tracers with better imaging properties and the improvement in quantitative data analysis should be of great importance for PET-guided neuroinflammation imaging in the future.

Tumor-Related Inflammation

Inflammation contributes to a tumor's immune escape phenomenon, creating a proper environment for neoplastic onset and continued growth. In fact, inflammatory cells and mediators are present in the microenvironment of virtually all tumors that are not epidemiologically related to inflammation. Recently, tumor-associated macrophages (TAMs) or tumor-infiltrating macrophages (TIMs) have been intensively investigated as a target for imaging and therapy. TAMs enhance tumor cell migration and invasion through their secretion of chemotactic and chemokinetic factors. Depletion of TAMs improved the effect of chemotherapy in some cancer models. Therefore, TAMs targeted imaging would have great value providing guidance for macrophage targeted cancer therapy and patient stratification for personalized treatment. Various molecular imaging techniques have been applied to study TAMs, including MRI, optical imaging, PET, SPEC, and hybrid molecular imaging modality, in which most of the imaging agents are nanomaterial based. Because FDG could also accumulate in nonneoplastic cells that infiltrate neoplasms, without histological validation, it remains unclear what percentage of FDG accumulation is caused by peritumoral and intratumoral inflammation. Hence, it is a consensus in clinical setting that cancer therapy evaluation using FDG PET should be carefully conducted, especially when effective treatment can lead to massive inflammation. Consequently, many studies focused on developing more tumor cell specific PET tracers beyond FDG. Some tumor proliferation markers such as lipid precursors, amino acids, nucleosides, and receptor ligands have been tested for this purpose. For example, ^{11}C-choline was developed to evaluate intracellular choline kinase activity, ^{11}C-methionine (MET) to image amino acid transporter, and ^{18}F-fluorothymidine (FLT) to determine thymidine kinase 1 activity. In some preclinical study, it has been demonstrated that ^{18}F-FET and ^{18}F-FLT selectively localized in tumor tissues but not inflammation. In some clinical studies also, it has been shown that ^{18}F-FLT is significantly better than ^{18}F-FDG as a measure of tumor proliferation and more specific than ^{18}F-FDG PET for cancer staging. However, tumor uptake of ^{18}F-FLT is much less than ^{18}F-FDG, resulting in a significantly lower sensitivity for ^{18}F-FLT PET than for ^{18}F-FDG PET. Several inflammation biomarkers may be promising in differentiating tumor from inflammation including VAP-1 and integrins. It has been shown that VAP-1 targeted peptidic tracer, ^{68}Ga-DOTAVAP-P1, showed accumulation in inflammation foci but not as much in tumors, making it a potential inflammation-targeting tracer ^{18}F-FPPRGD2, an integrin receptor targeting probe, was found to be superior to ^{18}F-FDG in monitoring tumor response to Abraxane treatment, possibly due to less uptake in TAM. It would indeed be very challenging to develop an imaging probe which can separate tumor and inflammation completely since inflammation is an inherent tumor microenvironment.

AIDS and Tuberculosis

FDG-PET plays a major role in the assessment and management of HIV-1-infected patients. FDG-PET data have shown that HIV-1 infection progresses by distinct anatomical steps, with involvement of the upper torso preceding involvement of the lower part of the body, and that the degree of FDG uptake is related to viral load. FDG uptake by the lymph nodes of HIV-positive patients was found to be inversely related to CD4 count, thereby supporting the theory of CD4 cell depletion through forced lymph node homing. In the clinical setting, FDG-PET has been shown to allow the differentiation of AIDS-related opportunistic infections and malignancies and to allow monitoring of the side effects of highly active antiretroviral treatment. Comparison of FDG uptake in patients who had received antiretroviral therapy (ART) and patients who were ART-naïve revealed different patterns of FDG uptake: All the ART-treated patients with either suppressed or high viraemia showed a normal pattern, while the ART-naïve patients with high viraemia displayed multiple foci of increased glucose metabolism in the lymph nodes. Together, this finding and the finding of a correlation between the well-established markers of progression to AIDS and positive FDG-PET in ART-naïve patients seem to confer prognostic value on FDG uptake.

Data on FDG-PET imaging in patients suffering from tuberculosis (TB) are very limited. Pulmonary TB commonly causes an increase in FDG uptake, whereas uptake is low in tuberculous pleural effusion. Sites of extra-pulmonary TB can be detected with FDG-PET, including disease involving the central nervous system and joint and bone TB.

Therapy Evaluation

In general, it is difficult in patients with high FDG uptake to differentiate between a malignancy, HIV infection, and TB. Several authors have shown that FDG uptake continues to increase over time in malignant lesions, whereas in inflammatory lesions uptake decreases or remains stable. However, studies conducted to assess the potential impact of double-phase FDG-PET versus routine staging in patients suffering from TB confirmed that it is extremely difficult to distinguish TB from malignant involvement. This was also borne out by the findings of a study in HIV-infected children, in whom FDG-PET scanning proved unable to discriminate reliably between malignant and inflammatory pathology.

In spite of many drawbacks, FDG-PET-CT could play a pivotal role in therapy assessment. The evolution of FDG uptake reflects the efficacy of the medical treatment, and its careful assessment can lead to better modulation of the drug dosage or prompt a radical modification of the therapeutic strategy. It has been shown to be a valuable imaging tool for assessing treatment efficacy in systemic sarcoidosis, large-vessel arteritis, tuberculosis, and aspergillosis. In vasculitis, FDG-PET-CT also has proven utility in assessing response to therapy. From this perspective, it is possible to conceive of a role for

FDG-PET-CT in monitoring therapy response and early relapses of inflammatory bowel diseases, even though only limited data are currently available in this regard. As regards the monitoring of antibiotic therapy in osteomyelitis, there is, as yet, no published evidence to support this indication.

However, differentiation between malignancies, AIDS, and TB is extremely important. In patients with HIV and TB, early identification and diagnosis are the keys to effective control of the disease. Correct identification of TB is crucially important in order to start anti-TB treatment and delay antiretroviral treatment for HIV. The presentation of TB in the HIV-infected patient is different from that observed in the HIV-negative patient: Apical predominance is less pronounced, while consolidation, cavitations, and hematogenic disseminations are less prevalent. In this context, FDG-PET-CT may aid in the diagnosis.

Aggressive lymphoid proliferations should be differentiated from generalized lymphadenopathy, which does not have a negative outcome, and here again FDG-PET may play a central role. FDG-PET scanning could also be indicated for monitoring the efficacy of TB treatment, while FDG activity may also be a useful tool for evaluating and excluding sites of active disease in the context of targeted screening for latent TB infection before immunosuppressive treatment.

Development of New Pet Agents

However, ^{18}F-FDG PET imaging of inflammation often gives false-positive results, especially in patients with cancer. Moreover, the high tracer accumulation in the heart and brain makes it difficult to detect inflammatory foci near those organs or tissues. Consequently, new imaging tracers and targets for more specific inflammation detection and therapy evaluation are under intensive investigation. PET imaging with these new tracers has greatly improved our understanding of the mechanism of inflammation and increased the diagnostic specificity and accuracy. Some of the newly developed agents and their applications are described below :

Choline Metabolism

Choline is an important precursor of phosphatidylcholine and sphingomyelin, two classes of phospholipids that are abundant in cell membranes. The phosphatidylcholine catabolism by many nucleated cells, mostly proliferative cells, serves as an imaging target both in cancers and in some inflammatory diseases. PET using radiolabeled choline has been used to image prostate cancer . By targeting the macrophages and monocytes in inflammatory diseases, choline is also used to image atherosclerosis and to evaluate necrosis after brain tumor radiation therapy. It has been documented that ^{18}F-choline had greater sensitivity in detecting atherosclerotic plaques than FDG (84 % versus 64 %). A major advantage offered by ^{18}F-choline imaging over $^{18F\text{-}FDG}$ is the lack of ^{18}F-choline uptake in the myocardium . Thus, choline may be superior to FDG in detecting coronary plaques.

Translocator Protein (TSPO)

Formerly known as peripheral benzodiazepine receptor (PBR), the 18 kDa translocator protein is located on the outer mitochondrial membrane and can bind with cholesterol and various classes of drug ligands. TSPO is ubiquitously expressed in peripheral tissues but is only minimally expressed in the healthy human brain. Previous studies found high TSPO expression in macrophages, neutrophils, lymphocytes, activated microglia, and astrocytes. Microglia have been found to contribute to neuroinflammation in many types of central nervous system (CNS) disorders, such as stroke, multiple sclerosis (MS), Alzheimer's disease (AD), Parkinson's disease (PD), amyotrophic lateral sclerosis (ALS), and epilepsy. Therefore, TPSO expressed on microglial cells in CNS emerges as a promising target for PET imaging of neuroinflammation. The most studied PET tracers binding to TPSO are ^{11}C or ^{18}F-labeled isoquinoline carboxamide PK11195 (1-(2-chlorophenyl)-*N*-methyl-*N*-(1-methylpropyl)-3-isoquinoline carboxamide) and more recently ^{11}C-PBR28 (*N*-(2-[^{11}C] methoxybenzyl)-*N*-(4-phenoxypyridin-3-yl)acetamide). Syntheses of these tracers are now mostly automated and are efficient, which guarantees the future application in the clinic.

Somatostatin Receptor

Somatostatin receptor (SSTR) has been investigated as a target for neuroendocrine tumor imaging. SPECT imaging of SSTR expression in neuroendocrine tumors has been well established for lesion detection and therapeutic monitoring. Since a high level of SSTR expression was found on activated lymphocytes and macrophages, this receptor has the potential to be used as a new target for inflammation imaging. Compared with tumor imaging, some new studies have reported using PET tracers to target SSTR in inflammatory disorders or diseases with mild/intense inflammatory infiltration, including atherosclerotic inflammation, inflamed pulmonary fibrosis, carcinoids, and inflammatory myofibroblastic tumor.

TATE (Tyr^3-octreotate) and TOC (Tyr^3-octreotide) are analogues of octreotide that bind to the somatostatin type 2 receptor (SSTR-2). 1,4,7,10-Tetraazacyclododecane-N,N′,N″,N‴-tetraacetic acid (DOTA) conjugation of these peptides allows for stable chelation to a variety of radio metals such as ^{111}In, ^{177}Lu, ^{90}Y, ^{68}Ga, and ^{64}Cu. In atherosclerosis imaging, clear plaque uptake of ^{68}Ga-DOTA-TATE or $^{68Ga\text{-}DOTA\text{-}TOC}$ in carotid arteries has been found, and the uptake has strong association with known risk factors of cardiovascular disease. Due to the much lower uptakes in myocardium, these tracers may provide clearer and more consistent detection of macrophage accumulation than FDG in coronary arteries plaques.

Prostaglandin H

Cyclooxygenase (COX) is an enzyme responsible for the conversion of arachidonic acid into prostaglandins. COX is the target of nonsteroidal anti-inflammatory drugs (NSAIDs). In addition, COX is an integral membrane glycoprotein, which can be induced by acute and chronic inflammatory stimulations. Thus far, three COX subtypes (COX-1, 2, and 3) have been identified. Among them, the inducible isoform COX-2 plays a pivotal role in cancer, cardiac/cerebral ischemia, Alzheimer's/Parkinson's disease, and response to inflammatory stimuli, especially neuroinflammation. Celecoxib (4-(5-*p*-tolyl-3-trifluoromethylpyrazol-1-yl)benzenesulfonamide) is broadly used as a selective COX-2 inhibitor to treat inflammatory diseases. Imaging tracers have also been developed using celecoxib and some other COX inhibitors by radiolabeling them with either ^{18}F or ^{11}C. They have been used to image neuroinflammations, tumors, or experimental skin inflammation. However, most of the tracers showed unsatisfactory *in vivo* properties due to either nonspecific bindings or low sensitivity in inflammatory foci.

Interleukin (IL)-2

Interleukin (IL)-2 is a small, single-chain glycoprotein (15.5 kDa) of 133 amino acids synthesized and secreted by activated T-lymphocytes, especially $CD4^+$ and $CD8^+$ Th1 lymphocytes. T-lymphocyte activation is seen in many types of inflammatory diseases, such as inflammatory degenerative diseases, graft rejection, tumor inflammation, organ-specific autoimmune diseases, and adipose inflammatory insulin resistance. IL-2 binds with high affinity to the cell membrane IL-2 receptor, which is mainly expressed on the cell surface of activated T-lymphocytes. PET imaging of activated T-lymphocytes by radiolabeled IL-2, therefore, provides an *in vivo*, dynamic approach in studying the immune-cell infiltration in these inflammatory diseases. Previously, ^{123}I and ^{99m}Tc labeled IL-2 have been used in many chronic inflammatory diseases, such as autoimmune diseases, celiac disease, and vulnerable atherosclerotic plaques using SPECT imaging. However, routine application of this technique was limited because the labeling procedures are complex and the spatial resolution of SPECT is not high enough.

Recently conducted several pilot studies, however, suggest that ^{18}F-FB-IL-2 is stable, biologically active, and allows for *in vivo* detection of activated T- lymphocytes.

Tumor Necrosis Factor (TNF-á)

Tumor necrosis factor-á (TNF-á) is a cytokine that can contribute to cell apoptosis and organ dysfunction. In the early phase of inflammation, TNF-á increases the transport of white blood cells to the inflammation sites. In the late phase, TNF-á level is lowered and can cause the apoptosis of inflammatory cells to terminate further unnecessary inflammation. Many studies show that TNF-á is important in acute immune response to infection, injury, autoimmune, and chronic inflammatory disorders such as rheumatoid arthritis and psoriasis. Some preliminary studies using a PET tracer ^{64}Cu-DOTA-etanercept, to image acute inflammatory process induced by 12-O-tetradecanoylphorbol-13-acetate (tetradecanoyl phorbol acetate, TPA) showed high ^{64}Cu-DOTA-etanercept uptake in the inflamed site only during the early acute inflammatory phase but not the chronic inflammation phase, indicating that TNF-á contributes to the onset of acute inflammation. This imaging trend was confirmed by *ex vivo* enzyme-linked immunosorbent assay (ELISA) of TNF-á levels in the inflamed ears. Recently synthesized ^{11}C-labeled tricyclic Nec-3 necroptosis inhibitor 3,3a,4,5-tetrahydro-2H-benz[*g*]indazoles appears to be a potential PET tracer for imaging TNF-á. However, many more studies are needed to confirm the observations.

Integrin Receptor

Integrin $Þ_vß_3$, a cell adhesion molecule, is over expressed in various cancer cells, endothelial cells of neo vessels, and also in some inflammatory cells such as macrophages. The study of integrin $á_vâ_3$ in cancers and tumor related angiogenesis has been extensively investigated in the past. RGD peptides containing the three amino acid sequence Arg-Gly-Asp, are ávâ3 specific ligands. Radiolabeled RGD peptides have been successfully tested in the clinical setting, complementing conventional FDG imaging. Some chronic inflammatory conditions with inflammatory angiogenesis, such as inflammatory bowel disease and rheumatoid arthritis, also show the participation of integrin ávâ3 in the inflammatory neo vessels in disease etiology and progression. Therefore, integrin ávâ3 emerges as a target for inflammation therapy as well as molecular imaging.

Vascular Adhesion Protein-1 (VAP-1)

Vascular adhesion protein 1 (VAP-1) is an endothelial adhesion protein stored in intracellular granules within endothelial cells. The expression of VAP-1 is quite low on the endothelial surface of normal tissues. Upon stimulation, VAP-1 is translocated onto the luminal surface of endothelial cells at sites of inflammation, causing the migration of leukocytes, especially $CD8^+$ T- lymphocytes, from the blood into the non-lymphoid inflammatory foci. VAP-1 is, therefore, a promising target for both anti-inflammation therapy and molecular imaging of inflammation. A number of studies using radio labeled synthetic peptides have been attempted to image VAP-1 expression. These ligands were either designed by molecular modeling based on the crystal structure of human VAP-1 or selected from phage display libraries and were labeled with ^{68}Ga to form ^{68}Ga-DOTA-Siglec-9, ^{68}Ga-DOTAVAP-P, ^{68}Ga-DOTAVAP-PEG-P1, or ^{68}Ga-DOTAVAP-PEG-P2. These VAP-1 targeted PET tracers have been tested in sterile/infectious inflammatory and tumor-bearing animal models. Comparing ^{68}Ga-DOTAVAP-P1 with ^{18}F-FDG showed intensive inflammatory foci uptake, concordant with the high VAP-1 expression examined by *ex vivo* studies. However, ^{68}Ga-DOTAVAP-P1 had very low uptake in BxPC3 tumors, suggesting the ability of this tracer to differentiate inflammation from tumor. In contrast, FDG showed high uptake in both inflammation foci and tumors, unable to discriminate one from the other. ^{68}Ga-DOTAVAP-P1 has also been used to image osteomyelitic bones and differentiate osteomyelitic bones from inflammation in healing bones. Using rat models ^{68}Ga-DOTAVAP-P1 PET studies indicated that the uptake of this tracer remained high in osteomyelitis whereas significant decrease was observed in healing cortical bone defects (representing the sterile inflammation process of bone healing). Future studies will reveal the potential of ^{68}Ga-DOTAVAP-P1 to differentiate bacterial infection from nonbacterial inflammation.

Concluding Remarks

FDG-PET-CT has emerged as a rapidly evolving diagnostic tool in infectious diseases. Its clinical impact on diagnosis, staging, and evaluation of therapy, already high, is destined to increase in the future. However, FDG-PET-CT is not always the best radionuclide imaging technique to choose in the presence of infectious diseases. The value of WBC/MoAb scanning in different infectious diseases has been well established and should not be overlooked. Larger prospective studies evaluating the role of FDG-PET-CT in the clinical management of patients with infectious diseases are needed, and in many cases it remains to be seen whether this modality will be able to record the high sensitivity and specificity values already shown by WBC/MoAb imaging. In the near future, the ongoing efforts to find better labeling agents may well result in the identification of the ideal PET-CT radiopharmaceutical for imaging infection and inflammation.

For Further Reading

1. Basu S, Zhuang H, Torigian DA, Rosenbaum J, Chen W, Alavi A. Functional imaging of inflammatory diseases using nuclear medicine techniques. Semin Nucl Med. 2009;39(2):124–45.
2. Gotthardt M, Bleeker-Rovers CP, Boerman OC, Oyen WJG. Imaging of inflammation by PET, conventional scintigraphy, and other imaging techniques. J Nucl Med. 2010;51(12):1937–49.
3. Jamar F, Buscombe J, Chiti A, et al. EANM/SNMMI guideline for 18F-FDG use in inflammation and infection. J Nucl Med. 2013;54:647–58.
4. Leone S, Borr'e S, Monforte AD, et al. Consensus document on controversial issues in the diagnosis and treatment of prosthetic joint infections. Int J Infect Dis. 2010;14 supplement 4:S67–77.
5. Wang G-L, Zhao K, Liu Z-F, Dong M-J, Yang S-Y. A meta-analysis of fluorodeoxyglucose-positron emission tomography versus scintigraphy in the evaluation of suspectedosteomyelitis. Nucl Med Commun. 2011;32(12):1134–42.

Application of PET (PET-CT) in Radiation Therapy Planning

18

Birendra Kishore Das

Introduction

The purpose of radiation therapy planning is to maximize dose delivery to the target while simultaneously decreasing radiation dose to the surrounding normal tissues. In the era of image-guided radiation therapy (IGRT), the greatest challenge is target delineation. Over the last two decades, technological advances in radiographic imaging, biochemistry, and molecular biology have played an increasing role in radiation treatment planning, delivery, and evaluation of response. In earlier times, fluoroscopy was the basis of radiation treatment planning. In the late 1980s, computed tomography (CT) became the basis for modern radiation treatment planning and delivery. Also multimodality anatomic imaging was found to be the solution to augment delineation of tumors and surrounding structures on CT-based treatment planning. Although these imaging modalities provide the customary anatomic details necessary for radiation treatment planning, they have limitations, including difficulty with identification of tumor extension, and distinction from scar tissues. To overcome these limitations, PET and, more recently, PET-CT have been innovative regarding the extent of disease appraisal, target delineation in the treatment planning, and assessment of therapy response. The use of multi-modality imaging fusion and the introduction of more sensitive and specific PET-CT tracers may further assist target definition. Novel markers of tumor hypoxia or proliferation have the potential to modify the delineation of target volumes, allowing for "dose painting" in selected subvolumes. Furthermore, the potential to predict early outcome or even detect early recurrence of tumor may allow for the tailoring of intervention in cancer patients. The implementation of three-dimensional radiotherapy and IMRT requires adequate selection and delineation of target volumes on the basis of anatomic or molecular imaging modalities, appropriate dose prescription and (dose) specification with regard to dose volume constraints, and quality control for both the clinical and the physical aspects of the entire procedure.

For target volume selection and delineation, anatomic imaging modalities, such as CT and, to a lesser extent, MRI, remain the most widely used modalities. CT is widely available, does not have geometric distortion, and provides intrinsic information on the electronic densities of various tissues—information that is used in dose calculation algorithms. As a limitation, CT lacks contrast resolution for normal soft-tissue structures and tumor extent. This limitation has led to significant inter- and intra-observer variations in delineation of the gross tumor volume (GTV) in head and neck, lung, esophageal, prostate, breast, cervical, and brain tumors.

B.K. Das, MD, ANM, FAR, FAMS
Director and Chief of Nuclear Medicine,
Utkal Institute of Medical Sciences,
Jaydev Vihar, Bhubaneswar, India
e-mail: bkishoredas@yahoo.com

B.K. Das (ed.), *Positron Emission Tomography: A Guide for Clinicians*,
DOI 10.1007/978-81-322-2098-5_18, © Springer India 2015

MRI with various sequences (e.g., unenhanced T1-weighted, contrast-enhanced T1-weighted, and T2-weighted sequences with or without the fat suppression option) is another anatomic imaging modality that can complement or sometimes replace CT. MRI has been shown to be more accurate than CT for evaluating the soft tissue or bone extent of nasopharynx, prostate, and brain tumors. However, for pharyngeal–laryngeal tumors, the advantage of MRI over CT has not been confirmed, either in terms of interobserver variability or in terms of target volume delineation.

Over the last few years, the use of molecular imaging, particularly the use of positron-labeled ^{18}F-FDG, has become increasingly popular in oncology. Given that adequate tracers are used, molecular imaging with PET enables visualization of the various molecular pathways of tumors, including metabolism, proliferation, oxygen delivery and consumption, and receptor or gene expression. Applied in the clinic, PET can be useful for tumor staging, for prediction of the tumor response, for selection or delineation of radiotherapy target volumes, for assessment of the tumor response to treatment, for the detection of early recurrence, or as a tool to evaluate modifications in organ function after treatment. The use of PET in general and of PET with FDG in particular for radiotherapy planning purposes has taken on increasing importance, so that more and more radiation oncologists believe that target volume selection and delineation cannot be adequately performed without the use of PET with FDG. It is important to discuss why should metabolic information be considered more important than the anatomic information provided by CT or MRI? What is the evidence supporting the use of FDG in the treatment planning process?

The ultimate goal of the planning process is to select and delineate target volumes (and organs at risk) on the basis of all of the available diagnostic information and on the knowledge of the physiology of the disease, that is, the probability of local and nodal infiltration. This goal is achieved in part through the use of various imaging modalities, which depict more or less accurately the true tumor extent. The difficulty with imaging modalities is that none of them has a sensitivity of 100 % (no false-negative examinations) or a specificity of 100 % (no false-positive examinations). Thus, false-negative and false-positive results for depicting neoplastic processes occur.

How the sensitivity and specificity of a particular imaging modality influence the radiation planning process depends on the underlying objective of the treatment. If, for a particular disease, the objective is to avoid missing a tumor at any expense, a highly sensitive approach needs to be selected. Such a selection will, of course, result in a lower specificity and in the inclusion of nonneoplastic tissue in the target volume. However, this approach reduces the likelihood of missing neoplastic cells. If, on the other hand, the aim is to avoid inclusion of nonneoplastic cells in the target volume to protect normal tissue, a highly specific approach needs to be adopted. However, such an approach reduces sensitivity and increases the risk for missing tumor cells.

When a novel imaging modality (e.g., PET with the tracer FDG) is introduced, its sensitivity and specificity need to be compared with those of the standard test which is for radiotherapy planning CT. Furthermore, its potential impact on treatment planning needs to be determined. For example, if an additional lymph node is visualized with a new imaging modality known to be more specific than the standard modality, it may be legitimate to increase the target volume(s) beyond what would have been used with a standard procedure; conversely, if fewer nodes are visualized with a new imaging modality known to be more sensitive than the standard modality, it may be legitimate to decrease the target volume(s) below what would have been used with a standard procedure.

Comparative analysis of using FDG-PET and CT to determine target volume has yielded different results in different cancers and locations. For example, compared with anatomic imaging modalities such as CT and MRI, FDG PET is not likely to be superior for the selection of the target volume in neck lymph nodes. In contrast, the sensitivity for the staging of lymph node involvement in lung cancer is significantly higher for FDG-PET than for CT. For esophageal cancer, the sensitivity of FDG-PET is similar to that of CT. However, FDG PET is more specific for the staging of lymph node involvement outside the mediastinum, like supraclavicular or celiac lymph node . For paraaortic lymph nodes in patients with cervical carcinoma, FDG-PET has been reported to be more specific than CT or MRI.

All these considerations have become redundant to some extent as more and more centers are using dual PET-CT systems. Selection of target volume using PET-CT systems has become more accurate than previous treatment planning using conventional imaging modalities.

The advent of dual-modality integrated PET-CT systems offers a unique opportunity of improving target localization and facilitating treatment planning for radiation therapy in contemporary oncologic practice.

Radiation Therapy Planning

Radiation therapy planning can be defined the process of image acquisition, volume delineation, dose-fractionation prescription, assigning of treatment fields and beam modifiers, evaluation of dose distribution, and quality assurance before final approval for treatment delivery. The standard imaging technique used in radiotherapy planning is CT as it provides both good anatomic detail for defining target volumes and the electron density data required for dose calculations. Over the last couple of decades, advances in radiation therapy planning and delivery have ushered in the era of high-precision conformal radiotherapy allowing generation of dose distributions that conform closely to the shape of the target volume while minimizing high-dose regions in the surrounding normal tissues. In general, anatomical cross-sectional CT images are used to delineate treatment volumes and design multiple uniform intensity fields that are shaped using multi-leaf collimators. Intensity modulated radiation therapy (IMRT) is an advanced form of conformal radiotherapy wherein the beam intensity is modulated to produce highly conformal dose distributions around irregular and complex-shaped target volumes. Modern radiotherapy departments are equipped with volumetric image-guidance for precise alignment of the patient with respect to the beam line. Rapid advances in technology allow highly sophisticated treatment planning coupled with extremely accurate localization and precise radiation dose delivery. However, the technology for target volume delineation, i.e., accurately defining what regions or tissues need to be targeted is still not very robust and continues to evolve. One distinct advantage of PET-CT in radiotherapy planning is its potential to improve tumor delineation, reducing intra-observer and inter-observer variability and making treatment volumes more standard across individuals and institutions.

PET for radiation therapy planning can be used in several ways: visual aid for target delineation, fusion of PET and CT images acquired from separate scanners, or a planning PET-CT scan done on an integrated PET-CT unit with the patient in treatment position. Positioning tools should include a firm flat couch top, immobilization devices, laser beams for patient alignment, and a wide-bore scanner (>70 cm). The PET and CT images thus acquired are complementary as well as supplementary. PET images can identify areas of disease not readily visible on CT alone. CT images can provide improved spatial resolution helping to anatomically localize sites of involvement. Also, the low-noise CT data can be used to generate patient-specific map of attenuation coefficients for correcting PET emission data for errors from photon attenuation, scattered radiation, and other physical degrading factors such as partial volume effect. Thus dual-modality PET-CT can improve both the visual quality and the quantitative accuracy of the correlated radiotracer data. It is now widely accepted and acknowledged that PET-CT impacts significantly on planning in the modern radiation therapy clinic. PET-CT not only has a direct impact on target volume delineation in a wide variety of cancers, but can also lead to a significant change in the therapeutic approach in 10–30 % of patients as compared to other reference imaging modalities (Figs. 18.1, 18.2, and 18.3).

Limitations of PET-Guided Planning

One of the main difficulties is the delineation of the treatment volume from noisy PET data. Identification of lesion edges in general is not a trivial problem in PET imaging. Major problems encountered in functional volume quantitation are image segmentation and imperfect system response function. The difficulty in image segmentation is compounded by the low spatial resolution and high-noise characteristics of PET images. Manual delineation of target volumes using different window-level settings and look-up tables is the most common and widely used technique in the clinic. However, the method is highly operator-dependent with wide interobserver variability. Semi-automated or fully automated delineation techniques offer an advantage over manual techniques by improving reproducibility. A collaborative effort between nuclear medicine physicians, radiologists, and radiation oncologists is desirable to fully exploit the potential of PET-CT-guided radiation therapy planning.

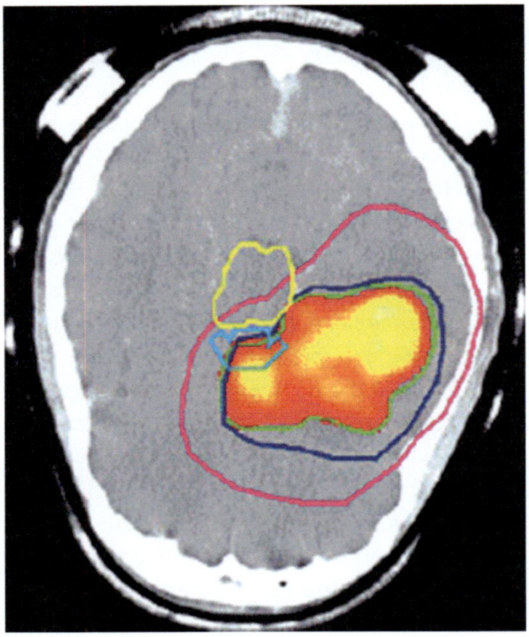

Fig. 18.1 PET-CT image of a brain tumor showing varying degree of tumor activity. Radiation therapy planning is represented by *colored* contours

Impact of PET-Guided Planning on Outcome

There have been recent reports of improved outcomes with PET-guided planning. The largest prospective dataset documenting improved outcomes comes out of a study consisting a total of 317 patients treated with a combination of

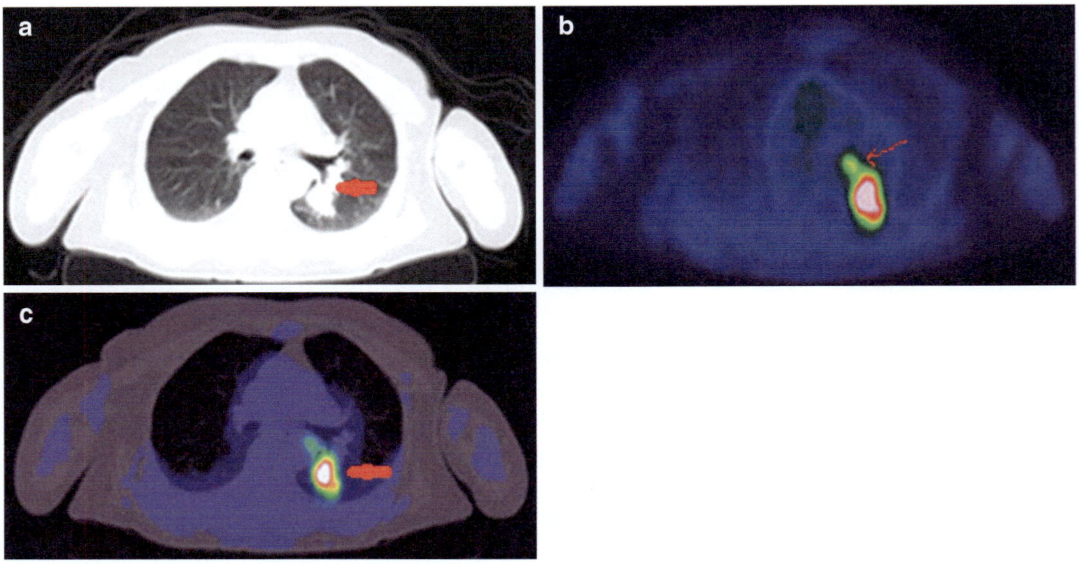

Fig. 18.2 CT lung showing suspect of several metastatic lymph nodes (**a**). PET and PET-CT fused images show one lymph node (see *arrow*) which is metabolically active and needs to be targeted for radiotherapy (**b**, **c**)

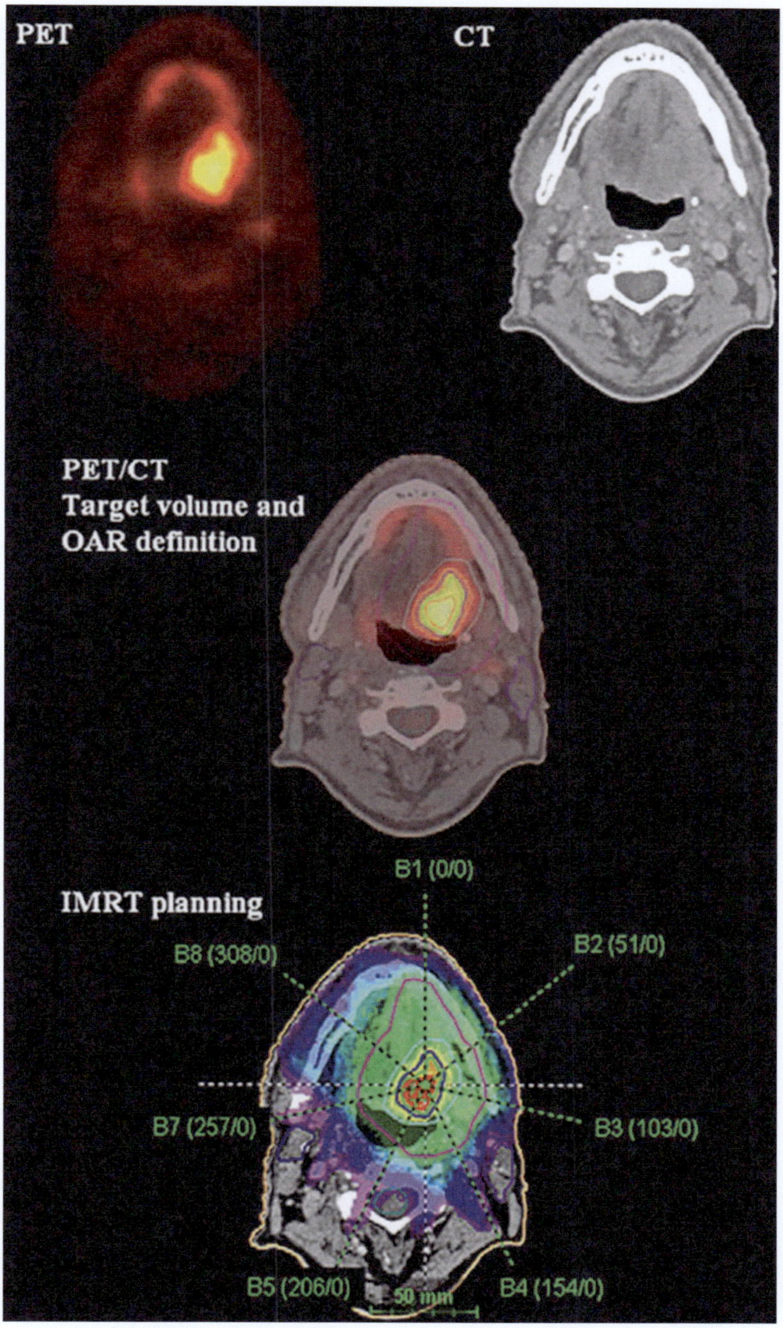

Fig. 18.3 Use of PET-CT in the radiotherapy planning of oropharyngeal tumors. *OAR* organ at risk, *IMRT* intensity modulated radiation therapy

whole-pelvis and split-field irradiation using an institutional step-wedge technique. Another 135 patients were treated with PET-CT-guided IMRT, using pseudo-step-wedge intensity modulation to match the target dose distribution of the conventional technique. Both groups had similar stage distribution, histology, brachytherapy, and concurrent chemotherapy. With a mean follow-up of 52 months for living patients, 178 patients (39IMRT, 139non-IMRT) had recurred. Patients in the PET-CT-guided IMRT group showed better overall and cause-specific survival ($P<0.0001$). Only eight patients in the PET-CT-guided IMRT group developed Grade 3 or worse large bowel or bladder complications which was significantly lesser compared to 54 patients in the non-IMRT group ($P = 0.0351$).

In another study of 115 patients with locally advanced NSCLC treated with definitive PET-guided conformal radiation therapy were analyzed for survival, local regional recurrence, and distant metastases. With a median follow-up of 18 months (range 3–44 months) for all patients, the median overall survival, 2-year actuarial overall survival, and disease-free survival were 19 months, 38, and 28 %, respectively. Majority of the patients died from distant metastases (overall rate of 36 %).

In a recent case control study, 45 patients with stage IVA pharyngeal carcinoma treated with definitive chemoradiation with PET-CT-guided IMRT were compared with 86 patients treated without PET-CT and 3D-conformal radiotherapy after matching with respect to gender, age, stage, grade, and tumor location. Median follow-up was 18 months (range, 6–49 months) for the PET-CT-IMRT group and 28 months (range, 1–168 months) for controls. PET-CT and treatment with IMRT improved cure rates compared to patients without PET-CT and IMRT. Overall survival of patients with PET-CT and IMRT was 97 and 91 % at 1 and 2 years respectively, compared to 74 and 54 % for patients without PET-CT or IMRT ($P = 0.002$). The event-free survival rate of PET-CT-IMRT group was 90 and 80 % at 1 and 2 years, respectively, compared to 72 and 56 % in the control group ($P = 0.005$). Thus more and more data is coming up showing distinct advantage of using PET-CT in radiation therapy planning.

Therapy Planning with Integrated MRI/PET

The combination of MRI and PET in a single gantry for simultaneous acquisition has been developed which has helped to bridge the gap between systems and molecular diagnosis. Both PET and MRI offer richly complementary information about disease. Their integration into a combined system has produced hybrid technology that is significantly better than the sum of its parts. The possibility of using this highly sophisticated hybrid technology for therapy planning is under way and may improve the results further.

Concluding Remarks

Radiation therapy planning has traditionally relied very heavily on CT imaging. Increasingly, FDG-PET-CT is being incorporated into the treatment planning process and promises to improve target volume delineation in a wide variety of cancers. The use of PET-CT for target volume selection should be considered within the framework of its sensitivity and specificity for various tumor types and also mandates specific tuning of parameters, such as image acquisition, processing, and segmentation. There is accumulating evidence that PET-CT guidance has significant impact on radiotherapy planning in many types of cancer. The potential benefits of improved staging and more accurate target localization can promote integrated PET-CT to become the gold standard for radiotherapy simulation and planning.

For Further Reading

1. Ford EC, Herman J, Yorke E, Wahl RL. 18F-FDG PET-CT for image-guided and intensity-modulated radiotherapy. J Nucl Med. 2009;50:1655–65.
2. Jarritt PH, Carson KJ, Hounsell AR, Visvikis D. The role of PET-CT scanning in radiotherapy planning. Br J Radiol. 2006;79:S27–35.
3. Rothschild S, Studer G, Seifert B, Huguenin P, Glanzmann C, Davis JB, et al. PET-CT staging followed by Intensity-Modulated Radiation Therapy (IMRT) improves treatment outcome of locally advanced pharyngeal carcinoma: a matched-pair comparison. Radiat Oncol. 2007;2:22.
4. Schlemmer HP, Pichler BJ, Krieg R, Heiss WD. An integrated MR/PET system: prospective applications. Abdom Imaging. 2009;34:668–74.
5. Vernon MR, Maheshwari M, Schultz CJ, Michel MA, Wong SJ, Campbell BH, et al. Clinical outcomes of patients receiving integrated PET-CT-guided radiotherapy for head and neck carcinoma. Int J Radiat Oncol Biol Phys. 2008;70:768–84.

Basic Principles of Cyclotron and Production of Positron-Emitting Isotopes

19

Birendra Kishore Das

What is a Cyclotron?

Cyclotron is a type of particle accelerator which accelerates charged particles using a high-frequency, alternating voltage (potential difference). A perpendicular magnetic field of constant magnitude and direction causes the particles to spiral almost in a circle so that they re-encounter the accelerating voltage many times. A cyclotron body consists of electrodes, called "dees" because of their shape, in a vacuum chamber. This vacuum chamber is flat and sits in a narrow gap between poles of a large magnet which creates a perpendicular magnetic field. A stream of charged particles is fed into the center of the chamber and a high-frequency alternating voltage is applied across the electrodes. This voltage alternately attracts and repels the charged particles causing them to accelerate. The magnetic field moves the particles in a circular path and, as they gain more energy from the accelerating voltage, they spiral outwards until they reach the outer edge of the chamber.

Inside the cyclotron, there are two D-shaped regions known as dees. In each dee, there is a magnetic field perpendicular to the plane of the page. In the gap separating the dees, there is a uniform electric field pointing from one dee to the other. When a charge is released from rest in the gap, it is accelerated by the electric field and carried into one of the dees. The magnetic field in the dee causes the charge to follow a half-circle that carries it back to the gap. While the charge is in the dee, the electric field in the gap is reversed, so the charge is once again accelerated across the gap. The cycle continues with the magnetic field in the dees continually bringing the charge back to the gap. Every time the charge crosses the gap, it picks up speed. This causes the half-circles in the dees to increase in radius, and eventually the charge emerges from the cyclotron at high speed (Fig. 19.1).

Historical Developments

The first cyclotron, built in 1930 by Ernest Lawrence and Stanley Livingston, was 4.5" (11 cm) in diameter and capable of accelerating protons to an energy of 80 keV. Lawrence soon went on to construct higher-energy and larger-diameter cyclotrons to provide particle beams for research in nuclear physics. Almost 80 years ago, he and Livingston published a seminal paper in which they described the production of light ions with kinetic energies in excess of 1 MeV using a device with magnetic pole pieces 28 cm across (Lawrence and Livingston 1932). By 1936, John Lawrence, Ernest's brother, had made the first recorded biomedical use of a cyclotron when

B.K. Das, MD, ANM, FAR, FAMS
Director and Chief of Nuclear Medicine,
Utkal Institute of Medical Sciences,
Jaydev Vihar, Bhubaneswar, India
e-mail: bkishoredas@yahoo.com

B.K. Das (ed.), *Positron Emission Tomography: A Guide for Clinicians*,
DOI 10.1007/978-81-322-2098-5_19, © Springer India 2015

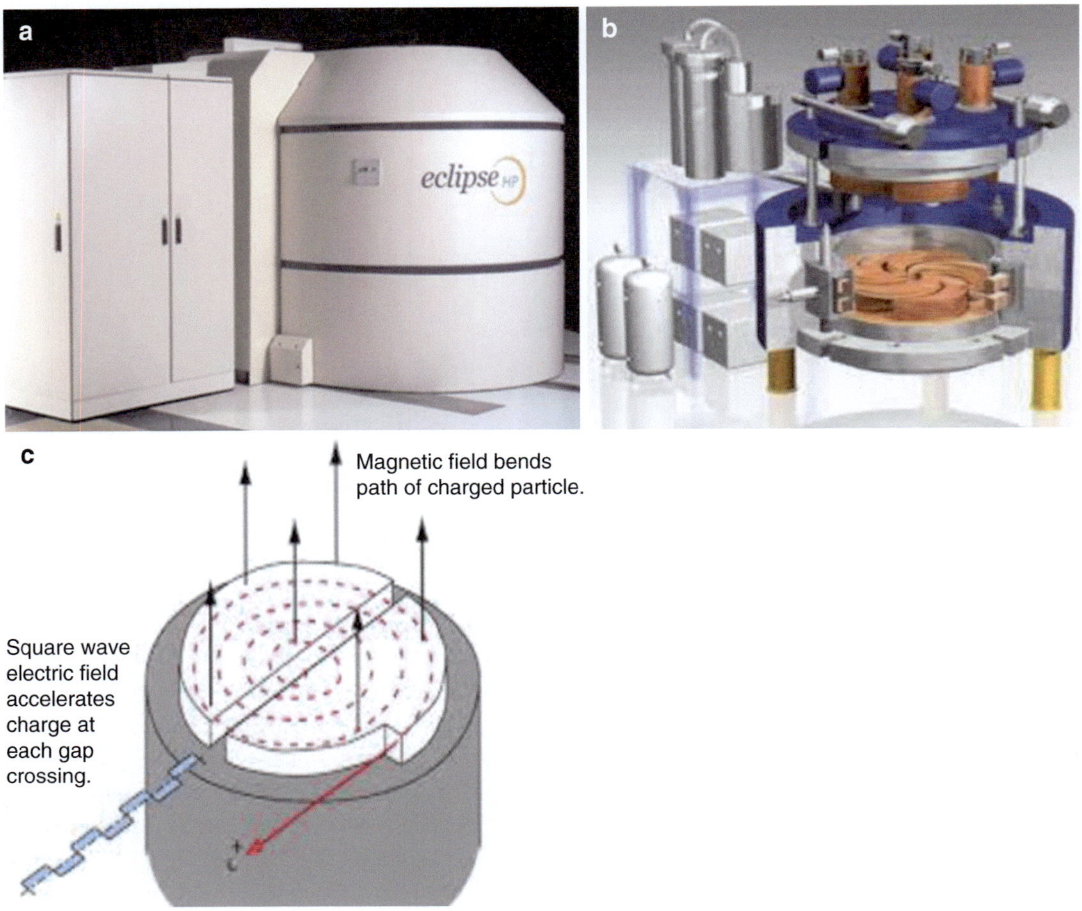

Fig. 19.1 Outward view of a cyclotron (**a**), inside view (**b**), and schematic diagram of functioning (**c**)

he used the 36" (91 cm) machine at Berkeley to produce ^{32}P for the treatment of leukemia. Since then, the physics design of the cyclotron has improved rapidly, with the introduction of alternating-gradient sector focusing, edge focusing, external ion-source injection, electron cyclotron-resonance sources, negative-ion acceleration, separated-sector technology, and the use of superconducting magnets.

However, other accelerator designs were evolving even faster, with the construction of the synchrocyclotron, the invention of the synchrotron, of linear accelerators and of particle colliders that were capable of generating the extremely high energies needed by the particle-physics community. The usefulness of the cyclotron appeared to diminish. But, in 1972, the TRIUMF laboratory in Canada turned on the world's largest cyclotron, at 2,000 tonnes with a beam-orbit diameter of 18 m and negative-ion acceleration. Two years later, in Switzerland, PSI brought into commission a large separated sector, 590 MeV proton cyclotron. Both of these machines have contributed to isotope-production programs. But the value of the cyclotron as a method for producing medical isotopes had come under further pressure due to the availability of numerous nuclear research reactors that had high neutron fluxes, large-volume irradiation positions, and considerable flexibility for isotope production. These attributes allowed the production of important radioisotopes such as ^{99}Mo, ^{131}I, ^{35}S, and even ^{32}P more easily and more cost-effectively.

How the Cyclotron Works

A cyclotron consists of two D-shaped regions known as dees. In each dee, there is a magnetic field perpendicular to the plane of the page. In the gap separating the dees, there is a uniform electric field pointing from one dee to the other. When a charge is released from rest in the gap it is accelerated by the electric field and carried into one of the dees. The magnetic field in the dee causes the charge to follow a half-circle that carries it back to the gap.

While the charge is in the dee the electric field in the gap is reversed, so the charge is once again accelerated across the gap. The cycle continues with the magnetic field in the dees continually bringing the charge back to the gap. Every time the charge crosses the gap it picks up speed. This causes the half-circles in the dees to increase in radius, and eventually the charge emerges from the cyclotron at high speed.

The electrodes would be in the vacuum chamber, which is flat, in a narrow gap between the two poles of a large magnet. In the cyclotron, a high-frequency alternating voltage applied across the "D" electrodes (also called "dees") alternately attracts and repels charged particles. The particles, injected near the center of the magnetic field, increase in speed (and therefore energy) only when passing through the gap between the electrodes. The perpendicular magnetic field (passing vertically through the "D" electrodes), combined with the increasing energy of the particles, forces the particles to travel in a spiral path. With no change in energy, the charged particles in a magnetic field will follow a circular path. In the cyclotron, energy is applied to the particles as they cross the gap between the dees and so they are accelerated (at the typical sub-relativistic speeds used) and will increase in mass as they approach the speed of light. Either of these effects (increased velocity or increased mass) will increase the radius of the circle and so the path will be a spiral. (The particles move in a spiral, because a current of electrons or ions, flowing perpendicular to a magnetic field, experiences a force perpendicular to its direction of motion. The charged particles move freely in a vacuum, so the particles follow a spiral path) (Fig. 19.2). The radius will increase until the particles hit a target at the perimeter of the vacuum chamber. Various materials may be used for the target, and the collisions will create secondary particles which may be guided outside of the cyclotron and into instruments for analysis. The results will enable the calculation of various properties, such as the mean spacing between atoms and the creation of various collision products. Subsequent chemical and particle analysis of the target material may give insight into nuclear transmutation of the elements used in the target.

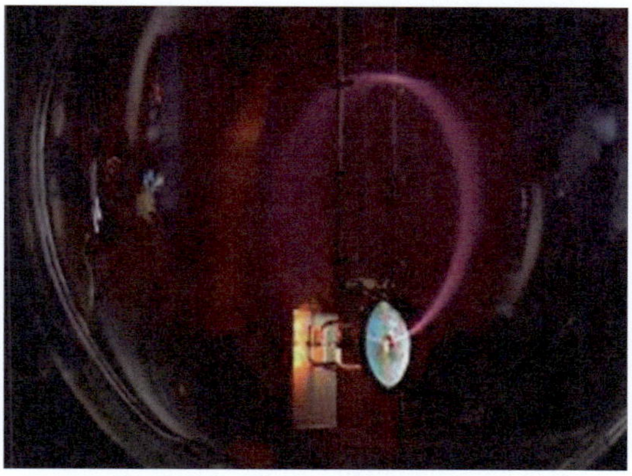

Fig. 19.2 Beam of electrons moving in a circle. Lighting is caused by excitation of gas atoms in a bulb

Production of Positron-Emitting Radiopharmaceuticals

Positron-emitting radioisotopes used in medicine are produced in cyclotrons. The cyclotron is an accelerator of subatomic particles. It produces a large quantity of protons (heavy particles with an electrical positive charge) and gets them moving at an accelerated rate along a circular orbit, inside a chamber controlled by powerful alternating electromagnetic fields. Thus, the particles gain energy and are smashed against a target at nearly the speed of light. The atoms of a substance placed in this target are transformed by this bombardment into radioactive, unstable isotopes, by means of a nuclear reaction.

As already described above, a cyclotron body consists of electrodes called "dees" because of their shape, in a vacuum chamber. This vacuum chamber is flat and sits in a narrow gap between poles of a large magnet which creates a perpendicular magnetic field. A stream of charged particles is fed into the center of the chamber and a high-frequency alternating voltage is applied across the electrodes. This voltage alternately attracts and repels the charged particles causing them to accelerate. The magnetic field moves the particles in a circular path and, as they gain more energy from the accelerating voltage, they spiral outwards until they reach the outer edge of the chamber.

Modern cyclotrons accelerate negative ions created in a plasma. When these negative ions reach the outer edge of the chamber, the excess electrons are stripped off the ions forming positive particles such as a proton or deuteron, which can then be extracted from the cyclotron as a beam. The size of the vacuum chamber determines the length of the spiral path and hence the amount of energy attained by the particle.

Medical Cyclotrons

Medical cyclotrons produce proton beams, which are used to manufacture radioisotopes used in medical diagnosis. Radioisotopes produced in a cyclotron decay by either positron emission or electron capture. Positron emission tomography (PET) and single photon emission computed tomography (SPECT), which utilizes the gamma rays associated with electron capture, are two imaging techniques that rely on cyclotron-produced radioisotopes.

Radionuclides used in PET scanning are typically isotopes with short half-lives such as

Carbon-11 (~20 min),
Nitrogen-13 (~10 min),
Oxygen-15 (~2 min),
Fluorine-18 (~110 min)., or
Rubidium-82 (~1.27 min).

These radionuclides are incorporated either into compounds normally used by the body such as glucose (or glucose analogues), water, or ammonia, or into molecules that bind to receptors or other sites of drug action. PET technology can be used to trace the biologic pathway of any compound in living humans (and many other species as well), provided it can be radiolabeled with a PET isotope. Thus, the specific processes that can be probed with PET are virtually limitless, At present, however, by far the most commonly used radiotracer in clinical PET scanning is fluorodeoxyglucose (also called FDG), an analogue of glucose that is labeled with Fluorine-18. This radiotracer is used in essentially all scans for oncology and most scans in neurology, and thus makes up the large majority of all of the radiotracer (>95 %) used in PET and PET-CT scanning.

Due to the short half-lives of most positron-emitting radioisotopes, the radiotracers have traditionally been produced using a cyclotron in close proximity to the PET imaging facility. The half-life of fluorine-18 is long enough that radiotracers labeled with fluorine-18 can be manufactured commercially at offsite locations and shipped to imaging centers. Rubidium-82 generators have also become commercially available. These contain strontium-82, which decays by electron capture to produce positron-emitting rubidium-82.

Limitations to the widespread use of PET arise from the high costs of cyclotrons needed to produce the short-lived radionuclides for PET scanning and the need for specially adapted on-site chemical synthesis apparatus to produce the radiopharmaceuticals after radioisotope preparation. Organic radiotracer molecules that will contain a positron-emitting radioisotope cannot be synthesized first and then the radioisotope prepared within them, because bombardment with a cyclotron to prepare the radioisotope destroys any organic carrier for it. Instead, the isotope must be prepared first, then afterward the chemistry to prepare any organic radiotracer (such as FDG) accomplished very quickly in the short time before the isotope decays. Few hospitals and universities are capable of maintaining such systems, and most clinical PET is supported by third-party suppliers of radiotracers that can supply many sites simultaneously. This limitation restricts clinical PET primarily to the use of tracers labeled with fluorine-18, which has a half-life of 110 min and can be transported a reasonable distance before use, or to rubidium-82 (used as rubidium-82 chloride) with a half-life of 1.27 min, which is created in a portable generator and is used for myocardial perfusion studies. In recent years, cyclotrons with integrated shielding and "hot labs" (automated chemistry labs that are able to work with radioisotopes) are available to accompany PET units to remote hospitals (Fig. 19.3 and 19.4).

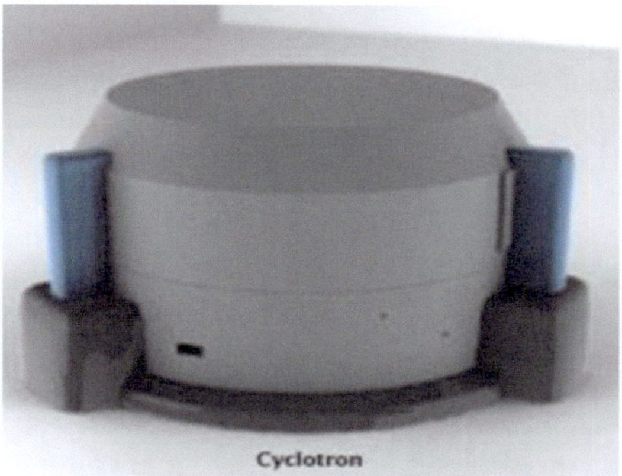

Fig. 19.3 View of a cyclotron used in many nuclear medicine centers to produce positron-emitting radioisotopes to be used locally

Because the half-life of fluorine-18 is about 2 h, the prepared dose of a radiopharmaceutical bearing this radionuclide will undergo multiple half-lives of decay during the working day. This necessitates frequent recalibration of the remaining dose (determination of activity per unit volume) and careful planning with respect to patient scheduling.

PET radiopharmaceuticals are formed by reactions of the organic compounds with radionuclides or positron emission isotopes. They are also called the labeled compounds. So-called tracers, required for the production of the radiopharmaceuticals, are usually produced by an accelerator of the elements, a cyclotron, possibly by a nuclear reactor. 11C, 13N, 15O, and 18F are the most used in positron emission tomography (next only PET) – scanning technology based on a detection of the gamma radiation, appeared because of the annihilation of the positrons emitted by the radioisotopes, made by PET scanners.

An advantage of the positron emission radionuclides is their short half-life time. For

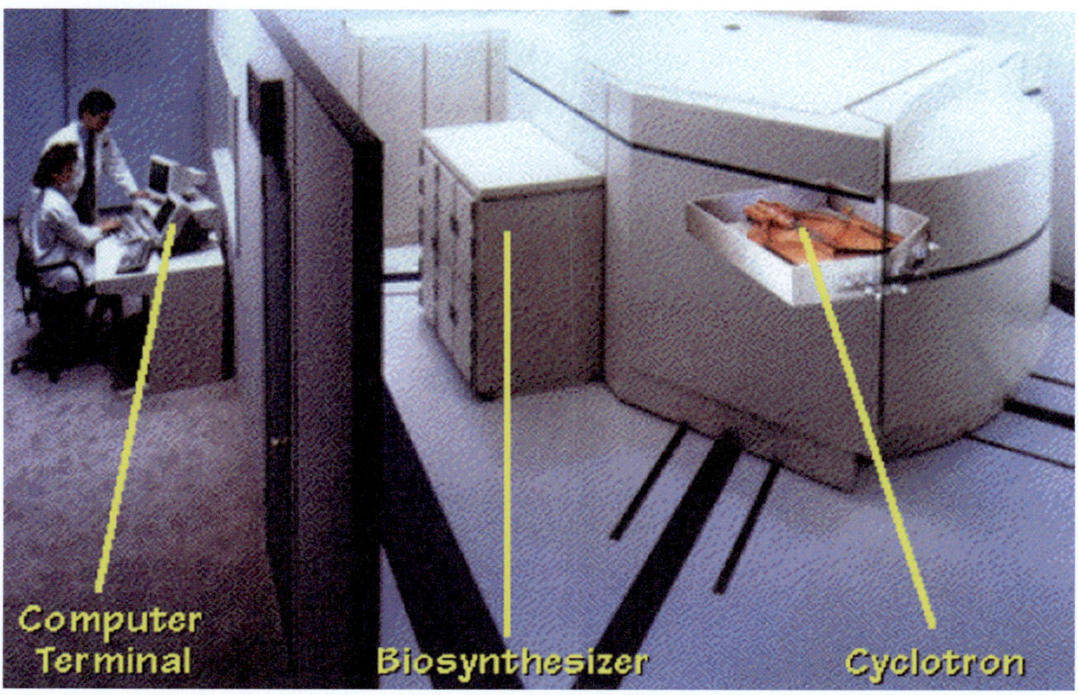

Fig. 19.4 Essential components of a cyclotron facility

that reason, a patient gets much smaller radiation dose then during the other similar medical examinations.

The targets for the preparation of the radiopharmaceuticals can be gases, liquids, and solid materials. The preparation of needed radionuclides precedes the radiopharmaceuticals synthesis. For the preparation of the artificial radionuclides in required amount for a study of chemical and biological processes is necessary to have a high intensity of the bombarding particles flow with the adequate energy.

The cyclotron produce from 1,014 to 1,015 accelerated particles per second. It can be protons, deuterons, hellions, and heavy nuclei. The targets mentioned previously are irradiating with a bunch of the accelerated particles.

Preparation of the [18F] FDG and the [18F] MISO

Irradiated water [18O] H_2O is evaporated in the presence of a cryptand (aminopolyether potassium carbonate complex – Kryptofix 222), which affect as a catalyst of stereospecific S_N2 substitution reaction. Dry evaporated mixture with developed 18F- is dissolved in waterless acetonitrile and leaves at 90 °C to react with prepared precursor, an analog of the mannose 1,3,4,6-tetra-O-acetyl-2-triflate-â-D-mannopyranose, so-called the triflate of mannose.

Formed 1,3,4,6-tetra-*O*-acetyl-2-[18F] fluoro-D-glucopyranose hydrolyzes at 110 °C with dilute hydrochloric acid (14 min) and a product [18F] FDG is clarified with ion exchange chromatography. The synthesis lasts for 30 min and radiochemical yield is 65 %. The product has got the molar activity higher than 400 GBqÅ"ìmol-1 [4]. Same procedure is used for the preparation of fluoromisonidazole ([18F] MISO). However, a precursor is used as ananalog of misonidazole. Radiochemical yield is 20 % and it is lower than in [18F] FDG preparation. The activity of a product [18F] MISO is 3.7 GBq (100 mCi).

For Further Reading

1. Lawrence EO. The Evolution of the cyclotron. Noblel Lecture. 1951.
2. Chao AW, Mess KH, Tigner M, Zimmermann F, editors. Handbook of accelerator physics and engineering. 2nd edn. World Scientific Publishing Company. 2013.
3. Pillai MRA. Cyclotron produced radionuclides: physical characteristics and production methods, IAEA technical report no. 468. Vienna: 2009.
4. Hichwa RD, et al. Design of target systems for production of PET nuclides. Nucl Instrum Methods Phys Res. 1989;B40/41:1110–3.
5. Jacobson MS, Hung JC, Mays TL, Mullan BP. The planning and design of a new PET radiochemistry facility. Mol Imaging Biol. 2002;4:119–27.
6. Ruth TJ, et al. Radionuclide production for the biosciences. Nucl Med Biol. 1989;16(4):323–36.

Basic Principles of CT Imaging 20

Birendra Kishore Das

CT uses X-rays to generate cross-sectional, two-dimensional images of the body. Images are acquired by rapid rotation of the X-ray tube 360° around the patient. The transmitted radiation is then measured by a ring of sensitive radiation detectors located on the gantry around the patient. The final image is generated from these measurements utilizing the basic principle that the internal structure of the body can be reconstructed from multiple X-ray projections.

The first CT scanner was developed in 1972 by Sir Godfrey Hounsfield. Since then, the modality has become established as an essential radiological technique applicable in a wide range of clinical situations (Fig. 20.1).

How is a CT Image Produced ?

Every acquired CT slice is subdivided into a matrix of up to 1,024 × 1,024 volume elements (voxels). Each voxel has been traversed during the scan by numerous X-ray photons and the intensity of the transmitted radiation measured by detectors. From these intensity readings, the density or attenuation value of the tissue at each point in the slice can be calculated. Specific attenuation values are assigned to each individual voxel. The viewed image is then reconstructed as a corresponding matrix of picture elements (pixels). Each pixel is assigned a numerical value (CT number), which is the average of all the attenuation values contained within the corresponding voxel. This number is compared to the attenuation value of water and displayed on a scale of arbitrary units named Hounsfield units (HU) after Sir Godfrey Hounsfield. In a simple term HU are displayed as various shades of gray to form an image.

Early CT scanners acquired images a single slice at a time (sequential scanning). However, during the 1980s significant advancements in technology heralded the development of slip ring technology, which enabled the X-ray tube to rotate continuously in one direction around the patient. This has contributed to the development of helical or spiral CT.

In spiral CT, the X-ray tube rotates continuously in one direction while the table on which the patient is lying is mechanically moved through the X-ray beam. The transmitted radiation thus takes on the form of a helix or spiral. Instead of acquiring data one slice at a time, information can be acquired as a continuous volume of contiguous slices. This allows larger anatomical regions of the body to be imaged during a single breath hold, thereby reducing the possibility of artifacts caused by patient movement.

B.K. Das, MD, ANM, FAR, FAMS
Director and Chief of Nuclear Medicine,
Utkal Institute of Medical Sciences,
Jaydev Vihar, Bhubaneswar, India
e-mail: bkishoredas@yahoo.com

B.K. Das (ed.), *Positron Emission Tomography: A Guide for Clinicians*,
DOI 10.1007/978-81-322-2098-5_20, © Springer India 2015

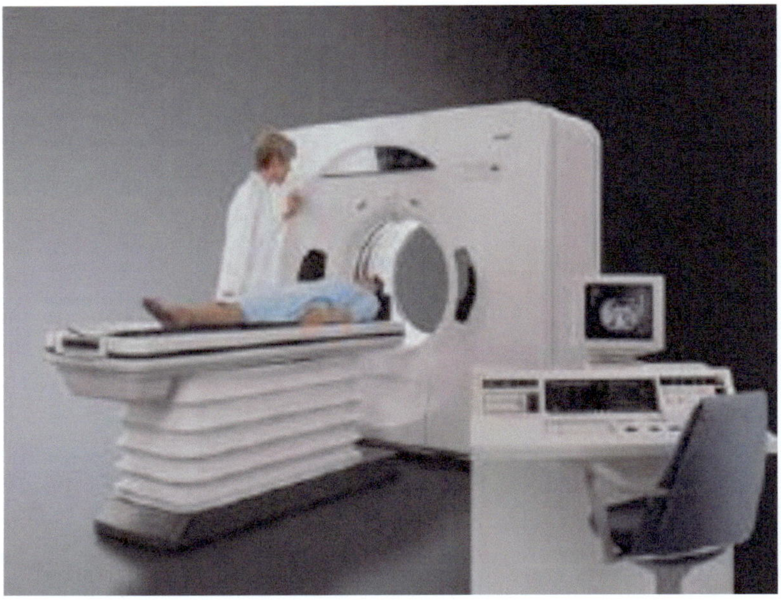

Fig. 20.1 A typical modern CT scan machine

The next generation of CT scanners is now commercially available. These multislice or multidetector machines utilize the principles of the helical scanner but incorporate multiple rows of detector rings. They can therefore acquire multiple slices per tube rotation, thereby increasing the area of the patient that can be covered in a given time by the X-ray beam.

Information Obtained from CT Scan

CT scans are used to image a wide variety of body structures and internal organs. Since the 1990s, CT equipment has become more affordable and available. In some diagnoses, CT scans have become the first imaging examination of choice. Because the computerized image is so sharp, focused, and three-dimensional, many tissues can be better differentiated than on standard X-rays. Common CT indications include:

- Sinus studies: The CT scan can show details of a *sinusitis* and bone *fractures*. Physicians may order CT of the sinuses to provide an accurate map for surgery.
- Brain studies: Brain scans can detect hematomas, tumors, and strokes. The introduction of CT scanning, especially spiral CT, has helped reduce the need for more invasive procedures such as cerebral *angiography* and more expensive MRI.
- Body scans: CT scans of the body will often be used to observe abdominal organs, such as the liver, kidneys, adrenal glands, spleen, and lymph nodes, and extremities (Fig. 20.2).
- Aorta scans: CT scans can focus on the thoracic or abdominal aorta to locate aneurysms and other possible aortic diseases.
- Chest scans: CT scans of the chest are useful in distinguishing tumors and in detailing accumulation of fluid in chest infections.

Information for Patient

Pregnant women or those who could possibly be pregnant should not have a CT scan unless the diagnostic benefits outweigh the risks. Pregnant patients should particularly avoid whole body or abdominal scans. If the exam is necessary for obstetrics purposes, technologists are instructed not to repeat films if there are errors. Pregnant patients receiving CT or any X-ray exam away from the abdominal area may be protected by a lead apron. However, most radiation, known as scatter, travels through the body and is not blocked by the apron.

Contrast agents are often used in CT exams and the use of these agents should be discussed

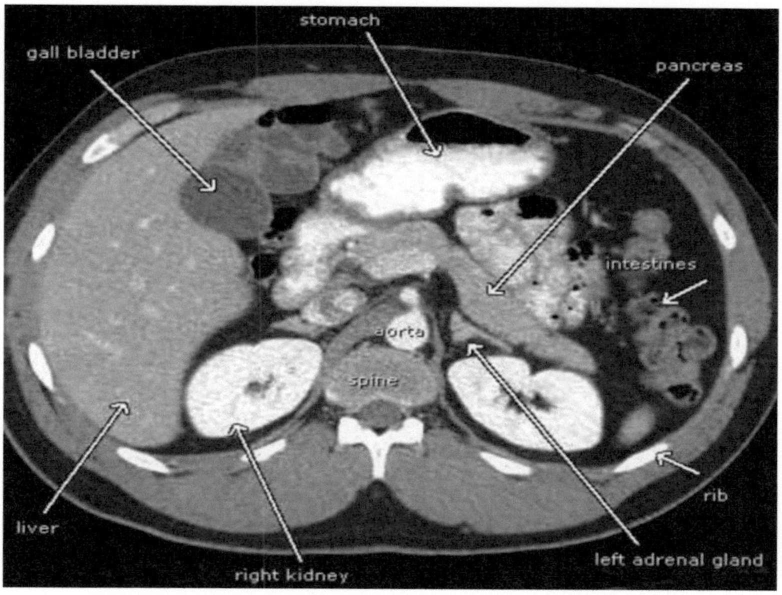

Fig. 20.2 A slice through the abdomen showing different anatomical structures in CT image

with the referring physician prior to the procedure. One of the common contrast agents, iodine, can cause allergic reactions. Patients who are known to be allergic to iodine have to inform the doctor.

Although the equipment looks large and intimidating, it is very sophisticated and fairly comfortable. The patient is asked to lie on a gantry, or narrow table, that slides into the center of the scanner. The scanner looks like a doughnut and is round in the middle, which allows the X-ray beam to rotate around the patient. The scanner section may also be tilted slightly to allow for certain cross-sectional angles.

CT Procedure

The patient will feel the gantry move very slightly as the precise adjustments for each sectional image is made. A technologist watches the procedure from a window and views the images on a computer screen.

It is essential that the patient lie very still during the procedure to prevent motion blurring. In some studies, such as chest CTs, the patient will be asked to hold his or her breath during image capture.

Following the procedure, films of the images are usually printed for the radiologist and referring physician to review. A radiologist can also interpret CT exams on a special computer screen. The procedure time will vary in length depending on the area being imaged. Average study times are from 30 to 60 min. Some patients may be concerned about claustrophobia, but the width of the "doughnut" portion of the scanner is such that many patients can be reassured of openness.

What Additional Information Is Obtained from CT Image ?

While traditional Xrays image organs in two dimensions, with the possibility that organs in the front of the body are superimposed over those in the back, CT scans allow for a more three-dimensional effect. Some have compared CT images to slices in a loaf of bread. Precise sections of the body can be located and imaged as cross-sectional views. The screen before the technologist shows a computer's analysis of each section detected by the X-ray beam. Thus, various densities of tissue can be easily distinguished.

Spiral CT

Spiral CT, also called helical CT, is a newer version of CT scanning which is continuous in motion and allows for three-dimensional recreation of images. For example, traditional CT allows the technologist to take slices at very small and precise intervals one after the other. Spiral CT allows for a continuous flow of images, without stopping the scanner to move to the next image slice. A major advantage of spiral CT is the ability to reconstruct images anywhere along the length of the study area. The procedure also speeds up the imaging process, meaning less time for the patient to lie still. The ability to image contrast more rapidly after it is injected, when it is at its highest level, is another advantage of spiral CT's high speed.

Radiation Exposure and Risks

Radiation exposure from a CT scan is similar to, though higher than, that of a conventional X-ray. Although this is a risk to pregnant women, the exposure to other adults is minimal and should produce no effects. Although severe contrast reactions are rare, they are a risk of many CT procedures, particularly to those who are allergic to iodine.

Normal Results

Normal findings on a CT exam show bone, the densest tissue, as white areas. Tissues and fat will show as various shades of gray, and fluids will be gray or black. Air will also look black. Intravenous, oral, and rectal contrast appears as white areas. The radiologist can determine if tissues and organs appear normal by the sensitivity of the gray shadows. In CT, the images that can cut through a section of tissue or organ provide three-dimensional viewing for the radiologist and referring physician.

Abnormal Results

Abnormal results may show different characteristics of tissues within organs. Accumulations of blood or other fluids where they do not belong may be detected. Radiologists can differentiate among types of tumors throughout the body by viewing details of their makeup.

PET scanner can be combined with sophisticated CT machine to produce hybrid PET-CT units. The CT component of a PET-CT unit can be used independently as a CT machine to provide diagnostic images. This is used by centers to generate additional revenue.

For Further Reading

1. Goldman LW. Principles of CT: multislice CT. J Nucl Med Technol. 2008;36:57–68.
2. Hendee WR. Cross sectional medical imaging: a history. Radiographics. 1989;9(6):1155–80.
3. Hounsfield GN. Computed medical imaging. Med Phys. 1980;7:277–82.
4. Ledley RS, Wilson JB, Golab T, Rotolo LS. The ACTA scanner: the whole body computerized transaxial tomograph. Comput Biol Med. 1974; 4:145–55.

21 Basic Principles of MR Imaging

Birendra Kishore Das

MRI stands for Magnetic Resonance Imaging, once called Nuclear Magnetic Resonance Imaging (NMR). The word "nuclear" was dropped off about 20 years ago because of fears that people would think there was something radioactive involved, which is not.

MRI is a way of getting pictures of various parts of the body without the use of x-rays, unlike regular x-rays pictures and CT scans. It is a non-invasive medical diagnostic technique that uses magnetism, radio waves, and a computer to produce images of body structures.

Magnetism is a property of matter that is a result of the orbiting electrons in atoms. The orbiting electrons cause the atoms to have a magnetic moment associated with an intrinsic angular momentum called "spin." Magnetic field strengths are measured in units of gauss (G) and Tesla (T). One Tesla is equal to 10,000 G. The earth's magnetic field is about 0.5 G. The strength of electromagnets used to pick up cars in junk yards is about the field strength of MRI machines (1.5–2.0 T).

The MRI scanner is a tube surrounded by a giant circular magnet. The patient is placed on a movable bed (Fig. 21.1) that is inserted into the magnet. The magnet creates a strong magnetic field that aligns the protons of hydrogen atoms, which are then exposed to a beam of radio waves. This spins the various protons of the body, and they produce a faint signal that is detected by the receiver portion of the MRI scanner. The receiver information is processed by a computer, and an image is produced.

Unlike CT where the image appearance is related primarily to the electron density of the material, MR images depend on many variables including proton density, T1 and T2 relaxation effects, flow effects, diffusion effects, and susceptibility effects. Image quality is generally described in terms of SNR, spatial resolution, and contrast. There is interdependence of the SNR, resolution, and time of acquisition.

The image and resolution produced by MRI is quite significant and can detect tiny changes of structures within the body. For some procedures,

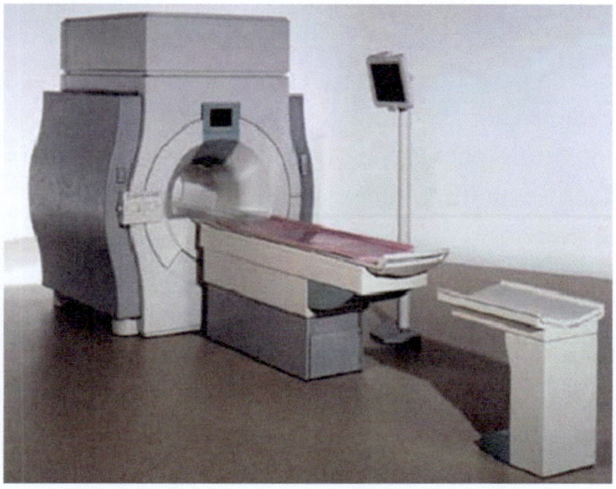

Fig. 21.1 A typical magnetic resonance imaging (MRI) equipment

B.K. Das, MD, ANM, FAR, FAMS
Director and Chief of Nuclear Medicine,
Utkal Institute of Medical Sciences,
Jaydev Vihar, Bhubaneswar, India
e-mail: bkishoredas@yahoo.com

B.K. Das (ed.), *Positron Emission Tomography: A Guide for Clinicians*,
DOI 10.1007/978-81-322-2098-5_21, © Springer India 2015

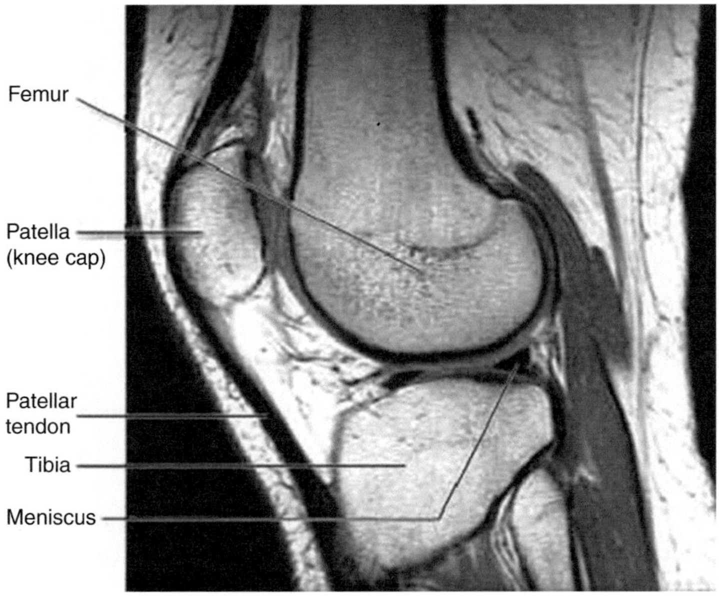

Fig. 21.2 MRI of the knee joint (side or lateral view) showing structural details of the bony and soft-tissue components

contrast agents, such as gadolinium, are used to increase the accuracy of the images (Fig. 21.2).

Magnetic resonance spectroscopy (*MRS*) of intact biological tissues was first reported by two groups: Moon and Richards using P-31 MRS to examine intact red blood cells in 1973, and Hoult et al. using P-31 MRS to examine excised leg muscle from the rat in 1974. Since then MRS has been applied to almost every organ of the body including brain, heart, liver, kidney, prostate, and extremities. MRS is useful for looking at disorders of metabolism, tumors, and certain inflammatory and ischemic diseases. Most of the work with in vivo MRS in humans has been in the brain. Abnormalities have been seen, sometimes with earlier detection than for any other diagnostic procedure short of biopsy, in primary brain tumors, infections such as AIDS, demyelinating disorders such as multiple sclerosis, stroke, and epilepsy.

Spectroscopic changes are documented in a variety of enzyme deficiencies, mitochondrial abnormalities, dystrophies, inflammatory myopathies, and thyroid disease. In muscle, these diseases include phosphofructokinase deficiency, amyloglucosidase deficiency, Duchene muscular dystrophy, Becker muscular dystrophy, dermatomyositis, polymyositis, inclusion body myositis, hypothyroidism, and congestive heart failure.

Unlike CT where the image appearance is related primarily to the electron density of the material, MR images depend on many variables including proton density, T1 and T2 relaxation effects, flow effects, diffusion effects, and susceptibility effects. Image quality is generally described in terms of SNR, spatial resolution, and contrast. There is interdependence of the SNR, resolution, and time of acquisition.

When Are MRI Scans Used?

An MRI scan can be used as an extremely accurate method of disease detection throughout the body. In the head, trauma to the brain can be seen as bleeding or swelling. Other abnormalities often found include brain aneurysms, stroke, tumors of the brain, as well as tumors or inflammation of the spine.

Neurosurgeons use an MRI scan not only in defining brain anatomy but in evaluating the integrity of the spinal cord after trauma. It is also used when considering problems associated with the vertebrae or intervertebral discs of the spine. An MRI scan can evaluate the structure of the heart and aorta, where it can detect aneurysms or tears.

It provides valuable information on glands and organs within the abdomen, and accurate information about the structure of the joints, soft tissues, and bones of the body. Often, surgery can be deferred or more accurately directed after knowing the results of an MRI scan.

What Are the Risks of an MRI Scan?

An MRI scan is a painless radiology technique that has the advantage of avoiding x-ray radiation exposure. There are no known side effects of an MRI scan. The benefits of an MRI scan relate to its precise accuracy in detecting structural abnormalities of the body.

Patients who have *any metallic materials within the body* must notify their physician prior to the examination or inform the MRI staff. Metallic chips, materials, surgical clips, or foreign material (artificial joints, metallic bone plates, or prosthetic devices, etc.) can significantly distort the images obtained by the MRI scanner. Patients who have heart pacemakers, metal implants, or metal chips or clips in or around the eyeballs cannot be scanned with an MRI because of the risk that the magnet may move the metal in these areas. Similarly, patients with artificial heart valves, metallic ear implants, bullet fragments, and chemotherapy or insulin pumps should not have MRI scanning.

During the MRI scan, patient lies in a closed area inside the magnetic tube. Some patients can experience a claustrophobic sensation during the procedure. Therefore, patients with any history of claustrophobia should relate this to the practitioner who is requesting the test, as well as the radiology staff. A mild sedative can be given prior to the MRI scan to help alleviate this feeling. It is customary that the MRI staff will be nearby during MRI scan. Furthermore, there is usually a means of communication with the staff (such as a buzzer held by the patient) which can be used for contact if the patient cannot tolerate the scan.

How Does a Patient Prepare for an MRI Scan and How Is It Performed?

All metallic objects on the body are removed prior to obtaining an MRI scan. Occasionally, patients will be given a sedative medication to decrease *anxiety* and relax the patient during the MRI scan. MRI scanning requires that the patient lie still for best accuracy. Patients lie within a closed environment inside the magnetic machine. Relaxation is important during the procedure and patients are asked to breathe normally. Interaction with the MRI technologist is maintained throughout the test. There are loud, repetitive clicking noises which occur during the test as the scanning proceeds. Occasionally, patients require injections of liquid intravenously to enhance the images which are obtained. The MRI scanning time depends on the exact area of the body studied, but ranges from half an hour to an hour and a half.

What Does the Patient Experience During the MRI Examination?

The patient will most likely be lying on a special table that moves into the center of the magnet. Prior to going into the magnet the patient will be offered earplugs to reduce the noise that he hears. He will then hear some "hammering" noises while the scanner is preparing for scanning and taking the pictures. During this hammering noise, it is important not to move, as this would blur the pictures. He may also feel some vibration during the hammering noise and some slight movement of the table during the examination. Some patients will be given an injection in their arm of a substance that improves certain types of pictures. This substance, called a "contrast agent", is very safe and is unrelated to the iodine used for CAT scans and kidney x-rays.

How Does a Patient Obtain the Results of the MRI Scan?

After the MRI scanning is completed, the computer generates visual images of the area of the body that was scanned. These images can be transferred to film (hard copy). A radiologist is a physician who is specially trained to interpret images of the body. The interpretation is transmitted in the form of a report to the practitioner who requested the MRI scan. The practitioner can then discuss the results with the patient and/or family.

How Safe Is MRI?

MRI is quite safe in the majority of patients. Certain patients may not be able to have an MRI. These include people who get nervous in small spaces (claustrophobic) and those with implanted medical devices such as aneurysm clips in the brain, heart pacemakers, and cochlear (inner ear) implants. Also, people with pieces of metal close to or in an important organ (such as the eye) may not be scanned. There are a few additional safety considerations and some exceptions based on individual circumstances.

Also, certain metal objects that we common have on our persons like watches, credit cards, hairpins, writing pens, etc., may be damaged by the MRI scanner or may be pulled away from our bodies if we go into an MRI room. Also, metal can sometimes cause poor pictures if it is close to the part being scanned. For these reasons, patients are asked to remove these objects before entering the MRI scanner.

What Are the Advantages of a MRI Scan Over Other Types of Scans?

MRI scans are good to study the non-bony parts or "soft tissues" of the body. In particular, the brain, spinal cord, and nerves are seen much more clearly with MRI than with regular x-rays and CAT scans. Also, muscles, ligaments, and tendons are seen quite well so that MRI scans are commonly used to look at knees and shoulders following injuries. A MRI scanner uses no x-rays or other radiation. A disadvantage of MRI is its higher cost compared to a regular x-ray or CAT scan. Also, CAT scans are often better at looking at the bones than MRI.

For Further Information

1. See Chapter 22 for Comparative advantages and disadvantages of PET-CT and PET-MRI units.

22 Comparison of PET-CT with PET-MRI

Birendra Kishore Das

Introduction

Hybrid imaging with positron emission tomography is an evolving technology. The idea to combine PET with CT was made in the early 1990s. A Swiss oncology surgeon suggested adding something useful such as a CT scanner that would provide anatomical information more familiar to surgeons at that time. Thus, the concept of PET-CT was born in 1991, in which the components of a CT scanner would be mounted in the gaps between the banks of BGO block detectors. However, it was soon evident from that such a concept would not be feasible owing to the density of x-ray components mounted on the rotating support. Thus, it took 7 years more before the first prototype combined PET-CT scanner was completed, installed, and became operational in 1998. The coregistered anatomy localized functional abnormalities and clarified equivocal situations, thus improving the accuracy and confidence of the scan interpretation. The use of a rapidly acquired, low-noise CT scan in place of a lengthy conventional PET transmission scan also reduced the overall scan duration. Positron emission tomography/computed tomography (PET-CT) is now established as the imaging modality of choice in many clinical conditions, particularly in oncology.

Although PET-CT has become a critical component of clinical diagnoses, it has certain limitations, largely related to the fact that the CT and PET scans are acquired sequentially rather than simultaneously. Especially problematic are artifacts caused by intra- and inter-scan patient and organ motion and differences between the breathing protocols used in PET and CT. These artifacts affect the accuracy of the registration and attenuation correction and compromise the accuracy of activity quantification. The excellent soft-tissue contrast and the fact that it does not use ionizing radiation are additional notable advantages of MRI.

Technical Issues

The idea to combine PET and MRI arose as early as the mid-1990s, even before PET-CT was introduced. It was felt necessary in small-animal imaging studies to add anatomic landmarks with high soft tissue contrast to the molecular information delivered by PET. Preclinical PET-MRI work was followed by immediate commercial interest to combine PET and MRI, which was partly driven by the limited sensitivity of MRI to trace biomarkers or to reveal metabolites.

There were many technical problems to be addressed before the two systems could be

B.K. Das, MD, ANM, FAR, FAMS
Director and Chief of Nuclear Medicine,
Utkal Institute of Medical Sciences,
Jaydev Vihar, Bhubaneswar, India
e-mail: bkishoredas@yahoo.com

B.K. Das (ed.), *Positron Emission Tomography: A Guide for Clinicians*,
DOI 10.1007/978-81-322-2098-5_22,

combined. First, the photomultiplier technology needed to be replaced with magnetic field–insensitive avalanche photodiodes, which were used in prototype animal PET systems that had a limited number of detectors and were temperature-controlled with costly cooling systems. Second, compact PET detectors were to be constructed which would be invisible to the MRI and not interfere with the field gradients or MR radiofrequency. Finally, the MRI scanner needed to be adapted to accommodate the PET detectors and to allow simultaneous data acquisition without mutual interference. The radiofrequency coils of the MRI scanner were to be integrated into the PET system. Dedicated coils needed to be built to minimize radiofrequency interference with the PET electronics and to avoid gamma-ray scatter and attenuation. Furthermore, completely new strategies for PET attenuation correction, based solely on MRI information, had to be developed. To overcome the foregoing problems, different methods have been proposed and implemented by various groups. Currently, three companies offer combined PET-MR systems: Philips, Siemens, and GE. Siemens was the first company which offered a fully integrated whole body and simultaneous acquisition PET-MRI system. This system received a CE mark and FDA approval for customer purchase in 2011.

Clinical Applications

Hybrid imaging has made great strides in the imaging evaluation of patients with a variety of diseases. PET-CT is now established as the imaging modality of choice in many clinical conditions, particularly in oncology. While the initial development of combined PET/MRI was in the preclinical arena, hybrid PET-MRI scanners are now available for clinical use. PET-MRI combines the unique features of MRI including excellent soft tissue contrast, diffusion-weighted imaging, dynamic contrast-enhanced imaging, fMRI, and other specialized sequences as well as MR spectroscopy with the quantitative physiologic information that is provided by PET. Most evidence for the potential clinical utility of PET-MRI is based on studies performed with side-by-side comparison or software-fused MRI and PET images. Data on distinctive utility of hybrid PET-MRI are rapidly emerging. There are potential competitive advantages of PET-MRI over PET-CT (Fig. 22.1). In general, PET-MRI may be preferred over PET-CT where the unique features of MRI provide more robust imaging evaluation in certain clinical settings (Fig. 22.2). The exact role and potential utility of simultaneous data acquisition in specific research and clinical settings will need to be defined. It may be that simultaneous PET-MRI will be best suited for clinical situations that are disease-specific, organ-specific, related to diseases of the children or in those patients undergoing repeated imaging for whom cumulative radiation dose must be kept as low as reasonably achievable.

PET and MRI data acquired simultaneously allows essentially perfect temporal correlation of dynamically acquired data sets from both modalities. This could be of special interest for brain imaging as well as in the fields of cardiology and oncology. Basically, MRI already provides a large variety of protocols which selectively enhance contrast and thus visual discrimination among different tissues in vivo, and which can be used for dynamic contrast-enhanced imaging, diffusion imaging, functional MRI (fMRI), etc. Thus, PET-MRI could provide MRI-derived information on both anatomy and function for correlation with PET-derived pathology-specific, quantitative information on other aspects of tissue function. Perhaps most importantly, MR is also capable of spectroscopy for detection of organ-specific abnormalities and pathologies by quantifying ratios of concentrations of molecules such as lactate or *N*-acetylaspartate (NAA). Functional MRI (fMRI) is of special interest in the fields of neurology and psychiatry, since it assesses brain function by detecting a contrast dependent on the blood oxygenation level (BOLD effect) and this on a combination of perfusion as well as oxygenation. To make use of the potential of combined PET and MRI to reveal such multifunctional information as well as anatomy in a single patient examination, any mutual interference between the two imaging modalities needs to be avoided.

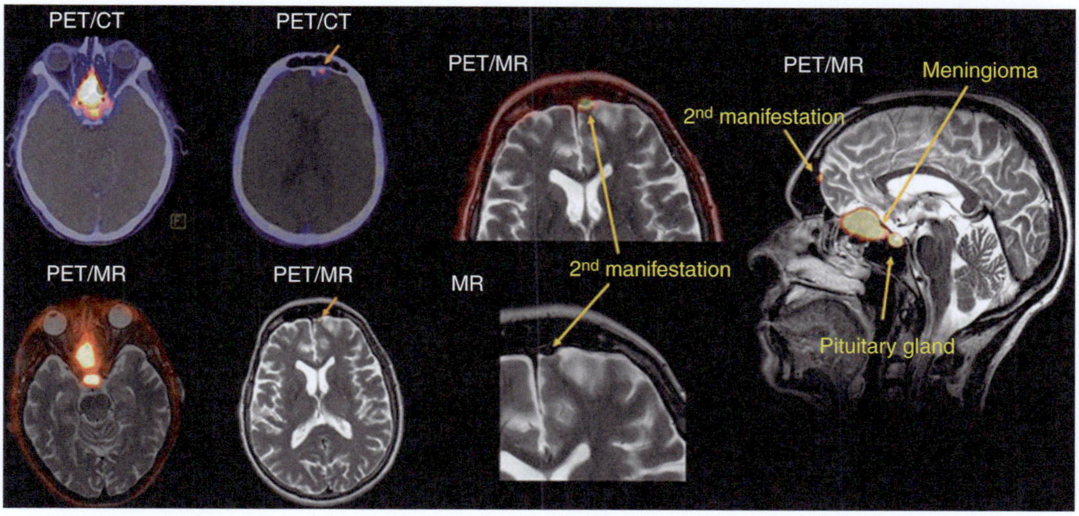

Fig. 22.1 PET-CT and PET/MRI with frontobasal meningioma in olfactory region. PET-CT images were acquired 20 min and PET/MR images 100 min after injection of 135 MBq of 68Ga-Octreotide. Tracer uptake in the tumor is seen on PET images. In addition, second smaller and previously unknown frontal meningioma was seen on PET and possibly corresponded to small mass demonstrated on T2-weighted turbo spin-echo MR images. This finding was not detected by CT

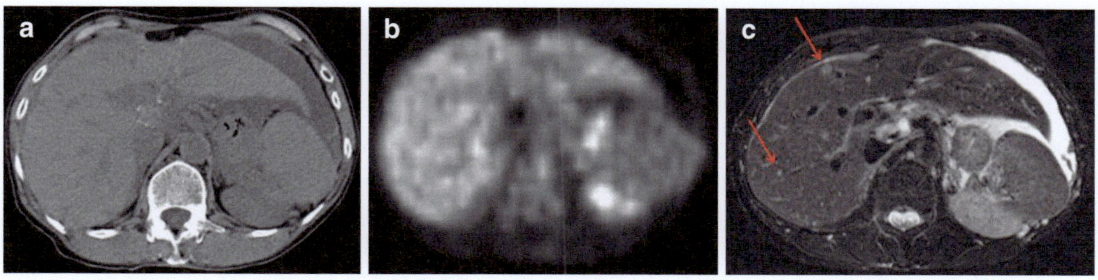

Fig. 22.2 Abdominal unenhanced 18F-FDG PET-CT and MRI of 55-year-old woman with ovarian cancer showing liver metastases detected by MRI (**c**, *arrows*) but neither by PET (**b**) nor native CT (**a**). MRI is most sensitive for detecting small liver lesions because of its superb soft-tissue contrast, whereas PET and CT are limited because of lower contrast and physiologic 18F-FDG liver uptake. PET/MRI provided better diagnostic confidence due to soft-tissue contrast and complementary information from different MRI sequences

Clinical Relevance

In terms of location and delineation of PET-positive bone lesions, the T1-weighted TSE MRI sequence performed significantly better than CT.

PET/MRI is more than adequate in characterizing liver lesions and provides greater lesion conspicuity than PET-CT, offering clinicians a powerful alternative for oncology imaging.

A small comparative study consisting of 31 patients were asked to undergo both PET-CT and PET/MRI scans to restage their recurrent prostate cancer after a single injection of ^{11}C-Choline. The PET-CT scans were carried out about 5 min after injection of the ^{11}C-Choline, and PET/MRI scans were given about 45 min later. Each patient's scans were interpreted separately and all evident lesions were categorized as definitely metastatic, probably metastatic, or indeterminate. The study revealed as follows:

PET/MR Found

- 17 areas of metastasis in 12 different patients
- 49 lymph node metastases
- 17 bone metastases in 5 patients

PET-CT Found

- 12 areas of metastasis in 8 patients
- 39 lymph node metastases
- 14 bone metastases in 4 patients

PET/MRI took longer to carry out than PET-CT but was still well tolerated by the patients.

Radiation exposure on PET/MRI was lower than on PET-CT.

Comparison of PET-CT with PET-MRI in Breast Cancer

Thirty six patients with carcinoma breast underwent a whole-body PET-CT scan 1 h after injection and an average of 62 min later a second scan using a hybrid PET/MRI system. PET/MRI and PET-CT were compared visually by rating anatomic allocation and image contrast. Regional tracer uptake in lesions was quantified using volumes of interest, and maximal and mean standardized uptake values (SUVmax and SUVmean, respectively) were calculated. Metabolic tumor volume (MTV) of each lesion was computed on PET/MRI and PET-CT. Tracer uptake in normal organ tissue was assessed as SUVmax and SUVmean in liver, spleen, left ventricular myocardium, lung, and muscle.

The Study Revealed

- Overall 74 FDG-positive lesions were visualized by both PET_CT and PET_MRI.
- No significant differences in anatomic allocation scores were found between PET-CT and PERT/MRI
- Contrast score of lesions on PET/MRI was significantly higher.
- Both SUVmax and SUVmean of lesions were significantly higher on PET-MRI than on PET-CT, with strong correlations between PET-MRI and PET-CT data ($\tilde{n}=0.71–0.88$).
- MTVs of all lesions were 4 % lower on PET-MRI than on PET-CT, but no statistically significant difference was observed, and an excellent correlation between measurements of MTV with PET-MRI and PET-CT was found ($\tilde{n}=0.95–0.97$; $p<0.0001$).
- Both SUVmax and SUVmean were significantly lower by PET-MRI than by PET-CT for lung, liver, and muscle, no significant difference was observed for spleen, while either SUV max and SUV mean of myocardium were significantly higher by PET/MRI.
- High correlations were found between PET-MRI and PET-CT for both SUV max and SUV mean of the left ventricular myocardium ($\tilde{n}=0.91$; $p<0.0001$), while moderate correlations were found for the other normal organ tissues ($\tilde{n}=0.36–0.61$; $p<0.05$).

In this rather very thorough study, PET-MRI showed equivalent performance in terms of qualitative lesion detection to PET-CT. Despite significant differences in tracer uptake quantification, due to either methodological and biological factors, PET-MRI and PET-CT measurements in lesions and normal organ tissues correlated well.

Radiation Therapy Planning

There are hardly any published study comparing head-to-head PET-CT and PET-MRI-based radiation therapy planning. When MRI images and/or PET images are used to define primary tumors and lymph nodes for gynecological malignancies, treatment planning on these structures is highly recommended. Because of a different size and shape of the target volumes on the MRI-CT and PET-CT scan, dose coverage to the PTV can be increased and mean dose to the organs at risk can be minimized. No major differences can be determined between CT, MRI-CT, and PET-CT scan in terms of maximum dose to the bladder and bowel area. This is because of a standard overlap of the bladder, the rectum, the anus, and the bowel area with the PTV, due to the large margin recipe used for gynecological malignancies.

For Further Reading

1. Pace L, Nicolai E, Luongo A, Aiello M, Catalano OA, Soricelli A, Salvatore M. Comparison of whole-body PET-CT and PET/MRI in breast cancer patients: lesion detection and quantitation of 18F-deoxyglucose uptake in lesions and in normal organ tissues. Eur J Radiol. 2014;83(2):289–96. doi:10.1016/j.ejrad.2013.11.002. Epub 2013 Nov 23.
2. Townsend DW. Dual-modality imaging: combining anatomy and function. J Nucl Med. 2008;49:938–55.
3. von Schulthess GK, Schlemmer HP. A look ahead: PET/MR versus PET-CT. Eur J Nucl Med Mol Imaging. 2009;36 Suppl 1:S3–9.

MIX
Papier aus verantwortungsvollen Quellen
Paper from responsible sources
FSC® C105338

If you have any concerns about our products,
you can contact us on
ProductSafety@springernature.com

In case Publisher is established outside the EU,
the EU authorized representative is:
Springer Nature Customer Service Center GmbH
Europaplatz 3, 69115 Heidelberg, Germany

Printed by Libri Plureos GmbH
in Hamburg, Germany